STATA MULTIPLE-IMPUTATION
REFERENCE MANUAL
RELEASE 12

A Stata Press Publication
StataCorp LP
College Station, Texas

D1262503

Table of contents

Cross-referencing the documentation

When reading this manual, you will find references to other Stata manuals. For example,

[U] **26 Overview of Stata estimation commands**

[R] **regress**

[XT] **xtreg**

The first example is a reference to chapter 26, *Overview of Stata estimation commands*, in the *User's Guide*; the second is a reference to the `regress` entry in the *Base Reference Manual*; and the third is a reference to the `xtreg` entry in the *Longitudinal-Data/Panel-Data Reference Manual*.

All the manuals in the Stata Documentation have a shorthand notation:

[GSM]	*Getting Started with Stata for Mac*
[GSU]	*Getting Started with Stata for Unix*
[GSW]	*Getting Started with Stata for Windows*
[U]	*Stata User's Guide*
[R]	*Stata Base Reference Manual*
[D]	*Stata Data-Management Reference Manual*
[G]	*Stata Graphics Reference Manual*
[XT]	*Stata Longitudinal-Data/Panel-Data Reference Manual*
[MI]	*Stata Multiple-Imputation Reference Manual*
[MV]	*Stata Multivariate Statistics Reference Manual*
[P]	*Stata Programming Reference Manual*
[SEM]	*Stata Structural Equation Modeling Reference Manual*
[SVY]	*Stata Survey Data Reference Manual*
[ST]	*Stata Survival Analysis and Epidemiological Tables Reference Manual*
[TS]	*Stata Time-Series Reference Manual*
[I]	*Stata Quick Reference and Index*
[M]	*Mata Reference Manual*

Detailed information about each of these manuals may be found online at

http://www.stata-press.com/manuals/

Title

Description

Missing data arise frequently. Various procedures have been suggested in the literature over the last several decades to deal with missing data (for example, Anderson [1957]; Hartley and Hocking [1971]; Rubin [1972, 1987]; and Dempster, Laird, and Rubin [1977]). The technique of multiple imputation, which originated in early 1970 in application to survey nonresponse (Rubin 1976), has gained popularity increasingly over the years as indicated by literature (for example, Rubin [1976, 1987, 1996]; Little [1992]; Meng [1994]; Schafer [1997]; van Buuren, Boshuizen, and Knook [1999]; Little and Rubin [2002]; Carlin et al. [2003]; Royston [2004, 2005a, 2005b, 2007, 2009]; Reiter and Raghunathan [2007]; Carlin, Galati, and Royston [2008]; Royston, Carlin, and White [2009]; and White, Royston, and Wood [2011]).

This entry presents a general introduction to multiple imputation and describes relevant statistical terminology used throughout the manual. The discussion here, as well as other statistical entries in this manual, is based on the concepts developed in Rubin (1987) and Schafer (1997).

Remarks

Remarks are presented under the following headings:

> Motivating example
> What is multiple imputation?
> Theory underlying multiple imputation
> How large should M be?
> Assumptions about missing data
> Patterns of missing data
> Proper imputation methods
> Analysis of multiply imputed data
> A brief introduction to MI using Stata
> Summary

We will use the following definitions and notation.

An imputation represents one set of plausible values for missing data, and so multiple imputations represent multiple sets of plausible values. With a slight abuse of the terminology, we will use the term *imputation* to mean the data where missing values are replaced with one set of plausible values.

We use M to refer to the number of imputations and m to refer to each individual imputation; that is, $m = 1$ means the first imputation, $m = 2$ means the second imputation, and so on.

Motivating example

Consider a fictional case–control study examining a relationship between smoking and heart attacks.

```
. use http://www.stata-press.com/data/r12/mheart0
(Fictional heart attack data; bmi missing)
. describe
Contains data from http://www.stata-press.com/data/r12/mheart0.dta
  obs:           154                          Fictional heart attack data;
                                                bmi missing
  vars:            9                          19 Jun 2011 10:50
  size:        2,310
```

variable name	storage type	display format	value label	variable label
attack	byte	%9.0g		Outcome (heart attack)
smokes	byte	%9.0g		Current smoker
age	float	%9.0g		Age, in years
bmi	float	%9.0g		Body Mass Index, kg/m^2
female	byte	%9.0g		Gender
hsgrad	byte	%9.0g		High school graduate
marstatus	byte	%9.0g	mar	Marital status: single, married, divorced
alcohol	byte	%24.0g	alc	Alcohol consumption: none, <2 drinks/day, >=2 drinks/day
hightar	byte	%9.0g		Smokes high tar cigarettes

```
Sorted by:
```

In addition to the primary variables attack and smokes, the dataset contains information about subjects' ages, body mass indexes (BMIs), genders, educational statuses, marital statuses, alcohol consumptions, and the types of cigarettes smoked (low/high tar).

We will use logistic regression to study the relationship between attack, recording heart attacks, and smokes:

```
. logit attack smokes age bmi hsgrad female

Iteration 0:   log likelihood = -91.359017
Iteration 1:   log likelihood = -79.374749
Iteration 2:   log likelihood = -79.342218
Iteration 3:   log likelihood =  -79.34221

Logistic regression                             Number of obs   =        132
                                                LR chi2(5)      =      24.03
                                                Prob > chi2     =     0.0002
Log likelihood =  -79.34221                     Pseudo R2       =     0.1315
```

attack	Coef.	Std. Err.	z	P>|z|	[95% Conf. Interval]	
smokes	1.544053	.3998329	3.86	0.000	.7603945	2.327711
age	.026112	.017042	1.53	0.125	-.0072898	.0595137
bmi	.1129938	.0500061	2.26	0.024	.0149837	.211004
hsgrad	.4048251	.4446019	0.91	0.363	-.4665786	1.276229
female	.2255301	.4527558	0.50	0.618	-.6618549	1.112915
_cons	-5.408398	1.810603	-2.99	0.003	-8.957115	-1.85968

The above analysis used 132 observations out of the available 154 because some of the covariates contain missing values. Let's examine the data for missing values, something we could have done first:

```
. misstable summarize
```

| | | | | Obs<. | | |
Variable	Obs=.	Obs>.	Obs<.	Unique values	Min	Max
bmi	22		132	132	17.22643	38.24214

We discover that `bmi` is missing in 22 observations. Our analysis ignored the information about the other covariates in these 22 observations. Can we somehow preserve this information in the analysis? The answer is yes, and one solution is to use multiple imputation.

What is multiple imputation?

Multiple imputation (MI) is a flexible, simulation-based statistical technique for handling missing data. Multiple imputation consists of three steps:

1. *Imputation step.* M imputations (completed datasets) are generated under some chosen imputation model.

2. *Completed-data analysis (estimation) step.* The desired analysis is performed separately on each imputation $m = 1, \ldots, M$. This is called completed-data analysis and is the primary analysis to be performed once missing data have been imputed.

3. *Pooling step.* The results obtained from M completed-data analyses are combined into a single multiple-imputation result.

The completed-data analysis step and the pooling step can be combined and thought of generally as the analysis step.

MI as a missing-data technique has two appealing main features: 1) the ability to perform a wide variety of completed-data analyses using existing statistical methods; and 2) separation of the imputation step from the analysis step. We discuss these two features in more detail in what follows.

Among other commonly used missing-data techniques that allow a variety of completed-data analyses are complete-case analysis or listwise (casewise) deletion, available-case analysis, and single-imputation methods. Although these procedures share one of MI's appealing properties, they lack some of MI's statistical properties.

For example, listwise deletion discards all observations with missing values and thus all information contained in the nonmissing values of these observations. With a large number of missing observations, this may lead to results that will be less efficient (larger standard errors, wider confidence intervals, less power) than MI results. In situations when the remaining complete cases are not representative of the population of interest, listwise deletion may also lead to biased parameter estimates.

In our opening logistic analysis of heart attacks, we used listwise deletion. The effect of age was not statistically significant based on the reduced sample. The MI analysis of these data (see *A brief introduction to MI using Stata* below) will reveal the statistical significance of age by using all available observations after imputing missing values for BMI.

Unlike listwise deletion, single-imputation methods do not discard missing values. They treat the imputed values as known in the analysis. This underestimates the variance of the estimates and so overstates precision and results in confidence intervals and significance tests that are too optimistic. MI rectifies this problem by creating multiple imputations and taking into account the sampling variability due to the missing data (between-imputation variability). See Little and Rubin (2002) and Allison (2001), among others, for a more detailed comparison of the methods.

The independence of the imputation step from the analysis step is the property MI shares with other imputation methods. The imputation step fills in missing values. The analysis step provides inference about multiply imputed results and does not require any information about the missing-data aspect of the problem.

The separation of the two steps allows different individuals, a data collector/imputer and a data analyst, to perform these steps independently of one another. The advantage is that the data collector/imputer usually has access to more information about the data than may be disclosed to the data analyst and thus can create more accurate imputations. The data analyst can use the imputed data released by the data collector in a number of different analyses. Of course, it is crucial that the imputer make the imputation model as general as possible to accommodate a wide variety of analyses that the data analyst might choose to perform; see *Proper imputation methods* below for details.

In our heart attack example, the imputer would create multiple imputations of missing values of BMI using, for example, a linear regression method, and then release the resulting data to the analyst. The analyst could then analyze these multiply imputed data using an ordinary logistic regression. That is, no adjustment is needed to the analysis model itself to account for missing BMI—the pooling portion of the analysis will account for the increased variability because of imputed missing data.

Theory underlying multiple imputation

MI was derived using the Bayesian paradigm yet was proved to be statistically valid from the frequentist (randomization-based) perspective. We use the definition from Rubin (1996) of statistical validity that implies approximately unbiased point estimates and implies confidence intervals achieving their nominal coverages when averaged over the randomization distributions induced by the known sampling and the posited missing-data mechanisms.

To explain the role the Bayesian and frequentist concepts play in MI, we need to consider the MI procedure in more detail. MI requires specification of two models—the imputation model and the analysis model. The imputation model is the model used to create imputations in the imputation step. The analysis model is the completed-data model used during the analysis step to obtain completed-data estimates, $\widehat{Q}$, of parameters of interest, Q, and the estimate, U, of sampling variability associated with $\widehat{Q}$. During the pooling step, the individual completed-data estimates $(\widehat{Q}, U)$ are combined into $(\widehat{Q}_{\mathrm{MI}}, T)$ to form one repeated-imputation inference. The statistical validity of the repeated-imputation inference is of interest.

Consider the case when both the imputation model and the analysis model are the same Bayesian models. Then the repeated imputations (multiple imputations) are repeated draws from the posterior predictive distribution of the missing data under a posited Bayesian model. The combined parameter estimates, $\widehat{Q}_{\mathrm{MI}}$, and their associated sampling variance estimate, $T = W + B$, are the approximations to the posterior mean and variance of Q. Here W represents the within-imputation variability (average of the completed-data variance estimates, U), and B represents the between-imputation variability (variance estimate of $\widehat{Q}_{\mathrm{MI}}$ over repeated imputations). Provided that the posterior mean and variance are adequate summaries of the posterior distribution, the repeated-imputation inference based on these combined estimates can be justified either from a purely Bayesian standpoint or from a purely frequentist standpoint. Thus a Bayesian apparatus is used to create imputations and also underlies the rules for combining parameter estimates.

In reality, the analysis model is rarely the same as the imputation model, and neither of them is an explicit Bayesian model. Repeated-imputation inference is still statistically valid in those cases. The rigorous justification is given in chapters 3 and 4 of Rubin (1987) from the frequentist perspective. Below we briefly summarize the conditions under which the repeated-imputation inference from the pooling step is statistically valid; also see Rubin (1987, 117–119) for more detail.

The repeated-imputation inference is statistically valid if 1) the multiple imputations from the imputation step are proper (see *Proper imputation methods* below) and 2) the completed-data inference based on $(\widehat{Q}, U)$ from the analysis step is randomization valid. Completed-data inference based on $(\widehat{Q}, U)$ is randomization valid if $\widehat{Q} \sim N\{Q, \mathrm{Var}(\widehat{Q})\}$ and U is a consistent estimate of $\mathrm{Var}(\widehat{Q})$ over the distribution of the sampling mechanism.

The randomization validity of MI was derived under the assumption of an infinite number of imputations. In practice, however, the number of imputations tends to be small and so the finite-M properties of the MI estimators must be explored. Rubin (1987) derives the fundamental result underlying the MI inference based on a finite M. We restate it below for a scalar Q:

$$T_M^{-1/2}(Q - \widehat{Q}_M) \sim t_{\nu_M}$$

where $\widehat{Q}_M$ is the average of M completed-data estimates of Q, $T_M = W + (1 + 1/M)B$, and t_{ν_M} is a Student's t distribution with degrees of freedom ν_M that depend on the number of imputations and rates of missing information (or the fraction of information missing because of nonresponse that measures the influence of the missing data on parameter estimates). Later, Li, Raghunathan, and Rubin (1991b) derived an improved procedure for multiple testing, and Barnard and Rubin (1999) and Reiter (2007) extended the MI inference to account for small samples. For computation details, see *Methods and formulas* in [MI] **mi estimate**.

How large should M be?

The theory underlying the validity of MI relies on an infinite number of imputations, M. The procedure is also known to have good statistical properties with finite M, but what values of M should we use in practice? Rubin (1987, 114) answers this question: the asymptotic relative efficiency (RE) of the MI procedure with finite M compared with infinite M is roughly 90% with only two imputations for a missing-information rate as high as 50%.

Most literature (for example, Rubin [1987] and van Buuren, Boshuizen, and Knook [1999]) suggests that $M = 5$ (corresponding to RE of 95% for 50% of information missing) should be sufficient to obtain valid inference. In general, however, the actual number of imputations necessary for MI to perform satisfactorily depends not only on the amount of information missing due to nonresponse but also on the analysis model and the data. Some analyses may require M to be 50 or more to obtain stable results (Kenward and Carpenter 2007; Horton and Lipsitz 2001).

Literature with formal recommendations on how to choose M is very sparse. Royston (2004), Royston, Carlin, and White (2009), and White, Royston, and Wood (2011) discuss the impact of the number of imputations on the precision of estimates and suggest ways of determining the required number of imputations by evaluating the sampling error of the MI estimates.

Because it is computationally feasible to obtain more imputations, we recommend using at least 20 imputations to reduce the sampling error due to imputations.

Assumptions about missing data

The theory underlying MI methodology makes no assumption about the missing-data mechanism. However, many imputation methods (including those provided by Stata) require that the missing-data mechanism be ignorable. Before we discuss the ignorability conditions, consider the following definitions.

Missing data are said to be missing completely at random (MCAR) if the probability that data are missing does not depend on observed or unobserved data. Under MCAR, the missing-data values are a simple random sample of all data values, and so any analysis that discards the missing values remains consistent, albeit perhaps inefficient.

Consider a hypothetical longitudinal study comparing different blood-pressure treatments. Suppose that the follow-up blood-pressure measurements were not collected from some subjects because they moved to a different area. These missing blood-pressure measurements can be viewed as MCAR as long as subjects' decisions to move were unrelated to any item in the study.

Missing data are said to be missing at random (MAR) if the probability that data are missing does not depend on unobserved data but may depend on observed data. Under MAR, the missing-data values do not contain any additional information given observed data about the missing-data mechanism. Note that MCAR can be viewed as a particular case of MAR. When missing data are MAR, listwise deletion may lead to biased results.

Suppose that some subjects decided to leave the study because of severe side effects from the assigned treatment of a high dosage of a medicine. Here it is unlikely that missing blood-pressure measurements are MCAR because the subjects who received a higher dosage of the medicine are more likely to suffer severe side effects than those who received a lower dosage and thus are more likely to drop out of the study. Missing blood-pressure measurements depend on the dosage of the received treatment and therefore are MAR.

On the other hand, if the subjects are withdrawn from the study for ethical reasons because of extremely high blood pressures, missing blood-pressure measurements would not be MAR. The measurements for the subjects with very high blood pressures will be missing and thus the reason for drop out will depend on the missing blood pressures. This type of missing-data mechanism is called missing not at random (MNAR). For such missing data, the reasons for its missingness must be accounted for in the model to obtain valid results.

Model parameters are said to be *distinct* from a Bayesian standpoint if their joint prior distribution can be factorized into independent marginal prior distributions.

The missing-data mechanism is said to be *ignorable* if missing data are MAR and the parameters of the data model and the parameters of the missing-data mechanism are distinct (Rubin 1976).

The ignorability assumption makes it possible to ignore the process that causes missing data in the imputation model—something not possible with MNAR—which simplifies the imputation step while still ensuring correct inference. The provided imputation methods assume that missing data are MAR.

In practice, it is difficult to test the ignorability assumption formally because the MAR mechanism can be distinguished from the MNAR mechanism only through the missing data that are not observed. Thus careful consideration is necessary before accepting this assumption. If in doubt, sensitivity analysis—analysis repeated under various missing-data models—needs to be performed to verify the stability of inference. In the context of MI, sensitivity analysis can be performed by modifying the imputation step to accommodate the nonignorable missing-data mechanism (for example, Kenward and Carpenter [2007] and van Buuren, Boshuizen, and Knook [1999]).

Patterns of missing data

Another issue we need to consider related to missing data is a pattern of missingness (or missing-data pattern).

Consider an $N \times p$ data matrix $Y = (Y_1, Y_2, \ldots, Y_p)'$ with p variables and N observations. Consider a permutation of column indices $(i_1, i_2, \ldots, i_p)$ such that Y_{i_1} is at least as observed as Y_{i_2}, which is at least as observed as Y_{i_3}, and so on. In other words, Y_{i_2} has missing values in the same observations (and possibly more) as Y_{i_1}, Y_{i_3} has missing values (and possibly more) in the same observations as Y_{i_2}, and so on. If such a permutation exists, then the pattern of missingness in Y is said to be monotone. If the pattern of missingness is not monotone, it is assumed to be arbitrary.

For example, consider the following indicator matrix recording the missing pattern in Y:

$$R_1 = \begin{pmatrix} 1 & 1 & 1 \\ 0 & 0 & 1 \\ 0 & 1 & 1 \\ 0 & 1 & 1 \end{pmatrix}$$

where R_1^{ij} is 1 if variable Y_j is observed (complete) in observation i and 0 otherwise. We can see that Y has a monotone-missing pattern if we interchange the first and the third columns of R_1. In fact, if we also rearrange the rows such that

$$R_1 = \begin{pmatrix} 1 & 1 & 1 \\ 1 & 1 & 0 \\ 1 & 1 & 0 \\ 1 & 0 & 0 \end{pmatrix}$$

then the monotonicity of missing values becomes even more evident. An example of a nonmonotone missing-value pattern is

$$R_2 = \begin{pmatrix} 1 & 1 & 1 \\ 1 & 1 & 0 \\ 0 & 1 & 0 \\ 1 & 0 & 0 \end{pmatrix}$$

There is no ordering of the first two columns of R_2 such that the missing values in one column imply missing values in the other column.

Why is it important to consider the monotone missing-value pattern? A monotone-missing pattern greatly simplifies the imputation task. Under a monotone-missing pattern, a multivariate imputation task can be formulated as a sequence of independent univariate (conditional) imputation tasks, which allows the creation of a flexible imputation model; see [MI] **mi impute monotone** for details, and see Rubin (1987, 174) for more technical conditions under which such a formulation is applicable.

Proper imputation methods

As we mentioned earlier, a key concept underlying the randomization-based evaluations of the repeated-imputation inference is proper multiple imputation.

A multiple-imputation method is said to be proper if it produces proper multiple imputations, which we are about to define. Rubin (1987, 118–119) gives a full technical definition for proper multiple imputations. Ignoring the more technical definition, Rubin (1996) states the following main conditions. The multiple imputations are said to be proper if

1. MI estimates $\widehat{Q}_{\mathrm{MI}}$ are asymptotically normal with mean $\widehat{Q}$ and a consistent variance–covariance estimate B.

2. The within-imputation variance estimate W is a consistent estimate of the variance–covariance estimate U with variability of a lower order than $\mathrm{Var}(\widehat{Q}_{\mathrm{MI}})$.

The above statements assume a large number of imputations and the randomization distribution induced by the missing-data mechanism.

In general, it is difficult to determine if an imputation method is proper using the above definition. Rubin (1987, sec. 4.3) and Binder and Sun (1996) describe several examples of proper and improper imputation methods. Rubin (1987, 125–127) recommends drawing imputations from a Bayesian posterior predictive distribution (or an appropriate approximation to it) of missing values under the chosen model for the data and the missing-data mechanism. The chosen imputation model must also be appropriate for the completed-data statistics likely to be used at the analysis stage. Schafer (1997, 145) points out that from a practical standpoint, it is more important that the chosen imputation model performs well over the repeated samples than that it be technically proper. This can be checked via simulation.

With the exception of predictive mean matching and chained equations, the imputation methods available in Stata obtain imputations by simulating from a Bayesian posterior predictive distribution of the missing data (or its approximation) under the conventional (or chosen) prior distribution; see *Imputation methods* in [MI] **mi impute** for details. To ensure that the multiple imputations are proper, you must choose an appropriate imputation model, which we briefly discuss next.

The imputation model must include all predictors relevant to the missing-data mechanism, and it must preserve all data characteristics likely to be explored at the analysis stage. For example, if the analysis model explores a correlation between two variables, then omitting either of those variables from the imputation model will lead to estimates of the correlation biased toward zero. Another common mistake that may lead to biased estimates is when an outcome variable of the analysis model is not used in the imputation model. In the survey context, all structural variables such as sampling weights, strata, and cluster identifiers (or at least main strata and main clusters) need to be included in the imputation model.

In general, any predictors involved in the definition of the completed-data estimators and the sampling design should be included in the imputation model. If you intend to use the multiply imputed data in an analysis involving a wide range of completed-data estimators, you should include as many variables as possible.

Using our heart attack data, if we were to release the multiply imputed version of it for general analyses, we would have included all available covariates as predictors in the regression model used to impute BMI and not only the subset of covariates (heart attacks, smoking status, age, gender, and educational status) used in our specific data analysis.

The severity of the effect of a misspecified imputation model will typically depend on the amount of imputed data relative to the observed data—a small number of observations with improperly imputed values may not affect the inference greatly if there is a large number of observations with complete data.

For more details about imputation modeling, see Rubin (1996), Schafer (1997, 139–144), Schafer and Olsen (1998), Allison (2001), Schafer and Graham (2002), Kenward and Carpenter (2007), Graham (2009), and White, Royston, and Wood (2011), among others. For imputation modeling of large surveys, see, for example, Schafer, Khare, and Ezzati-Rice (1993) and Ezzati-Rice et al. (1995).

Analysis of multiply imputed data

Once we have multiply imputed data, we perform our primary analysis on each completed dataset and then use Rubin's combination rules to form one set of results. Assuming that the underlying imputation model is properly specified (see, for example, Abayomi, Gelman, and Levy [2008] and Gelman et al. [2005] for multiple-imputation diagnostics), we can choose from a variety of statistical methods. For example, the methods can include maximum-likelihood methods, survey methods, nonparametric methods, and any other method appropriate for the type of data we have.

Each of the methods have certain concepts associated with them. For example, maximum-likelihood methods use a likelihood function, whereas a deviance is associated with generalized linear models. While these concepts are well defined within each individual completed-data analysis, they may not have a clear interpretation when the individual analyses are combined in the pooling step. (Only in the special case when the imputation and analysis models are compatible Bayesian models can the estimated parameters be viewed as approximations to the mode of the posterior distribution.)

As a result, various statistical (postestimation) procedures based on these concepts, such as likelihood-ratio tests, goodness-of-fit tests, etc., are not directly applicable to MI results. Instead, their "MI" versions are being studied in the literature (Li et al. 1991a; Meng and Rubin 1992). Another concept that is not uniquely defined within MI is that of prediction; see Carlin, Galati, and Royston (2008) and White, Royston, and Wood (2011) for one definition.

Donald Bruce Rubin (1943–) was born in Washington, DC. He entered Princeton intending to become a physicist but ended up majoring in psychology. He entered Harvard intending to continue as a psychologist, but in the event, gained further degrees in computer science and statistics. After periods at the Educational Testing Service and the University of Chicago, Rubin returned to Harvard in 1984. He has had many visiting appointments and has carried out extensive consultancy work. Rubin has long been a leader in research on causal inference in experiments and observational studies, and problems of nonresponse and missing data. Among many major contributions is his formalization of the expectation-maximization algorithm with Arthur Dempster and Nan Laird. Rubin's work ranges over a wide variety of sciences and is often Bayesian in style. Rubin was elected a member of the National Academy of Sciences in 2010.

A brief introduction to MI using Stata

Stata offers full support for MI analysis from the imputation step to the pooling step.

The imputation step can be performed for one variable or multiple variables. A number of imputation methods, including flexible methods accommodating variables of different types and an iterative Markov chain Monte Carlo method based on multivariate normal, are available; see [MI] **mi impute** for details.

The analysis and pooling steps are combined into one step and performed by `mi estimate`; see [MI] **mi estimate**. You can fit many commonly used models and obtain combined estimates of coefficients (or transformed coefficients) (see [MI] **estimation** for a list of supported estimation commands), or you can create your own estimation command and use it with the `mi estimate` prefix.

In addition to the conventional estimation steps, Stata facilitates many data-manipulation routines for managing your multiply imputed data and verifying its integrity over the imputations; see [MI] **intro** for a full list of commands.

As a short demonstration of `mi`, let's analyze the heart attack data introduced earlier using MI; see [MI] **workflow** for more thorough guidelines.

The goals are 1) to fill in missing values of `bmi` using, for example, a linear regression imputation method (`mi impute regress`) to obtain multiply imputed data and 2) to analyze the multiply imputed data using logistic regression, which we will do using `mi estimate`. Before we can accomplish these two steps, we need to prepare the data so they can be used with `mi`. First, we declare the data to be `mi` data:

```
. use http://www.stata-press.com/data/r12/mheart0
(Fictional heart attack data; bmi missing)
. mi set mlong
```

We choose to use the data in the marginal long style (mlong) because it is a memory-efficient style; see [MI] **styles** for details.

To use mi impute, we must first register imputation variables. In general, we recommend that you register all variables relevant to the analysis as imputed, passive, or regular with mi register (see [MI] **mi set**), especially if you plan on doing any data management of your multiply imputed data.

```
. mi register imputed bmi
(22 m=0 obs. now marked as incomplete)
. mi register regular attack smokes age hsgrad female
```

We are now ready to use mi impute. To lessen the simulation (Monte Carlo) error, we arbitrarily choose to create 20 imputations (add(20) option). We also specify the rseed() option for reproducibility:

```
. mi impute regress bmi attack smokes age hsgrad female, add(20) rseed(2232)
Univariate imputation              Imputations =        20
Linear regression                        added =        20
Imputed: m=1 through m=20              updated =         0
```

	Observations per *m*			
Variable	Complete	Incomplete	Imputed	Total
bmi	132	22	22	154

```
(complete + incomplete = total; imputed is the minimum across m
 of the number of filled-in observations.)
```

From the output, we see that all 22 incomplete values of bmi were successfully imputed. You may want to examine your imputations to verify that nothing abnormal occurred during imputation. For example, as a quick check, we can compare main descriptive statistics from some imputations (say, the first and the last one) to those from the observed data. We use mi xeq (see [MI] **mi xeq**) to execute Stata's summarize command on the original data ($m = 0$), the first imputation ($m = 1$), and the last imputation ($m = 20$):

```
. mi xeq 0 1 20: summarize bmi
m=0 data:
-> summarize bmi
```

Variable	Obs	Mean	Std. Dev.	Min	Max
bmi	132	25.24136	4.027137	17.22643	38.24214

```
m=1 data:
-> summarize bmi
```

Variable	Obs	Mean	Std. Dev.	Min	Max
bmi	154	25.11855	3.990918	15.47331	38.24214

```
m=20 data:
-> summarize bmi
```

Variable	Obs	Mean	Std. Dev.	Min	Max
bmi	154	25.37117	4.051929	15.4505	38.24214

The summary statistics of the imputed datasets look reasonable.

We now fit the logistic regression using the `mi estimate` prefix command:

```
. mi estimate, dots: logit attack smokes age bmi hsgrad female
Imputations (20):
.........10.........20 done
```

Multiple-imputation estimates				Imputations	=	20
Logistic regression				Number of obs	=	154
				Average RVI	=	0.0404
				Largest FMI	=	0.1678
DF adjustment: Large sample				DF: min	=	694.17
				avg	=	115477.35
				max	=	287682.23
Model F test: Equal FMI				F(5,43531.9)	=	3.74
Within VCE type: OIM				Prob > F	=	0.0022

attack	Coef.	Std. Err.	t	P>|t|	[95% Conf. Interval]	
smokes	1.239172	.3630877	3.41	0.001	.5275236	1.950821
age	.0354929	.0154972	2.29	0.022	.0051187	.065867
bmi	.1184188	.0495676	2.39	0.017	.0210985	.2157391
hsgrad	.185709	.4075301	0.46	0.649	-.6130435	.9844615
female	-.0996102	.4193583	-0.24	0.812	-.9215408	.7223204
_cons	-5.845855	1.72309	-3.39	0.001	-9.225542	-2.466168

Compared with the earlier `logit` analysis (using listwise deletion), we detect the significance of `age`, whose effect was apparently disguised by the missing data. See [MI] **mi estimate** for details.

We will be using variations of these data throughout the `mi` documentation.

Summary

- MI is a simulation-based procedure. Its purpose is not to re-create the individual missing values as close as possible to the true ones but to handle missing data in a way resulting in valid statistical inference (Rubin 1987, 1996).

- MI yields valid inference if 1) the imputation method is proper with respect to the posited missing-data mechanism (see *Proper imputation methods* above) and 2) completed-data analysis is valid in the absence of missing data.

- A small number of imputations (5 to 20) may be sufficient when fractions of missing data are low. High fractions of missing data as well as particular data structures may require up to 100 (or more) imputations. Whenever feasible to do so, we recommend that you vary the number of imputations to see if this affects your results.

- With a small number of imputations, the reference distribution for the MI inference is Student's t (or F in multiple-hypothesis testing). The residual degrees of freedom depend on M and the rates of missing information and thus are different for each parameter of interest.

- With a large number of imputations, the reference distribution for MI inference is approximately normal (or χ^2 in multiple-hypothesis testing).

- When the imputer's model is more restrictive than the analyst's model, the MI inference can be invalid if the imputer's assumptions are not true. On the other hand, when the analyst's model is more restrictive than the imputer's model, the MI results will be valid but somewhat conservative if the analyst's assumptions are true. If the analyst's assumptions are false, the results can be biased; see, for example, Schafer (1997) for details.

- MI is relatively robust to departures from the correct specification of the imputation model, provided the rates of missing information are low and the correct completed-data model is used in the analysis.

- Certain concepts, for example, likelihood and deviance, do not have clear interpretation within the MI framework. As such, various statistical (postestimation) procedures based on these concepts (for example, likelihood-ratio tests, goodness-of-fit tests) are not directly applicable to MI results.

References

Abayomi, K., A. Gelman, and M. Levy. 2008. Diagnostics for multivariate imputations. *Journal of the Royal Statistical Society, Series C* 57: 273–291.

Allison, P. D. 2001. *Missing Data.* Thousand Oaks, CA: Sage.

Anderson, T. W. 1957. Maximum likelihood estimates for a multivariate normal distribution when some observations are missing. *Journal of the American Statistical Association* 52: 200–203.

Arnold, B. C., E. Castillo, and J. M. Sarabia. 1999. *Conditional Specification of Statistical Models.* New York: Springer.

———. 2001. Conditionally specified distributions: An introduction. *Statistical Science* 16: 249–274.

Barnard, J., and D. B. Rubin. 1999. Small-sample degrees of freedom with multiple imputation. *Biometrika* 86: 948–955.

Binder, D. A., and W. Sun. 1996. Frequency valid multiple imputation for surveys with a complex design. *Proceedings of the Survey Research Methods Section, American Statistical Association* 281–286.

Carlin, J. B., J. C. Galati, and P. Royston. 2008. A new framework for managing and analyzing multiply imputed data in Stata. *Stata Journal* 8: 49–67.

Carlin, J. B., N. Li, P. Greenwood, and C. Coffey. 2003. Tools for analyzing multiple imputed datasets. *Stata Journal* 3: 226–244.

Dempster, A. P., N. M. Laird, and D. B. Rubin. 1977. Maximum likelihood from incomplete data via the EM algorithm. *Journal of the Royal Statistical Society, Series B* 39: 1–38.

Ezzati-Rice, T. M., W. Johnson, M. Khare, R. J. A. Little, D. B. Rubin, and J. L. Schafer. 1995. A simulation study to evaluate the performance of model-based multiple imputations in NCHS health examination surveys. *Proceedings of the Annual Research Conference*, 257–266. U.S. Bureau of the Census: Washington, DC.

Gelman, A., J. B. Carlin, H. S. Stern, and D. B. Rubin. 2004. *Bayesian Data Analysis.* 2nd ed. London: Chapman & Hall/CRC.

Gelman, A., and D. B. Rubin. 1992. Inference from iterative simulation using multiple sequences. *Statistical Science* 7: 457–472.

Gelman, A., I. Van Mechelen, G. Verbeke, D. F. Heitjan, and M. Meulders. 2005. Multiple imputation for model checking: Completed-data plots with missing and latent data. *Biometrics* 61: 74–85.

Graham, J. W. 2009. Missing data analysis: Making it work in the real world. *Annual Review of Psychology* 60: 549–576.

Hartley, H. O., and R. R. Hocking. 1971. The analysis of incomplete data (with discussion). *Biometrics* 27: 783–823.

Horton, N. J., and K. P. Kleinman. 2007. Much ado about nothing: A comparison of missing data methods and software to fit incomplete data regression models. *American Statistician* 61: 79–90.

Horton, N. J., and S. R. Lipsitz. 2001. Multiple imputation in practice: Comparison of software packages for regression models with missing variables. *American Statistician* 55: 244–254.

Jenkins, S. P., R. V. Burkhauser, S. Feng, and J. Larrimore. 2011. Measuring inequality using censored data: a multiple-imputation approach to estimation and inference. *Journal of the Royal Statistical Society, Series A* 174: 63–81.

Kenward, M. G., and J. R. Carpenter. 2007. Multiple imputation: Current perspectives. *Statistical Methods in Medical Research* 16: 199–218.

Lee, K. J., and J. B. Carlin. 2010. Multiple imputation for missing data: Fully conditional specification versus multivariate normal imputation. *American Journal of Epidemiology* 171: 624–632.

Li, K.-H. 1988. Imputation using Markov chains. *Journal of Statistical Computation and Simulation* 30: 57–79.

Li, K.-H., X.-L. Meng, T. E. Raghunathan, and D. B. Rubin. 1991a. Significance levels from repeated p-values with multiply-imputed data. *Statistica Sinica* 1: 65–92.

Li, K.-H., T. E. Raghunathan, and D. B. Rubin. 1991b. Large-sample significance levels from multiply imputed data using moment-based statistics and an F reference distribution. *Journal of the American Statistical Association* 86: 1065–1073.

Little, R. J. A. 1988. Missing-data adjustments in large surveys. *Journal of Business and Economic Statistics* 6: 287–296.

———. 1992. Regression with missing X's: A review. *Journal of the American Statistical Association* 87: 1227–1237.

Little, R. J. A., and D. B. Rubin. 2002. *Statistical Analysis with Missing Data.* 2nd ed. Hoboken, NJ: Wiley.

Marchenko, Y. V., and J. P. Reiter. 2009. Improved degrees of freedom for multivariate significance tests obtained from multiply imputed, small-sample data. *Stata Journal* 9: 388–397.

Meng, X.-L. 1994. Multiple-imputation inferences with uncongenial sources of input (with discussion). *Statistical Science* 9: 538–573.

Meng, X.-L., and D. B. Rubin. 1992. Performing likelihood ratio tests with multiply-imputed data sets. *Biometrika* 79: 103–111.

Raghunathan, T. E., J. M. Lepkowski, J. Van Hoewyk, and P. Solenberger. 2001. A multivariate technique for multiply imputing missing values using a sequence of regression models. *Survey Methodology* 27: 85–95.

Reiter, J. P. 2007. Small-sample degrees of freedom for multi-component significance tests with multiple imputation for missing data. *Biometrika* 94: 502–508.

———. 2008. Multiple imputation when records used for imputation are not used or disseminated for analysis. *Biometrika* 95: 933–946.

Reiter, J. P., and T. E. Raghunathan. 2007. The multiple adaptations of multiple imputation. *Journal of the American Statistical Association* 102: 1462–1471.

Royston, P. 2004. Multiple imputation of missing values. *Stata Journal* 4: 227–241.

———. 2005a. Multiple imputation of missing values: Update. *Stata Journal* 5: 188–201.

———. 2005b. Multiple imputation of missing values: Update of ice. *Stata Journal* 5: 527–536.

———. 2007. Multiple imputation of missing values: Further update of ice, with an emphasis on interval censoring. *Stata Journal* 7: 445–464.

———. 2009. Multiple imputation of missing values: Further update of ice, with an emphasis on categorical variables. *Stata Journal* 9: 466–477.

Royston, P., J. B. Carlin, and I. R. White. 2009. Multiple imputation of missing values: New features for mim. *Stata Journal* 9: 252–264.

Rubin, D. B. 1972. A non-iterative algorithm for least squares estimation of missing values in any analysis of variance design. *Journal of the Royal Statistical Society, Series C* 21: 136–141.

———. 1976. Inference and missing data. *Biometrika* 63: 581–592.

———. 1986. Statistical matching using file concatenation with adjusted weights and multiple imputations. *Journal of Business and Economic Statistics* 4: 87–94.

———. 1987. *Multiple Imputation for Nonresponse in Surveys.* New York: Wiley.

———. 1996. Multiple imputation after 18+ years. *Journal of the American Statistical Association* 91: 473–489.

Schafer, J. L. 1997. *Analysis of Incomplete Multivariate Data.* Boca Raton, FL: Chapman & Hall/CRC.

Schafer, J. L., and J. W. Graham. 2002. Missing data: Our view of the state of the art. *Psychological Methods* 7: 147–177.

Schafer, J. L., M. Khare, and T. M. Ezzati-Rice. 1993. Multiple imputation of missing data in NHANES III. *Proceedings of the Annual Research Conference,* 459–487. U.S. Bureau of the Census: Washington, DC.

Schafer, J. L., and M. K. Olsen. 1998. Multiple imputation for multivariate missing-data problems: A data analyst's perspective. *Multivariate Behavioral Research* 33: 545–571.

Schenker, N., and J. M. G. Taylor. 1996. Partially parametric techniques for multiple imputation. *Computational Statistics & Data Analysis* 22: 425–446.

Tanner, M. A., and W. H. Wong. 1987. The calculation of posterior distributions by data augmentation (with discussion). *Journal of the American Statistical Association* 82: 528–550.

van Buuren, S. 2007. Multiple imputation of discrete and continuous data by fully conditional specification. *Statistical Methods in Medical Research* 16: 219–242.

van Buuren, S., H. C. Boshuizen, and D. L. Knook. 1999. Multiple imputation of missing blood pressure covariates in survival analysis. *Statistics in Medicine* 18: 681–694.

van Buuren, S., J. P. L. Brand, C. G. M. Groothuis-Oudshoorn, and D. B. Rubin. 2006. Fully conditional specification in multivariate imputation. *Journal of Statistical Computation and Simulation* 76: 1049–1064.

White, I. R., R. Daniel, and P. Royston. 2010. Avoiding bias due to perfect prediction in multiple imputation of incomplete categorical data. *Computational Statistics & Data Analysis* 54: 2267–2275.

White, I. R., P. Royston, and A. M. Wood. 2011. Multiple imputation using chained equations: Issues and guidance for practice. *Statistics in Medicine* 30: 377–399.

Also see

[MI] **intro** — Introduction to mi

[MI] **workflow** — Suggested workflow

[MI] **mi impute** — Impute missing values

[MI] **estimation** — Estimation commands for use with mi estimate

[MI] **mi estimate** — Estimation using multiple imputations

[MI] **Glossary**

Title

intro — Introduction to mi

Syntax

To become familiar with mi as quickly as possible, do the following:

1. See *A simple example* under *Remarks* below.
2. If you have data that require imputing, see [MI] **mi set** and [MI] **mi impute**.
3. Alternatively, if you have already imputed data, see [MI] **mi import**.
4. To fit your model, see [MI] **mi estimate**.

To create mi data from original data

mi set	declare data to be mi data
mi register	register imputed, passive, or regular variables
mi unregister	unregister previously registered variables
mi unset	return data to unset status (rarely used)

See *Description* below for a summary of mi data and these commands.
See [MI] **Glossary** for a definition of terms.

To import data that already have imputations for the missing values (do not mi set the data)

mi import	import mi data
mi export	export mi data to non-Stata application

Once data are mi set or mi imported

mi query	query whether and how mi set
mi describe	describe mi data
mi varying	identify variables that vary over m
mi misstable	tabulate missing values
mi passive	create passive variable and register it

To perform estimation on mi data

mi impute	impute missing values
mi estimate	perform and combine estimation on $m > 0$
mi ptrace	check stability of MCMC
mi test	perform tests on coefficients
mi testtransform	perform tests on transformed coefficients
mi predict	obtain linear predictions
mi predictnl	obtain nonlinear predictions

To stset, svyset, tsset, or xtset any mi data that were not set at the time they were mi set

mi fvset	fvset for mi data
mi svyset	svyset for mi data
mi xtset	xtset for mi data
mi tsset	tsset for mi data
mi stset	stset for mi data
mi streset	streset for mi data
mi st	st for mi data

To perform data management on mi data

mi rename	rename variable
mi append	append for mi data
mi merge	merge for mi data
mi expand	expand for mi data
mi reshape	reshape for mi data
mi stsplit	stsplit for mi data
mi stjoin	stjoin for mi data
mi add	add imputations from one mi dataset to another

To perform data management for which no mi prefix command exists

mi extract	extract $m = 0$ data
...	perform data management the usual way
mi replace0	replace $m = 0$ data in mi data

To perform the same data-management or data-reporting command(s) on $m = 0$, $m = 1$, ...

mi xeq: ...	execute commands on $m = 0$, $m = 1$, $m = 2$, ..., $m = M$
mi xeq #: ...	execute commands on $m = \#$
mi xeq # # ...: ...	execute commands on specified values of m

Useful utility commands

mi convert	convert mi data from one style to another
mi extract #	extract $m = \#$ from mi data
mi select #	programmer's command similar to mi extract
mi copy	copy mi data
mi erase	erase files containing mi data
mi update	verify/make mi data consistent
mi reset	reset imputed or passive variable

For programmers interested in extending mi

[MI] **technical**	Detail for programmers

Summary of styles

There are four styles or formats in which mi data are stored: flongsep, flong, mlong, and wide.

1. Flongsep: $m = 0$, $m = 1$, ..., $m = M$ are each separate .dta datasets. If $m = 0$ data are stored in pat.dta, then $m = 1$ data are stored in _1_pat.dta, $m = 2$ in _2_pat.dta, and so on. Flongsep stands for *full long and separate*.

2. Flong: $m = 0$, $m = 1$, ..., $m = M$ are stored in one dataset with $\_N = N + M \times N$ observations, where N is the number of observations in $m = 0$. Flong stands for *full long*.

3. Mlong: $m = 0$, $m = 1$, ..., $m = M$ are stored in one dataset with $\_N = N + M \times n$ observations, where n is the number of incomplete observations in $m = 0$. Mlong stands for *marginal long*.

4. Wide: $m = 0$, $m = 1$, ..., $m = M$ are stored in one dataset with $\_N = N$ observations. Each imputed and passive variable has M additional variables associated with it. If variable bp contains the values in $m = 0$, then values for $m = 1$ are contained in variable _1_bp, values for $m = 2$ in _2_bp, and so on. Wide stands for *wide*.

See *style* in [MI] **Glossary** and see [MI] **styles** for examples. See [MI] **technical** for programmer's details.

Description

The `mi` suite of commands deals with multiple-imputation data, abbreviated as `mi` data.

In summary,

1. `mi` data may be stored in one of four formats—flongsep, flong, mlong, and wide—known as styles. Descriptions are provided in *Summary of styles* directly above.

2. `mi` data contain M imputations numbered $m = 1, 2, \ldots, M$, and contain $m = 0$, the original data with missing values.

3. Each variable in `mi` data is registered as imputed, passive, or regular, or it is unregistered.

   a. Unregistered variables are mostly treated like regular variables.

   b. Regular variables usually do not contain missing, or if they do, the missing values are not imputed in $m > 0$.

   c. Imputed variables contain missing in $m = 0$, and those values are imputed, or are to be imputed, in $m > 0$.

   d. Passive variables are algebraic combinations of imputed, regular, or other passive variables.

4. If an imputed variable contains a value greater than . in $m = 0$—it contains .a, .b, ..., .z—then that value is considered a hard missing and the missing value persists in $m > 0$.

See [MI] **Glossary** for a more thorough description of terms used throughout this manual.

All `mi` commands are implemented as ado-files.

Remarks

Remarks are presented under the following headings:

> *A simple example*
> *Suggested reading order*

A simple example

We are about to type six commands:

```
. use http://www.stata-press.com/data/r12/mheart5            (1)
. mi set mlong                                               (2)
. mi register imputed age bmi                                (3)
. set seed 29390                                             (4)
. mi impute mvn age bmi = attack smokes hsgrad female, add(10)  (5)
. mi estimate: logistic attack smokes age bmi hsgrad female  (6)
```

The story is that we want to fit

```
. logistic attack smokes age bmi hsgrad female
```

but the `age` and `bmi` variables contain missing values. Fitting the model by typing `logistic ...` would ignore some of the information in our data. Multiple imputation (MI) attempts to recover that information. The method imputes M values to fill in each of the missing values. After that, statistics are performed on the M imputed datasets separately and the results combined. The goal is to obtain better estimates of parameters and their standard errors.

In the solution shown above,

1. We load the data.

2. We set our data for use with mi.

3. We inform mi which variables contain missing values for which we want to impute values.

4. We impute values in command 5; we prefer that our results be reproducible, so we set the random-number seed in command 4. This step is optional.

5. We create $M = 10$ imputations for each missing value in the variables we registered in command 3.

6. We fit the desired model separately on each of the 10 imputed datasets and combine the results.

The results of running the six-command solution are

```
. use http://www.stata-press.com/data/r12/mheart5
(Fictional heart attack data; bmi and age missing)
. mi set mlong
. mi register imputed age bmi
(28 m=0 obs. now marked as incomplete)
. set seed 29390
. mi impute mvn age bmi = attack smokes hsgrad female, add(10)

Performing EM optimization:
note: 12 observations omitted from EM estimation because of all imputation
      variables missing
  observed log likelihood = -651.75868 at iteration 7

Performing MCMC data augmentation ...

Multivariate imputation                     Imputations =        10
Multivariate normal regression                    added =        10
Imputed: m=1 through m=10                        updated =         0

Prior: uniform                               Iterations =      1000
                                                burn-in =       100
                                                between =       100
```

	Observations per m			
Variable	Complete	Incomplete	Imputed	Total
age	142	12	12	154
bmi	126	28	28	154

```
(complete + incomplete = total; imputed is the minimum across m
 of the number of filled-in observations.)
```

```
. mi estimate: logistic attack smokes age bmi hsgrad female
Multiple-imputation estimates          Imputations       =          10
Logistic regression                    Number of obs     =         154
                                       Average RVI       =      0.1031
                                       Largest FMI       =      0.3256
DF adjustment:    Large sample         DF:       min     =       92.90
                                                 avg     =    25990.98
                                                 max     =    77778.66
Model F test:      Equal FMI           F(   5, 3279.8) =        3.27
Within VCE type:         OIM           Prob > F          =      0.0060
```

attack	Coef.	Std. Err.	t	P>|t|	[95% Conf. Interval]
smokes	1.18324	.3605462	3.28	0.001	.4765251	1.889954
age	.0321028	.016145	1.99	0.047	.0004071	.0637984
bmi	.1100667	.0546424	2.01	0.047	.0015561	.2185772
hsgrad	.1413171	.4043884	0.35	0.727	-.6512819	.933916
female	-.0759589	.416927	-0.18	0.855	-.8931367	.7412189
_cons	-5.38815	1.85184	-2.91	0.004	-9.047656	-1.728644

Note that the output from the last command,

```
. mi estimate: logistic attack smokes age bmi hsgrad female
```

reported coefficients rather than odds ratios, which logistic would usually report. That is because the estimation command is not logistic, it is mi estimate, and mi estimate happened to use logistic to obtain results that mi estimate combined into its own estimation results.

mi estimate by default displays coefficients. If we now wanted to see odds ratios, we could type

```
. mi estimate, or
```
(*output showing odds ratios would appear*)

Note carefully: We replay results by typing mi estimate, not by typing logistic. If we had wanted to see the odds ratios from the outset, we would have typed

```
. mi estimate, or: logistic attack smokes age bmi hsgrad female
```

Suggested reading order

The order of suggested reading of this manual is

[MI] **intro substantive**
[MI] **intro**
[MI] **Glossary**
[MI] **workflow**

[MI] **mi set**
[MI] **mi import**
[MI] **mi describe**
[MI] **mi misstable**

[MI] **mi impute**
[MI] **mi estimate**
[MI] **mi estimate postestimation**

[MI] **styles**
[MI] **mi convert**
[MI] **mi update**

[MI] **mi rename**
[MI] **mi copy**
[MI] **mi erase**
[MI] **mi XXXset**

[MI] **mi extract**
[MI] **mi replace0**

[MI] **mi append**
[MI] **mi add**
[MI] **mi merge**
[MI] **mi reshape**
[MI] **mi stsplit**
[MI] **mi varying**

Programmers will want to see [MI] **technical**.

What's new

This section is intended for previous Stata users. If you are new to Stata, you may as well skip it.

1. **Chained equations**, which is to say, fully conditional specifications for imputing missing values given arbitrary patterns for continuous, binary, ordinal, cardinal, or count variables. See [MI] **mi impute chained**.

2. **Four new imputation methods**. You can impute

    1) truncated data,

    2) interval-censored data,

    3) count data, and

    4) overdispersed count data.

    See [MI] **mi impute truncreg**, [MI] **mi impute intreg**, [MI] **mi impute poisson**, and [MI] **mi impute nbreg**.

3. **Conditional imputation** is now supported by all univariate imputation methods, which is to say, you can impute values for variables with restrictions, such as the number of pregnancies being imputed only for females, even if female itself is imputed. See *Conditional imputation* in [MI] **mi impute** and new option `conditional()` in the univariate imputation entries such as [MI] **mi impute regress**.

4. **Panel-data and multilevel models** are now supported by mi estimate. Included are xtcloglog, xtgee, xtlogit, xtmelogit, xtmepoisson, xtmixed, xtnbreg, xtpoisson, xtprobit, xtrc, and xtreg. See [MI] **estimation**.

5. **Linear and nonlinear predictions after MI estimation** using new commands mi predict and mi predictnl. See [MI] **mi predict**.

6. **Imputation by groups**, which is to say, imputations can be made separately for different groups of the data. See new option by() in [MI] **mi impute**.

7. **Imputation by drawing posterior estimates from bootstrapped samples**. See new option bootstrap in the univariate imputation entries such as [MI] **mi impute regress**.

8. **Handling of perfect prediction** during imputation of categorical data using logit, ologit, and mlogit. See *The issue of perfect prediction during imputation of categorical data* in [MI] **mi impute** and see new option augment in [MI] **mi impute logit**, [MI] **mi impute ologit**, and [MI] **mi impute mlogit**.

9. **Faster imputation.** `mi impute` no longer secretly converts to `flongsep` and back again.

10. **mi estimate now supports total.** See [MI] **estimation**.

11. **Monte Carlo jackknife error estimates** obtained by omitting one imputation at a time and reapplying the combination rules. See new option `mcerror` in [MI] **mi estimate**.

12. **Estimation output improved.**

    a. **Implied zero coefficients now shown.** When a coefficient is omitted, it is now shown as being zero and the reason it was omitted—collinearity, base, empty—is shown in the standard-error column. (The word "omitted" is shown if the coefficient was omitted because of collinearity.)

    b. **You can set displayed precision for all values in coefficient tables** using `set cformat`, `set pformat`, and `set sformat`. Or you may use options `cformat()`, `pformat()`, and `sformat()` now allowed on all estimation commands. See [R] **set cformat** and [R] **estimation options**.

    c. **Estimation commands now respect the width of the Results window.** This feature may be turned off by new display option `nolstretch`. See [R] **estimation options**.

    d. **You can now set whether base levels, empty cells, and omitted are shown** using `set showbaselevels`, `set showemptycells`, and `set showomitted`. See [R] **set showbaselevels**.

13. **misstable summarize will now create summary variables** recording the missing-values pattern. See new option `generate()` for `summarize` in [R] **misstable**. Note that `mi misstable` does not have this new option. The new option is useful before data are imputed.

For a complete list of all the new features in Stata 12, see [U] **1.3 What's new**.

Acknowledgments

We thank Jerry (Jerome) Reiter of Duke University, Patrick Royston of the MRC Clinical Trials Unit, and Ian White of the MRC Biostatistics Unit for their comments and assistance in the development of mi. We also thank James Carpenter of the London School of Hygiene and Tropical Medicine and Jonathan Sterne of the University of Bristol for their comments.

Previous and still ongoing work on multiple imputation in Stata influenced the design of mi. For their past and current contributions, we thank Patrick Royston and Ian White again for ice; John Carlin and John Galati, both of the Murdoch Children's Research Institute and University of Melbourne, and Patrick Royston and Ian White (yet again) for mim; John Galati for inorm; and Rodrigo Alfaro of the Banco Central de Chile for mira.

Also see

[MI] **intro substantive** — Introduction to multiple-imputation analysis

[MI] **Glossary**

[MI] **styles** — Dataset styles

[MI] **workflow** — Suggested workflow

[U] **1.3 What's new**

Title

> **estimation** — Estimation commands for use with mi estimate

Description

Multiple-imputation data analysis in Stata is similar to standard data analysis. The standard syntax applies, but you need to remember the following for MI data analysis:

1. The data must be declared as mi data.

   If you already have multiply imputed data (saved in Stata format), use mi import to import it into mi; see [MI] **mi import**.

   If you do not have multiply imputed data, use mi set (see [MI] **mi set**) to declare your original data to be mi data and use mi impute (see [MI] **mi impute**) to fill in missing values.

2. After you have declared mi data, commands such as svyset, stset, and xtset cannot be used. Instead use mi svyset to declare survey data, use mi stset to declare survival data, and use mi xtset to declare panel data. See [MI] **mi XXXset**.

3. Prefix the estimation commands with mi estimate: (see [MI] **mi estimate**).

The following estimation commands support the `mi estimate` prefix.

Command	Entry	Description
Linear regression models		
regress	[R] **regress**	Linear regression
cnsreg	[R] **cnsreg**	Constrained linear regression
mvreg	[R] **mvreg**	Multivariate regression
Binary-response regression models		
logistic	[R] **logistic**	Logistic regression, reporting odds ratios
logit	[R] **logit**	Logistic regression, reporting coefficients
probit	[R] **probit**	Probit regression
cloglog	[R] **cloglog**	Complementary log-log regression
binreg	[R] **binreg**	GLM for the binomial family
Count-response regression models		
poisson	[R] **poisson**	Poisson regression
nbreg	[R] **nbreg**	Negative binomial regression
gnbreg	[R] **nbreg**	Generalized negative binomial regression
Ordinal-response regression models		
ologit	[R] **ologit**	Ordered logistic regression
oprobit	[R] **oprobit**	Ordered probit regression
Categorical-response regression models		
mlogit	[R] **mlogit**	Multinomial (polytomous) logistic regression
mprobit	[R] **mprobit**	Multinomial probit regression
clogit	[R] **clogit**	Conditional (fixed-effects) logistic regression
Quantile regression models		
qreg	[R] **qreg**	Quantile regression
iqreg	[R] **qreg**	Interquantile range regression
sqreg	[R] **qreg**	Simultaneous-quantile regression
bsqreg	[R] **qreg**	Bootstrapped quantile regression
Survival regression models		
stcox	[ST] **stcox**	Cox proportional hazards model
streg	[ST] **streg**	Parametric survival models
stcrreg	[ST] **stcrreg**	Competing-risks regression
Other regression models		
glm	[R] **glm**	Generalized linear models
areg	[R] **areg**	Linear regression with a large dummy-variable set
rreg	[R] **rreg**	Robust regression
truncreg	[R] **truncreg**	Truncated regression
Descriptive statistics		
mean	[R] **mean**	Estimate means
proportion	[R] **proportion**	Estimate proportions
ratio	[R] **ratio**	Estimate ratios
total	[R] **total**	Estimate totals

Panel-data models

xtreg	[XT] **xtreg**	Fixed-, between- and random-effects, and population-averaged linear models
xtmixed	[XT] **xtmixed**	Multilevel mixed-effects linear regression
xtrc	[XT] **xtrc**	Random-coefficients regression
xtlogit	[XT] **xtlogit**	Fixed-effects, random-effects, and population-averaged logit models
xtprobit	[XT] **xtprobit**	Random-effects and population-averaged probit models
xtcloglog	[XT] **xtcloglog**	Random-effects and population-averaged cloglog models
xtpoisson	[XT] **xtpoisson**	Fixed-effects, random-effects, and population-averaged Poisson models
xtnbreg	[XT] **xtnbreg**	Fixed-effects, random-effects, and population-averaged negative binomial models
xtmelogit	[XT] **xtmelogit**	Multilevel mixed-effects logistic regression
xtmepoisson	[XT] **xtmepoisson**	Multilevel mixed-effects Poisson regression
xtgee	[XT] **xtgee**	Fit population-averaged panel-data models by using GEE

Survey regression models

svy:	[SVY] **svy**	Estimation commands for survey data (excluding commands that are not listed above)

Also see

[MI] **mi estimate** — Estimation using multiple imputations

[MI] **mi estimate postestimation** — Postestimation tools for mi estimate

[MI] **mi set** — Declare multiple-imputation data

[MI] **mi import** — Import data into mi

[MI] **mi impute** — Impute missing values

[MI] **workflow** — Suggested workflow

[MI] **intro substantive** — Introduction to multiple-imputation analysis

[MI] **intro** — Introduction to mi

[MI] **Glossary**

Title

---

mi add — Add imputations from another mi dataset

Syntax

mi add *varlist* using *filename* [, *options*]

options	Description
assert(master)	assert all observations found in master
assert(match)	assert all observations found in master and in using
noupdate	see [MI] **noupdate option**

Notes:

1. Jargon:

    match variables = *varlist*, variables on which match performed

    master = data in memory

    using = data on disk (*filename*)

2. Master must be mi set.

3. Using must be mi set.

4. *filename* must be enclosed in double quotes if *filename* contains blanks or other special characters.

Menu

Statistics > Multiple imputation

Description

mi add adds the imputations from the using to the end of the master.

Options

assert(*results*) specifies how observations are expected to match. If results are not as you expect, an error message will be issued and the master data left unchanged.

assert(master) specifies that you expect a match for every observation in the master, although there may be extra observations in the using that mi add is to ignore.

assert(match) specifies that you expect every observation in the master to match an observation in the using and vice versa.

The default is that the master may have observations that are missing from the using and vice versa. Only observations in common are used by mi add.

noupdate in some cases suppresses the automatic mi update this command might perform; see [MI] **noupdate option**.

26

Remarks

Think of the result produced by `mi add` as being

Result	Source
$m = 0$	$m = 0$ from master
$m = 1$	$m = 1$ from master
$m = 2$	$m = 2$ from master
.	.
.	.
.	.
$m = M_{\text{master}}$	$m = M_{\text{master}}$ from master
$m = M_{\text{master}} + 1$	$m = 1$ from using
$m = M_{\text{master}} + 2$	$m = 2$ from using
.	.
.	.
.	.
$m = M_{\text{master}} + M_{\text{using}}$	$m = M_{\text{using}}$ from using

That is, the original data in the master remain unchanged. All that happens is the imputed data from the using are added to the end of the master as additional imputations.

For instance, say you discover that you and a coworker have been working on the same data. You have added $M = 20$ imputations to your data. Your coworker has separately added $M = 17$. To combine the data, type something like

```
. use mydata

. mi add patientid using karensdata
(17 imputations added; M=37)
```

The only thing changed in your data is M. If your coworker's data have additional variables, they are ignored. If your coworker has variables registered differently from how you have them registered, that is ignored. If your coworker has not yet registered as imputed a variable that you have registered as imputed, that is noted in the output. You might see

```
. use mydata

. mi add patientid using karensdata
(17 imputations added; M=37)
(imputed variable grade not found in using data;
    added imputations contain m=0 values for that variable)
```

Saved results

`mi add` saves the following in `r()`:

Scalars
`r(m)`	number of added imputations
`r(unmatched_m)`	number of unmatched master observations
`r(unmatched_u)`	number of unmatched using observations

Macros
`r(imputed_f)`	variables for which imputed found
`r(imputed_nf)`	variables for which imputed not found

Also see

[MI] **intro** — Introduction to mi

[MI] **mi append** — Append mi data

[MI] **mi merge** — Merge mi data

Title

> **mi append** — Append mi data

Syntax

mi append using *filename* $\left[\, , \, options \right]$

options	Description
generate(*newvar*)	create *newvar*; 0 = master, 1 = using
nolabel	do not copy value labels from using
nonotes	do not copy notes from using
force	string $\leftrightarrow$ numeric not type mismatch error
noupdate	see [MI] **noupdate option**

Notes:

1. Jargon:
   master = data in memory
     using = data on disk (*filename*)

2. Master must be mi set; using may be mi set.

3. mi append is syntactically and logically equivalent to append; see [D] **append**. The resulting data have $M = \max(M_{\text{master}}, M_{\text{using}})$, not their sum. See [MI] **mi add** to append imputations holding $m = 0$ constant.

4. mi append syntactically differs from append in that multiple using files may not be specified and the keep(*varlist*) option is not allowed.

5. *filename* must be enclosed in double quotes if *filename* contains blanks or other special characters.

Menu

Statistics > Multiple imputation

Description

mi append is append for mi data; see [D] **append** for a description of appending datasets.

Options

generate(*newvar*) specifies that new variable *newvar* be created containing 0 for observations from the master and 1 for observations from the using.

nolabel prevents copying the value-label definitions from the using data to the master. Even if you do not specify this option, label definitions from the using never replace those of the master.

nonotes prevents any notes in the using from being incorporated into the master; see [D] **notes**.

force allows string variables to be appended to numeric variables and vice versa. The results of such type mismatches are, of course, missing values. Default behavior is to issue an error message rather than append datasets with such violently differing types.

noupdate in some cases suppresses the automatic mi update this command might perform; see [MI] **noupdate option**.

Remarks

Use mi append when you would use append if the data were not mi.

Remarks are presented under the following headings:

> Adding new observations
> Adding new observations and imputations
> Adding new observations and imputations, M unequal
> Treatment of registered variables

Adding new observations

Assume that file mymi.dta contains data on three-quarters of the patients in the ICU. The data are mi set and $M = 5$. File remaining.dta arrives containing the remaining patients. The data are not mi set. To combine the datasets, you type

    . use mymi, clear
    . mi append using remaining

The original mi data had $M = 5$ imputations, and so do the resulting data. If the new data contain no missing values of the imputed variables, you are ready to go. Otherwise, you will need to impute values for the new data.

Adding new observations and imputations

Assume that file westwing.dta contains data on patients in the west wing of the ICU. File eastwing.dta contains data on patients in the east wing of the ICU. Both datasets are mi set with $M = 5$. You originally intended to analyze the datasets separately, but you now wish to combine them. You type

    . use westwing, clear
    . mi append using eastwing

The original data had $M = 5$ imputations, and so do the resulting data.

The data for $m = 0$ are the result of running an ordinary append on the two $m = 0$ datasets.

The data for $m = 1$ are also the result of running an ordinary append, this time on the two $m = 1$ datasets. Thus the result is a combination of observations of westwing.dta and eastwing.dta in the same way that $m = 0$ is. Imputations for observations that previously existed are obtained from westwing.dta, and imputations for the newly appended observations are obtained from eastwing.dta.

Adding new observations and imputations, M unequal

Consider the same situation as above, but this time assume $M = 5$ in westwing.dta and $M = 4$ in eastwing.dta. The combined result will still have $M = 5$. Imputed values in $m = 5$ will be missing for imputed variables.

Treatment of registered variables

It is possible that the two datasets will have variables registered inconsistently.

Variables registered as imputed in either dataset will be registered as imputed in the final result regardless of how they were registered (or unregistered) in the other dataset.

Barring that, variables registered as passive in either dataset will be registered as passive in the final result.

Barring that, variables registered as regular in either dataset will be registered as regular in the final result.

Saved results

mi append saves the following in r():

Scalars
    r(N_master)   number of observations in $m=0$ in master
    r(N_using)   number of observations in $m=0$ in using
    r(M_master)   number of imputations (M) in master
    r(M_using)   number of imputations (M) in using

Macros
    r(newvars)   new variables added

Thus values in the resulting data are

$$N = \text{\# of observations in } m = 0$$
$$= \texttt{r(N\_master) + r(N\_using)}$$

$$k = \text{\# of variables}$$
$$= k\_master + \text{`:word count `r(newvars)''}$$

$$M = \text{\# of imputations}$$
$$= \texttt{max(r(M\_master), r(M\_using))}$$

Also see

[MI] **intro** — Introduction to mi

[D] **append** — Append datasets

[MI] **mi add** — Add imputations from another mi dataset

[MI] **mi merge** — Merge mi data

Title

| **mi convert** — Change style of mi data |

Syntax

mi convert <u>w</u>ide   [, *options*]

mi convert <u>ml</u>ong   [, *options*]

mi convert <u>fl</u>ong   [, *options*]

mi convert <u>flongsep</u> *name*   [, *options*]

options	Description
clear	okay to convert if data not saved
<u>noup</u>date	see [MI] **noupdate option**

Menu

Statistics > Multiple imputation

Description

mi convert converts mi data from one style to another.

Options

clear specifies that it is okay to convert the data even if the data have not been saved to disk since they were last changed.

noupdate in some cases suppresses the automatic mi update this command might perform; see [MI] **noupdate option**.

Remarks

Remarks are presented under the following headings:

> Using mi convert as a convenience tool
> Converting from flongsep
> Converting to flongsep

32

Using mi convert as a convenience tool

Some tasks are easier in one style than another. `mi convert` allows you to switch to the more convenient style. It would not be unreasonable for a snippet of a session to read

```
. mi convert wide
. drop if sex=="male"
. mi convert mlong, clear
. replace age2 = age^2
```

This user is obviously exploiting his or her knowledge of [MI] **styles**. The official way to do the above would be

```
. drop if sex=="male"
. mi update
. mi passive: replace age2 = age^2
```

It does not matter which approach you choose.

Converting from flongsep

If you have flongsep data, it is worth finding out whether you can convert it to one of the other styles. The other styles are more convenient than flongsep, and `mi` commands run faster on them. With your flongsep data in memory, type

```
. mi convert mlong
```

The result will be either success or an insufficient-memory error.

If you wish, you can make a crude guess as to how much memory is required as follows:

1. Use your flongsep data. Type `mi describe`. Write down M, the number of imputations, and write down the number of complete observations, which we will call N, and the number of incomplete observations, which we will call n.

2. With your flongsep data still in memory, type `memory`. Write down the sum of the numbers reported as "data" and "overhead" under the "used" column. We will call this sum S for size.

3. Calculate $T = S + M \times S \times (n/N)$. T is an approximation of the memory your `mi` data would consume in the mlong style. To that, we need to add a bit to account for extra memory used by Stata commands and for variables or observations you might want to add. How much to add is always debatable. For large datasets, add 10% or 5 MB, whichever is smaller.

   For instance, you might have

$$
\begin{aligned}
M &= & 30 \\
N &= & 10{,}000 \\
n &= & 1{,}500 \\
S &= 8{,}040{,}000 = 8 \text{ MB}
\end{aligned}
$$

and thus we would calculate $T = 8 + 30 \times 8 \times (1500/10000) = 44$ MB, to which we would add another 4 or 5 MB, to obtain 48 or 49 MB.

Converting to flongsep

Note that `mi convert`'s syntax for converting to flongsep is

`mi convert flongsep` *name*

You must specify a name, and that name will become the basis for the names of the datasets that comprise the collection of flongsep data. Data for $m = 0$ will be stored in *name*.`dta`; data for $m = 1$, in _1_*name*.`dta`; data for $m = 2$, in _2_*name*.`dta`; and so on. The files will be stored in the current directory; see the `pwd` command in [D] **cd**.

If you are going to use flongsep data, see *Advice for using flongsep* in [MI] **styles**. Also see [MI] **mi copy** and [MI] **mi erase**.

Also see

[MI] **intro** — Introduction to mi

[MI] **styles** — Dataset styles

Title

> **mi copy** — Copy mi flongsep data

Syntax

> mi copy *newname* [, replace]

Menu

Statistics > Multiple imputation

Description

> mi copy *newname* copies flongsep data in memory to *newname* and sets it so that you are working with that copy. *newname* may not be specified with the .dta suffix.
>
> In detail, mi copy *newname* 1) completes saving the flongsep data to its current name if that is necessary; 2) copies the data to *newname*.dta, _1_*newname*.dta, _2_*newname*.dta, ..., _*M*_*newname*.dta; and 3) tells mi that you are now working with *newname*.dta in memory.
>
> mi copy can also be used with wide, mlong, or flong data, although there is no reason you would want to do so. The data are not saved to the original filename as flongsep data would be, but otherwise actions are the same: the data in memory are copied to *newname*.dta, and *newname*.dta is loaded into memory.

Option

> replace specifies that it is okay to overwrite *newname*.dta, _1_*newname*.dta, _2_*newname*.dta, ..., if they already exist.

Remarks

> In Stata, one usually works with a copy of the data in memory. Changes you make to the data are not saved in the underlying disk file until and unless you explicitly save your data. That is not true when working with flongsep data.
>
> Flongsep data are a matched set of datasets, one containing $m = 0$, another containing $m = 1$, and so on. You work with one of them in memory, namely, $m = 0$, but as you work, the other datasets are automatically updated; as you make changes, the datasets on disk change.
>
> Therefore, it is best to work with a copy of your flongsep data and then periodically save the data to the real files, thus mimicking how you work with ordinary Stata datasets. mi copy is for just that purpose. After loading your flongsep data, type, for example,

> . use myflongsep

and immediately make a copy,

> . mi copy *newname*

35

You are now working with the same data but under a new name. Your original data are safe.

When you reach a point where you would ordinarily save your data, whether under the original name or a different one, type

. mi copy *original_name_or_different_name*, replace

. use *newname*

Later, when you are done with *newname*, you can erase it by typing

. mi erase *newname*

Concerning erasure, you will discover that mi erase will not let you erase the files when you have one of the files to be erased in memory. Then you will have to type

. mi erase *newname*, clear

See [MI] **mi erase** for more information.

For more information on flongsep data, see *Advice for using flongsep* in [MI] **styles**.

Also see

[MI] **intro** — Introduction to mi

[MI] **styles** — Dataset styles

[MI] **mi erase** — Erase mi datasets

Title

> **mi describe** — Describe mi data

Syntax

mi query

mi describe [, *describe_options*]

describe_options	Description
detail	show missing-value counts for $m = 1$, $m = 2$, ...
noupdate	see [MI] **noupdate option**

Menu

Statistics > Multiple imputation

Description

mi query reports whether the data in memory are mi data and, if they are, reports the style in which they are set.

mi describe provides a more detailed report on mi data.

Options

detail reports the number of missing values in $m = 1$, $m = 2$, ..., $m = M$ in the imputed and passive variables, along with the number of missing values in $m = 0$.

noupdate in some cases suppresses the automatic mi update this command might perform; see [MI] **noupdate option**.

Remarks

Remarks are presented under the following headings:

> *mi query*
> *mi describe*

mi query

mi query without mi data in memory reports

```
. mi query
(data not mi set)
```

37

With mi data in memory, you see something like

```
. mi query
data mi set wide, M = 15
last mi update 30mar2011 12:46:49, approximately 5 minutes ago
```

mi query does not burden you with unnecessary information. It mentions when mi update was last run because you should run it periodically; see [MI] **mi update**.

mi describe

mi describe more fully describes mi data:

```
. mi describe
Style:  mlong
        last mi update 30mar2011 10:21:07, approximately 2 minutes ago
Obs.:   complete      90
        incomplete    10   (M = 20 imputations)
        ─────────────────
        total         100
Vars.:  imputed:  2; smokes(10) age(5)
        passive: 1; agesq(5)
        regular: 0
        system:  3; _mi_m _mi_id _mi_miss
        (there are 3 unregistered variables; gender race chd)
```

mi describe lists the style of the data, the number of complete and incomplete observations, M (the number of imputations), the registered variables, and the number of missing values in $m = 0$ of the imputed and passive variables. In the output, the line

```
Vars.:  imputed:  2; smokes(10) age(5)
```

means that the smokes variable contains 10 missing values in $m = 0$ and that age contains 5. Those values are soft missings and thus eligible to be imputed. If one of smokes' missing values in $m = 0$ were hard, the line would read

```
Vars.:  imputed:  2; smokes(9+1) age(5)
```

mi describe reports information about $m = 0$. To obtain information about all m's, use mi describe, detail:

```
. mi describe, detail
Style:  mlong
        last mi update 30mar2011 10:36:50, approximately 3 minutes ago
Obs.:   complete      90
        incomplete    10   (M = 20 imputations)
        ─────────────────
        total         100
Vars.:  imputed:  2; smokes(10; 20*0) age(5; 20*0)
        passive: 1; agesq(5; 20*0)
        regular: 0
        system:  3; _mi_m _mi_id _mi_miss
        (there are 3 unregistered variables; gender race chd)
```

In this example, all imputed values are nonmissing. We can see that from

```
Vars.:  imputed:  2; smokes(10; 20*0) age(5; 20*0)
```

Note the 20*0 after the semicolons. That is the number of missing values in $m = 1$, $m = 2$, ..., $m = 20$. In the smokes variable, there are 10 missing values in $m = 0$, then 0 in $m = 1$, then 0 in $m = 2$, and so on. If $m = 17$ had two missing imputed values, the line would read

```
Vars.:  imputed:  2; smokes(10; 16*0, 2, 3*0) age(5; 20*0)
```

16*0, 2, 3*0 means that for $m = 1$, $m = 2$, ..., $m = 20$, the first 16 have 0 missing values, the next has 2, and the last 3 have 0.

If smokes had $9 + 1$ missing values rather than 10—that is, 9 soft missing values plus 1 hard missing rather than all 10 being soft missing—and all 9 soft missings were filled in, the line would read

```
Vars.:  imputed:  2; smokes(9+1; 20*0) age(5; 20*0)
```

The 20 imputations are shown as having no soft missing values. It goes without saying that they have 1 hard missing. Think of 20*0 as meaning 20*(0+1).

If smokes had $9 + 1$ missing values and two of the soft missings in $m = 18$ were still missing, the line would read

```
Vars.:  imputed:  2; smokes(9+1; 16*0, 2, 3*0) age(5; 20*0)
```

Saved results

mi query saves the following in r():

Scalars
r(update)	seconds since last mi update
r(m)	m if r(style)=="flongsep"
r(M)	M if r(style)!="flongsep"

Macros
r(style)	*style*
r(name)	*name* if r(style)=="flongsep"

Note that mi query issues a return code of 0 even if the data are not mi. In that case, r(style) is "".

mi describe saves the following in r():

Scalars
r(update)	seconds since last mi update
r(N)	number of observations in $m=0$
r(N_incomplete)	number of incomplete observations in $m=0$
r(N_complete)	number of complete observations in $m=0$
r(M)	M

Macros
r(style)	*style*
r(ivars)	names of imputed variables
r(_0_miss_ivars)	$\#=.$ in each r(ivars) in $m=0$
r(_0_hard_ivars)	$\#>.$ in each r(ivars) in $m=0$
r(pvars)	names of passive variables
r(_0_miss_pvars)	$\#\geq.$ in each r(pvars) in $m=0$
r(rvars)	names of regular variables

If the detail option is specified, for each m, $m = 1, 2, ..., M$, also saved are

Macros
r(_m_miss_ivars)	$\#=.$ in each r(ivars) in m
r(_m_miss_pvars)	$\#\geq.$ in each r(pvars) in m

Also see

[MI] **intro** — Introduction to mi

Title

> **mi erase** — Erase mi datasets

Syntax

> mi erase *name* [, clear]

Menu

Statistics > Multiple imputation

Description

mi erase erases mi .dta datasets.

Option

clear specifies that it is okay to erase the files even if one of the files is currently in memory. If clear is specified, the data are dropped from memory and the files are erased.

Remarks

Stata's ordinary erase (see [D] **erase**) is not sufficient for erasing mi datasets because an mi dataset might be flongsep, in which case the single name would refer to a collection of files, one containing $m = 0$, another containing $m = 1$, and so on. mi erase deletes all the files associated with mi dataset *name*.dta, which is to say, it erases *name*.dta, _1_*name*.dta, _2_*name*.dta, and so on:

```
. mi erase mysep
(files mysep.dta, _1_mysep.dta _2_mysep.dta _3_mysep.dta erased)
```

Also see

[MI] **intro** — Introduction to mi

[MI] **styles** — Dataset styles

[MI] **mi copy** — Copy mi flongsep data

Title

> **mi estimate** — Estimation using multiple imputations

Syntax

Compute MI estimates of coefficients by fitting estimation command to mi data

mi estimate [, *options*] : *estimation_command* ...

Compute MI estimates of transformed coefficients by fitting estimation command to mi data

mi estimate [*spec*] [, *options*] : *estimation_command* ...

where *spec* may be one or more terms of the form ([*name*:] *exp*). *exp* is any function of the parameter estimates allowed by nlcom; see [R] **nlcom**.

options	Description
Options	
nimputations(#)	specify number of imputations to use; default is to use all existing imputations
imputations(*numlist*)	specify which imputations to use
mcerror	compute Monte Carlo error estimates
ufmitest	perform unrestricted FMI model test
nosmall	do not apply small-sample correction to degrees of freedom
saving(*miestfile*[, replace])	save individual estimation results to *miestfile*.ster
Tables	
[no]citable	suppress/display standard estimation table containing parameter-specific confidence intervals; default is citable
dftable	display degrees-of-freedom table; dftable implies nocitable
vartable	display variance information about estimates; vartable implies citable
table_options	control table output
display_options	control column formats, row spacing, and display of omitted variables and base and empty cells
Reporting	
level(#)	set confidence level; default is level(95)
dots	display dots as estimations are performed
noisily	display any output from *estimation_command* (and from nlcom if transformations specified)
trace	trace *estimation_command* (and nlcom if transformations specified); implies noisily
nogroup	suppress summary about groups displayed for xt commands
xtme_options	control output from mixed-effects commands

Advanced

esample(*newvar*)	store estimation sample in variable *newvar*; available only in the flong and flongsep styles
errorok	allow estimation even when *estimation_command* (or nlcom) errors out; such imputations are discarded from the analysis
esampvaryok	allow estimation when estimation sample varies across imputations
cmdok	allow estimation when *estimation_command* is not one of the supported estimation commands
coeflegend	display legend instead of statistics
nowarning	suppress the warning about varying estimation samples
eform_option	display coefficient table in exponentiated form
post	post estimated coefficients and VCE to e(b) and e(V)
noupdate	do not perform mi update; see [MI] **noupdate option**

You must mi set your data before using mi estimate; see [MI] **mi set**.

coeflegend, nowarning, *eform_option*, post, and noupdate do not appear in the dialog box.

table_options	Description
noheader	suppress table header(s)
notable	suppress table(s)
nocoef	suppress table output related to coefficients
nocmdlegend	suppress command legend that appears in the presence of transformed coefficients when nocoef is used
notrcoef	suppress table output related to transformed coefficients
nolegend	suppress table legend(s)
nocnsreport	do not display constraints

See [MI] **mi estimate postestimation** for features available after estimation. mi estimate is its own estimation command. The postestimation features for mi estimate do not include by default the postestimation features for *estimation_command*. To replay results, type mi estimate without arguments.

Menu

Statistics > Multiple imputation

Description

mi estimate: *estimation_command* runs *estimation_command* on the imputed mi data, and adjusts coefficients and standard errors for the variability between imputations according to the combination rules by Rubin (1987).

Options

    ⌐ Options ⌐

nimputations(#) specifies that the first # imputations be used; # must be $M_{\min} \leq \# \leq M$, where $M_{\min} = 3$ if mcerror is specified and $M_{\min} = 2$, otherwise. The default is to use all imputations, M. Only one of nimputations() or imputations() may be specified.

imputations(*numlist*) specifies which imputations to use. The default is to use all of them. *numlist* must contain at least two numbers. If mcerror is specified, *numlist* must contain at least three numbers. Only one of nimputations() or imputations() may be specified.

mcerror specifies to compute Monte Carlo error (MCE) estimates for the results displayed in the estimation, degrees-of-freedom, and variance-information tables. MCE estimates reflect variability of MI results across repeated uses of the same imputation procedure and are useful for determining an adequate number of imputations to obtain stable MI results; see White, Royston, and Wood (2011) for details and guidelines.

MCE estimates are obtained by applying the jackknife procedure to multiple-imputation results. That is, the jackknife pseudovalues of MI results are obtained by omitting one imputation at a time; see [R] **jackknife** for details about the jackknife procedure. As such, the MCE computation requires at least three imputations.

If level() is specified during estimation, MCE estimates are obtained for confidence intervals using the specified confidence level instead of using the default 95% confidence level. If any of the options described in [R] *eform_option* is specified during estimation, MCE estimates for the coefficients, standard errors, and confidence intervals in the exponentiated form are also computed. mcerror can also be used upon replay to display MCE estimates. Otherwise, MCE estimates are not reported upon replay even if they were previously computed.

ufmitest specifies that the unrestricted fraction missing information (FMI) model test be used. The default test performed assumes equal fractions of information missing due to nonresponse for all coefficients. This is equivalent to the assumption that the between-imputation and within-imputation variances are proportional. The unrestricted test may be preferable when this assumption is suspect provided the number of imputations is large relative to the number of estimated coefficients.

nosmall specifies that no small-sample correction be made to the degrees of freedom. The small-sample correction is made by default to estimation commands that account for small samples. If the command saves residual degrees of freedom in e(df_r), individual tests of coefficients (and transformed coefficients) use the small-sample correction of Barnard and Rubin (1999) and the overall model test uses the small-sample correction of Reiter (2007). If the command does not save residual degrees of freedom, the large-sample test is used and the nosmall option has no effect.

saving(*miestfile*[, replace]) saves estimation results from each model fit in *miestfile*.ster. The replace suboption specifies to overwrite *miestfile*.ster if it exists. *miestfile*.ster can later be used by mi estimate using (see [MI] **mi estimate using**) to obtain MI estimates of coefficients or of transformed coefficients without refitting the completed-data models. This file is written in the format used by estimates use; see [R] **estimates save**.

    ⌐ Tables ⌐

All table options below may be specified at estimation time or when redisplaying previously estimated results. Table options must be specified as options to mi estimate, not to *estimation_command*.

citable and nocitable specify whether the standard estimation table containing parameter-specific confidence intervals is displayed. The default is citable. nocitable can be used with vartable to suppress the confidence-interval table.

dftable displays a table containing parameter-specific degrees of freedom and percentages of increase in standard errors due to nonresponse. dftable implies nocitable.

vartable displays a table reporting variance information about MI estimates. The table contains estimates of within-imputation variances, between-imputation variances, total variances, relative increases in variance due to nonresponse, fractions of information about parameter estimates missing due to nonresponse, and relative efficiencies for using finite M rather than a hypothetically infinite number of imputations. vartable implies citable.

table_options control the appearance of all displayed table output:

noheader suppresses all header information from the output. The table output is still displayed.

notable suppresses all tables from the output. The header information is still displayed.

nocoef suppresses the display of tables containing coefficient estimates. This option affects the table output produced by citable, dftable, and vartable.

nocmdlegend suppresses the table legend showing the specified command line, *estimation_command*, from the output. This legend appears above the tables containing transformed coefficients (or above the variance-information table if vartable is used) when nocoef is specified.

notrcoef suppresses the display of tables containing estimates of transformed coefficients (if specified). This option affects the table output produced by citable, dftable, and vartable.

nolegend suppresses all table legends from the output.

nocnsreport; see [R] **estimation options**.

display_options: noomitted, vsquish, noemptycells, baselevels, allbaselevels, cformat(*%fmt*), pformat(*%fmt*), and sformat(*%fmt*); see [R] **estimation options**.

⌐ Reporting ⌐

Reporting options must be specified as options to mi estimate and not as options to *estimation_command*.

level(#); see [R] **estimation options**.

dots specifies that dots be displayed as estimations are successfully completed. An x is displayed if the *estimation_command* returns an error, if the model fails to converge, or if nlcom fails to estimate one of the transformed coefficients specified in *spec*.

noisily specifies that any output from *estimation_command* and nlcom, used to obtain the estimates of transformed coefficients if transformations are specified, be displayed.

trace traces the execution of *estimation_command* and traces nlcom if transformations are specified. trace implies noisily.

nogroup suppresses the display of group summary information (number of groups, average group size, minimum, and maximum) as well as other command-specific information displayed for xt commands; see the list of commands under *Panel-data models* in [MI] **estimation**.

xtme_options: variance, noretable, nofetable, and estmetric. These options are relevant only with the mixed-effects commands such as xtmixed (see [XT] **xtmixed**), xtmelogit (see [XT] **xtmelogit**), and xtmepoisson (see [XT] **xtmepoisson**). The estmetric option is implied when vartable or dftable is used.

⌐   ⌐Advanced ⌐
---

esample(*newvar*) creates *newvar* containing e(sample). This option is useful to identify which observations were used in the estimation, especially when the estimation sample varies across imputations (see *Potential problems that can arise when using mi estimate* for details). *newvar* is zero in the original data ($m = 0$) and in any imputations ($m > 0$) in which the estimation failed or that were not used in the computation. esample() may be specified only if the data are flong or flongsep; see [MI] **mi convert** to convert to one of those styles. The variable created will be super varying and therefore must not be registered; see [MI] **mi varying** for more explanation. The saved estimation sample *newvar* may be used later with mi extract (see [MI] **mi extract**) to set the estimation sample.

errorok specifies that estimations that fail be skipped and the combined results be based on the successful individual estimation results. The default is that mi estimate stops if an individual estimation fails. If errorok is specified with saving(), all estimation results, including failed, are saved to a file.

esampvaryok allows estimation to continue even if the estimation sample varies across imputations. mi estimate stops if the estimation sample varies. If esampvaryok is specified, results from all imputations are used to compute MI estimates and a warning message is displayed at the bottom of the table. Also see the esample() option. See *Potential problems that can arise when using mi estimate* for more information.

cmdok allows unsupported estimation commands to be used with mi estimate; see [MI] **estimation** for a list of supported estimation commands. Alternatively, if you want mi estimate to work with your estimation command, add the property mi to the program properties; see [P] **program properties**.

The following options are available with mi estimate but are not shown in the dialog box:

coeflegend; see [R] **estimation options**. coeflegend implies nocitable and cannot be combined with citable or dftable.

nowarning suppresses the warning message at the bottom of table output that occurs if the estimation sample varies and esampvaryok is specified. See *Potential problems that can arise when using mi estimate* for details.

eform_option; see [R] **eform_option**. Regardless of the *estimation_command* specified, mi estimate reports results in the coefficient metric under which the combination rules are applied. You may use the appropriate *eform_option* to redisplay results in exponentiated form, if desired. If dftable is also specified, the reported degrees of freedom and percentage increases in standard errors are not adjusted and correspond to the original coefficient metric.

post requests that MI estimates of coefficients and their respective VCEs be posted in the usual way. This allows the use of *estimation_command*-specific postestimation tools with MI estimates. There are issues; see *Using the command-specific postestimation tools* in [MI] **mi estimate postestimation**. post may be specified at estimation time or when redisplaying previously estimated results.

noupdate in some cases suppresses the automatic mi update this command might perform; see [MI] **noupdate option**. This option is seldom used.

Remarks

mi estimate requires that imputations be already formed; see [MI] **mi impute**. To import existing multiply imputed data, see [MI] **mi import**.

Remarks are presented under the following headings:

Using mi estimate
Example 1: Completed-data logistic analysis
Example 2: Completed-data linear regression analysis
Example 3: Completed-data survival analysis
Example 4: Panel data and multilevel models
Example 5: Estimating transformations
Example 6: Monte Carlo error estimates
Potential problems that can arise when using mi estimate

Using mi estimate

`mi estimate` estimates model parameters from multiply imputed data and adjusts coefficients and standard errors for the variability between imputations. It runs the specified *estimation_command* on each of the M imputed datasets to obtain the M completed-data estimates of coefficients and their VCEs. It then computes MI estimates of coefficients and standard errors by applying combination rules (Rubin 1987, 77) to the M completed-data estimates. See [MI] **intro substantive** for a discussion of MI analysis and see *Methods and formulas* for computational details.

To use `mi estimate`, your data must contain at least two imputations. The basic syntax of `mi estimate` is

. mi estimate: *estimation_command* ...

estimation_command is any estimation command from the list of supported estimation commands; see [MI] **estimation**.

If you wish to estimate on survey data, type

. mi estimate: svy: *estimation_command* ...

If you want to vary the number of imputations or select which imputations to use in the computations, use the `nimputations()` or the `imputations()` option, respectively.

. mi estimate, nimputations(9): *estimation_command* ...

Doing so is useful to evaluate the stability of MI results. MCE estimates of the parameters are also useful for determining the stability of MI results. You can use the `mcerror` option to obtain these estimates. Your data must contain at least three imputations to use `mcerror`.

You can obtain more-detailed information about imputation results by specifying the `dftable` and `vartable` options.

You can additionally obtain estimates of transformed coefficients by specifying expressions with `mi estimate`; see *Example 5: Estimating transformations* for details.

When using `mi estimate`, keep in mind that

1. `mi estimate` is its own estimation command.

2. `mi estimate` uses different degrees of freedom for each estimated parameter when computing its significance level and confidence interval.

3. `mi estimate` reports results in the coefficient metric under which combination rules are applied regardless of the default reporting metric of the specified *estimation_command*. Use *eform_option* with `mi estimate` to report results in the exponentiated metric, if you wish.

4. `mi estimate` has its own reporting options and does not respect command-specific reporting options. The reporting options specified with *estimation_command* affect only the output of the command that is displayed when `mi estimate`'s `noisily` option is specified. Specify `mi estimate`'s options immediately after the `mi estimate` command:

. mi estimate, *options*: *estimation_command* ...

Example 1: Completed-data logistic analysis

Recall the logistic analysis of the heart attack data from [MI] **intro substantive**. The goal of the analysis was to explore the relationship between heart attacks and smoking adjusted for other factors such as age, body mass index (BMI), gender, and educational status. The original data contain missing values of BMI. The listwise-deletion analysis on the original data determined that smoking and BMI have significant impact on a heart attack. After imputing missing values of BMI, age was determined to be a significant factor as well. See *A brief introduction to MI using Stata* in [MI] **intro substantive** for details. The data we used are stored in mheart1s20.dta.

Below we refit the logistic model using the imputed data. We also specify the dots option so that dots will be displayed as estimations are completed.

```
. use http://www.stata-press.com/data/r12/mheart1s20
(Fictional heart attack data; bmi missing)

. mi estimate, dots: logit attack smokes age bmi hsgrad female

Imputations (20):
.........10.........20 done

Multiple-imputation estimates          Imputations       =         20
Logistic regression                    Number of obs     =        154
                                        Average RVI       =     0.0312
                                        Largest FMI       =     0.1355
DF adjustment:    Large sample          DF:      min      =    1060.38
                                                 avg      =  223362.56
                                                 max      =  493335.88
Model F test:       Equal FMI           F(   5,71379.3) =       3.59
Within VCE type:          OIM           Prob > F          =     0.0030
```

attack	Coef.	Std. Err.	t	P>|t|	[95% Conf. Interval]	
smokes	1.198595	.3578195	3.35	0.001	.4972789	1.899911
age	.0360159	.0154399	2.33	0.020	.0057541	.0662776
bmi	.1039416	.0476136	2.18	0.029	.010514	.1973692
hsgrad	.1578992	.4049257	0.39	0.697	-.6357464	.9515449
female	-.1067433	.4164735	-0.26	0.798	-.9230191	.7095326
_cons	-5.478143	1.685075	-3.25	0.001	-8.782394	-2.173892

The left header column reports information about the fitted MI model. The right header column reports the number of imputations and the number of observations used, the average relative variance increase (RVI) due to nonresponse, the largest fraction of missing information (FMI), a summary about parameter-specific degrees of freedom (DF), and the overall model test that all coefficients, excluding the constant, are equal to zero.

Notice first that mi estimate reports Student's t and F statistics for inference although logit would usually report Z and χ^2 statistics.

mi estimate: logit is not logit. mi estimate uses Rubin's combination rules to obtain the estimates from multiply imputed data. The variability of the MI estimates consists of two components: variability within imputations and variability between imputations. Therefore, the precision of the MI estimates is governed not only by the number of observations in the sample but also by the number of imputations. As such, even if the number of observations is large, if the number of imputations is small and the FMI are not low, the reference distribution used for inference will deviate from the normal distribution. Because in practice the number of imputations tends to be small, mi estimate uses a reference t distribution.

Returning to the output, average RVI reports the average relative increase (averaged over all coefficients) in variance of the estimates because of the missing bmi values. A relative variance increase is an increase in the variance of the estimate because of the loss of information about the parameter due to nonresponse relative to the variance of the estimate with no information lost. The closer this number is to zero, the less effect missing data have on the variance of the estimate. Note that the reported RVI will be zero if you use `mi estimate` with the complete data or with missing data that have not been imputed. In our case, average RVI is small: 0.0312.

`Largest FMI` reports the largest of all the FMI about coefficient estimates due to nonresponse. This number can be used to get an idea of whether the specified number of imputations is sufficient for the analysis. A rule of thumb is that $M \geq 100 \times$ FMI provides an adequate level of reproducibility of MI analysis. In our example, the largest FMI is 0.14 and the number of imputations, 20, exceeds the required number of imputations: 14 ($= 100 \times 0.14$) according to this rule.

The coefficient-specific degrees of freedom (DF) averaging 223,363 are large. They are large because the MI degrees of freedom depends not only on the number of imputations but also on the RVI due to nonresponse. Specifically, the degrees of freedom is inversely related to RVI. The closer RVI is to zero, the larger the degrees of freedom regardless of the number of imputations.

To the left of the DF, we see that the degrees of freedom is obtained under a large-sample assumption. The alternative is to use a small-sample adjustment. Whether the small-sample adjustment is applied is determined by the type of the reference distribution used for inference by the specified estimation command. For the commands that use a large-sample (normal) approximation for inference, a large-sample approximation is used when computing the MI degrees of freedom. For the commands that use a small-sample (Student's t) approximation for inference, a small-sample approximation is used when computing the MI degrees of freedom. See *Methods and formulas* for details. As we already mentioned, `logit` assumes large samples for inference, and thus the MI degrees of freedom is computed assuming a large sample.

The model F test rejects the hypothesis that all coefficients are equal to zero and thus rules out a constant-only model for heart attacks. By default, the model test uses the assumption that the fractions of missing information of all coefficients are equal (as noted by `Equal FMI` to the left). Although this assumption may not be supported by the data, it is used to circumvent the difficulties arising with the estimation of the between-imputation variance matrix based on a small number of imputations. See *Methods and formulas* and [MI] **mi test** for details.

`mi estimate` also reports the type of variance estimation used by the estimation command to compute variance estimates in the individual completed-data analysis. These completed-data variance estimates are then used to compute the within-imputation variance. In our example, the observed-information-matrix (OIM) method, the default variance-estimation method used by maximum likelihood estimation, is used to compute completed-data VCEs. This is labeled as `Within VCE type: OIM` in the output.

Finally, `mi estimate` reports a coefficient table containing the combined estimates. Unlike all other Stata estimation commands, the reported significance levels and confidence intervals in this table are based on degrees of freedom that is specific to each coefficient. Remember that the degrees of freedom depends on the relative variance increases and thus on how much information is lost about the estimated parameter because of missing data. How much information is lost is specific to each parameter and so is the degrees of freedom.

As we already saw, a summary of the coefficient-specific degrees of freedom (minimum, average, and maximum) was reported in the header. We can obtain a table containing coefficient-specific degrees of freedom by replaying the results with the `dftable` option:

```
. mi estimate, dftable
```

| Multiple-imputation estimates | | | | | | Imputations | = | 20 |
| Logistic regression | | | | | | Number of obs | = | 154 |

Multiple-imputation estimates   Imputations   =   20
Logistic regression   Number of obs =   154
                                                   Average RVI   =   0.0312
                                                   Largest FMI   =   0.1355
DF adjustment: Large sample   DF: min   =   1060.38
                                                           avg   = 223362.56
                                                           max   = 493335.88
Model F test: Equal FMI   F(5,71379.3) =   3.59
Within VCE type: OIM   Prob > F   =   0.0030

attack	Coef.	Std. Err.	t	P>|t|	DF	% Increase Std. Err.
smokes	1.198595	.3578195	3.35	0.001	320019.4	0.39
age	.0360159	.0154399	2.33	0.020	493335.9	0.31
bmi	.1039416	.0476136	2.18	0.029	1060.4	7.45
hsgrad	.1578992	.4049257	0.39	0.697	165126.7	0.54
female	-.1067433	.4164735	-0.26	0.798	358078.3	0.37
_cons	-5.478143	1.685075	-3.25	0.001	2554.8	4.61

Notice that we type `mi estimate` to replay the results, not `logit`.

The header information remains the same. In particular, degrees of freedom ranges from 1,060 to 493,336 and averages 223,363. In the table output, the columns for the confidence intervals are replaced with the `DF` and `% Increase Std. Err.` columns. We now see that the smallest degrees of freedom corresponds to the coefficient for `bmi`. We should have anticipated this because `bmi` is the only variable containing missing values in this example. The largest degrees of freedom is observed for the coefficient for `age`, which suggests that the loss of information due to nonresponse is the smallest for the estimation of this coefficient.

The last column displays as a percentage the increase in standard errors of the parameters due to nonresponse. We observe a 7% increase in the standard error for the coefficient of `bmi` and a 4% increase in the standard error for the constant. Increases in standard errors of other coefficients are negligible.

In this example, we displayed a degrees-of-freedom table on replay by specifying the `dftable` option. We could also obtain this table if we specified this option at estimation time. Alternatively, if desired, we could display both tables by specifying the `citable` and `dftable` options together.

We can obtain more detail about imputation results by specifying the `vartable` option. We specify this option on replay and also use the `nocitable` option to suppress the default confidence-interval table:

```
. mi estimate, vartable nocitable
```

Multiple-imputation estimates   Imputations   =   20
Logistic regression
Variance information

| | Imputation variance | | | | | Relative |
attack	Within	Between	Total	RVI	FMI	efficiency
smokes	.127048	.00094	.128035	.007765	.007711	.999615
age	.000237	1.4e-06	.000238	.006245	.00621	.99969
bmi	.001964	.000289	.002267	.154545	.135487	.993271
hsgrad	.162206	.001675	.163965	.010843	.010739	.999463
female	.172187	.001203	.17345	.007338	.00729	.999636
_cons	2.5946	.233211	2.83948	.094377	.086953	.995671

The first three columns of the table provide the variance information specific to each parameter. As we already discussed, MI variance contains two sources of variation: within imputation and between imputation. The first two columns provide estimates for the within-imputation and between-imputation variances. The third column is a total variance that is the sum of the two variances plus an adjustment for using a finite number of imputations. The next two columns are individual RVIs and fractions of missing information (FMIs) due to nonresponse. The last column records relative efficiencies for using a finite number of imputations (20 in our example) versus the theoretically optimal infinite number of imputations.

We notice that the coefficient for age has the smallest within-imputation and between-imputation variances. The between-imputation variability is very small relative to the within-imputation variability, which is why age had such a large estimate of the degrees of freedom we observed earlier. Correspondingly, this coefficient has the smallest values for RVI and FMI. As expected, the coefficient for bmi has the highest RVI and FMI.

The reported relative efficiencies are high for all coefficient estimates, with the smallest relative efficiency, again, corresponding to bmi. These estimates, however, are only approximations and thus should not be used exclusively to determine the required number of imputations. See Royston, Carlin, and White (2009) and White, Royston, and Wood (2011) for other ways of determining a suitable number of imputations.

Example 2: Completed-data linear regression analysis

Recall the data on house resale prices from example 3 of [MI] **mi impute mvn**. We use the imputed data stored in mhouses1993s30.dta to examine the relationship of various predictors on price via linear regression:

```
. use http://www.stata-press.com/data/r12/mhouses1993s30
(Albuquerque Home Prices Feb15-Apr30, 1993)

. mi estimate, ni(5): regress price tax sqft age nfeatures ne custom corner
```

Multiple-imputation estimates				Imputations	=	5
Linear regression				Number of obs	=	117
				Average RVI	=	0.0685
				Largest FMI	=	0.2075
				Complete DF	=	109
DF adjustment:	Small sample			DF: min	=	48.59
				avg	=	85.22
				max	=	104.79
Model F test:	Equal FMI			F(7, 103.9)	=	67.50
Within VCE type:	OLS			Prob > F	=	0.0000

price	Coef.	Std. Err.	t	P>|t|	[95% Conf. Interval]	
tax	.6631356	.122443	5.42	0.000	.4195447	.9067265
sqft	.2185884	.0670182	3.26	0.002	.0856051	.3515718
age	-.0395402	1.613185	-0.02	0.981	-3.28205	3.202969
nfeatures	8.735622	13.42251	0.65	0.517	-18.01198	35.48323
ne	4.069381	36.94491	0.11	0.913	-69.4355	77.57426
custom	130.4925	42.93286	3.04	0.003	45.36257	215.6225
corner	-71.25406	40.06697	-1.78	0.078	-150.7152	8.207084
_cons	130.2002	70.38012	1.85	0.068	-9.624642	270.025

By default, mi estimate uses all available imputations in the analysis. For the purpose of illustration, we use only the first 5 imputations out of the available 30 by specifying the nimputations(5) option, which we abbreviated as ni(5).

Compared with the output from the previous example, an additional result, Complete DF, is reported. Also notice that the adjustment for the degrees of freedom is now labeled as Small sample. Remember that mi estimate determines what adjustment to use based on the reference distribution used for inference by the specified estimation command.

regress uses a reference t distribution with $117 - 8 = 109$ residual degrees of freedom. Thus a small-sample adjustment is used by mi estimate for the MI degrees of freedom.

Complete DF contains the degrees of freedom used for inference with complete data. It corresponds to the completed-data residual degrees of freedom saved by the command in e(df_r). In most applications, the completed-data residual degrees of freedom will be the same, and so Complete DF will correspond to the complete degrees of freedom, the degrees of freedom that would have been used for inference if the data were complete. In the case when the completed-data residual degrees of freedom varies across imputations (as may happen when the estimation sample varies; see *Potential problems that can arise when using mi estimate*), Complete DF reports the smallest of them.

In our example, all completed-data residual degrees of freedom are equal, and Complete DF is equal to 109, the completed-data residual degrees of freedom obtained from regress. mi estimate uses the complete degrees of freedom to adjust the MI degrees of freedom for a small sample (Barnard and Rubin 1999).

Example 3: Completed-data survival analysis

Consider survival data on 48 participants in a cancer drug trial. The dataset contains information about participants' ages, treatments received (drug or placebo), times to death measured in months, and a censoring indicator. The data are described in more detail in *Cox regression with censored data* of [ST] **stcox**. We consider a version of these data containing missing values for age. The imputed data are saved in mdrugtrs25.dta:

```
. use http://www.stata-press.com/data/r12/mdrugtrs25
(Patient Survival in Drug Trial)

. mi describe
  Style:  mlong
          last mi update 30mar2011 12:46:48, 1 day ago
  Obs.:   complete           40
          incomplete          8   (M = 25 imputations)
          _____
          total              48
  Vars.:  imputed:  1; age(8)
          passive:  0
          regular:  3; studytime died drug
          system:   3; _mi_m _mi_id _mi_miss
          (there are no unregistered variables)
```

The dataset contains 25 imputations for 8 missing values of age. Missing values were imputed following guidelines in White and Royston (2009).

We analyze these data using stcox with mi estimate. These data have not yet been stset, so we use mi stset (see [MI] **mi XXXset**) to set them and then perform the analysis using mi estimate: stcox:

```
. mi stset studytime, failure(died)

     failure event:  died != 0 & died < .
obs. time interval:  (0, studytime]
 exit on or before:  failure


      48  total obs.
       0  exclusions


      48  obs. remaining, representing
      31  failures in single record/single failure data
     744  total analysis time at risk, at risk from t =          0
                          earliest observed entry t =            0
                              last observed exit t =            39

. mi estimate, dots: stcox drug age

Imputations (25):
    .........10.........20..... done

Multiple-imputation estimates              Imputations     =          25
Cox regression: Breslow method for ties    Number of obs   =          48
                                           Average RVI     =      0.1059
                                           Largest FMI     =      0.1567
DF adjustment:   Large sample              DF:      min    =      998.63
                                                    avg    =    11621.53
                                                    max    =    22244.42
Model F test:       Equal FMI              F(   2, 4448.6) =       13.43
Within VCE type:        OIM                Prob > F        =      0.0000
```

_t	Coef.	Std. Err.	t	P>|t|	[95% Conf. Interval]	
drug	-2.204572	.4589047	-4.80	0.000	-3.104057	-1.305086
age	.1242711	.040261	3.09	0.002	.0452652	.2032771

Notice that `mi estimate` displays the results in the coefficient metric and not in the hazard-ratio metric. By default, `mi estimate` reports results in the metric under which the combination rules were applied. To obtain the results as hazard ratios, we can use the `hr` option with `mi estimate`:

```
. mi estimate, hr
Multiple-imputation estimates              Imputations     =          25
Cox regression: Breslow method for ties    Number of obs   =          48
                                           Average RVI     =      0.1059
                                           Largest FMI     =      0.1567
DF adjustment:   Large sample              DF:      min    =      998.63
                                                    avg    =    11621.53
                                                    max    =    22244.42
Model F test:       Equal FMI              F(   2, 4448.6) =       13.43
Within VCE type:        OIM                Prob > F        =      0.0000
```

_t	Haz. Ratio	Std. Err.	t	P>|t|	[95% Conf. Interval]	
drug	.1102977	.0506161	-4.80	0.000	.0448668	.2711491
age	1.132323	.0455885	3.09	0.002	1.046305	1.225412

We obtain results similar to those from the corresponding example in [ST] **stcox**.

We specified the `hr` option above on replay. We can also specify it at estimation time:

```
. mi estimate, hr: stcox drug age
  (output omitted )
```

Notice that the `hr` option must be specified with `mi estimate` to obtain hazard ratios. Specifying it with the command itself,

```
. mi estimate: stcox drug age, hr
  (output omitted )
```

will not affect the output from `mi estimate` but only that of the command, `stcox`. You see `stcox`'s output only if you specify `mi estimate`'s `noisily` option.

See Cleves et al. (2010, sec. 9.6) for more information on Cox regression with multiply imputed data.

Example 4: Panel data and multilevel models

We have data on the math scores of students in their third and fifth years of education. There are 887 students from 48 schools in inner London; see Mortimore et al. (1988) for more information on the study. We would like to fit a random-effects model to the fifth-year score, `math5`, on the third-year score, `math3`, using a random effect at the school level.

We created a version of the data that contains missing values for `math3` and then performed imputation following guidelines from the Stata FAQ "How can I account for clustering when creating imputations with mi impute?"; see http://www.stata.com/support/faqs/stat/impute_cluster.html. The resulting imputed data are saved in `mjsps5.dta`:

```
. use http://www.stata-press.com/data/r12/mjsps5, clear
(LEA Junior School Project data (Mortimore et al., 1988) with missing values)
. mi describe
  Style:  mlong
          last mi update 30mar2011 12:46:49, 1 day ago
  Obs.:   complete          705
          incomplete        182   (M = 5 imputations)
          ─────────────────────────
          total             887
  Vars.:  imputed:  1; math3(182)
          passive:  0
          regular:  2; school math5
          system:   3; _mi_m _mi_id _mi_miss
          (there are no unregistered variables)
```

There are five imputations for 182 missing values of the third-year score, `math3`. Variable `math3` is an imputed variable, whereas variable `math5` and variable `school`, recording school identifiers, are complete and are registered as regular.

Our random-effects model includes only a random intercept, the school effect, so we can use the `xtreg` command, or more specifically `mi estimate: xtreg`, for our primary analysis.

Without imputed data, to use `xtreg` or any other panel-data command, we must first declare data to be panel (xt) data by using `xtset`. With imputed data, we should use the `mi xtset` command instead. We declare `school` as our panel variable:

```
. mi xtset school
       panel variable:  school (unbalanced)
```

Next we use `mi estimate: xtreg` to regress the fifth-year math score on the third-year score.

```
. mi estimate: xtreg math5 math3
Multiple-imputation estimates            Imputations      =          5
Random-effects GLS regression            Number of obs    =        887

Group variable: school                   Number of groups =         48
                                         Obs per group: min =         5
                                                        avg =      18.5
                                                        max =        62

                                         Average RVI      =     0.0595
                                         Largest FMI      =     0.1071
DF adjustment:    Large sample           DF:    min       =     381.40
                                                avg       =   85771.71
                                                max       =  171162.01
Model F test:         Equal FMI          F(   1,  381.4)  =     305.71
Within VCE type: Conventional            Prob > F         =     0.0000

─────────────┬─────────────────────────────────────────────────────────
       math5 │     Coef.    Std. Err.      t    P>|t|    [95% Conf. Interval]
─────────────┼─────────────────────────────────────────────────────────
       math3 │   .6101277   .0348951    17.48   0.000    .5415168   .6787385
       _cons │   30.48295   .3576417    85.23   0.000    29.78198   31.18392
─────────────┼─────────────────────────────────────────────────────────
     sigma_u │  2.0684286
     sigma_e │  5.3206673
         rho │  .13128791    (fraction of variance due to u_i)
─────────────┴─────────────────────────────────────────────────────────
Note: sigma_u and sigma_e are combined in the original metric.
```

Third-year math scores are positively associated with fifth-year math scores. Because we use a random-effects model, the coefficient on `math3` is for comparison of students from the same school or from different schools.

In the above results, multiple-imputation estimates of variance components `sigma_u` and `sigma_e` are obtained by applying Rubin's combination rules to the completed-data estimates in the original, standard-deviation metric.

Alternatively, we can use the `xtmixed` command to fit our two-level random-effects model and to obtain variance-component estimates of the school effect. `xtmixed` can be used to fit more complicated multilevel models; see [XT] **xtmixed** for details.

We fit a two-level linear model with `mi estimate: xtmixed` and specify `school` as our second-level variable. `xtmixed` does not require prior declaration of the data, so we do not need to use `mi xtset` with `mi estimate: xtmixed`:

```
. mi estimate: xtmixed math5 math3 || school: , reml
```

Multiple-imputation estimates   Imputations   =   5
Mixed-effects REML regression   Number of obs   =   887

Group variable: school   Number of groups   =   48
                                               Obs per group: min =   5
                                                              avg =   18.5
                                                              max =   62

                                               Average RVI   =   0.0574
                                               Largest FMI   =   0.1079
DF adjustment: Large sample   DF: min   =   376.05
                                                         avg   =   44112.02
                                                         max   = 167428.86
Model F test: Equal FMI   F(1,   376.0)   =   305.41
                                               Prob > F   =   0.0000

math5	Coef.	Std. Err.	t	P>|t|	[95% Conf. Interval]	
math3	.6100335	.0349069	17.48	0.000	.5413963	.6786708
_cons	30.48217	.3536049	86.20	0.000	29.78911	31.17522

Random-effects Parameters	Estimate	Std. Err.	[95% Conf. Interval]	
school: Identity				
sd(_cons)	2.033826	.3069989	1.512894	2.734129
sd(Residual)	5.321503	.1355669	5.061821	5.594508

The estimated coefficients, error standard deviations, and other statistics are similar to those from mi estimate: xtreg. Unlike mi estimate: xtreg, the mi estimate: xtmixed command combines variance components in the estimation metric described in [XT] **xtmixed** and then back-transforms the estimates to display results in the original metric. In our example, the reported standard deviations are exponentiated multiple-imputation estimates of the log-standard deviations.

The random-effects parameters are displayed as standard deviations. We can display variances instead by replaying the mi estimate command with the variance option:

```
. mi estimate, variance
Multiple-imputation estimates              Imputations      =          5
Mixed-effects REML regression              Number of obs    =        887

Group variable: school                     Number of groups =         48
                                           Obs per group: min =         5
                                                          avg =       18.5
                                                          max =         62

                                           Average RVI      =     0.0574
                                           Largest FMI      =     0.1079
DF adjustment:    Large sample             DF:     min      =     376.05
                                                   avg      =   44112.02
                                                   max      =  167428.86
Model F test:       Equal FMI              F(   1,   376.0) =     305.41
                                           Prob > F         =     0.0000
```

math5	Coef.	Std. Err.	t	P>|t|	[95% Conf. Interval]	
math3	.6100335	.0349069	17.48	0.000	.5413963	.6786708
_cons	30.48217	.3536049	86.20	0.000	29.78911	31.17522

Random-effects Parameters	Estimate	Std. Err.	[95% Conf. Interval]	
school: Identity				
var(_cons)	4.136447	1.248765	2.288848	7.475462
var(Residual)	28.3184	1.442839	25.62204	31.29852

Although the random-effects parameters are now displayed as variances, they are still combined and stored in the log–standard-deviation metric.

To obtain variance components, we should specify the variance option with mi estimate,

```
. mi estimate, variance: xtmixed ...
```

and not with xtmixed:

```
. mi estimate: xtmixed ..., variance
```

Example 5: Estimating transformations

Stata estimation commands usually support lincom and nlcom (see [R] **lincom** and [R] **nlcom**) to obtain estimates of the transformed coefficients after estimation by using the delta method. Because MI estimates based on a small number of imputations may not yield a valid VCE, this approach is not generally viable. Also, transformations applied to the combined coefficients are only asymptotically equivalent to the combined transformed coefficients. With a small number of imputations, these two ways of obtaining transformed coefficients can differ significantly.

Thus mi estimate provides its own way of combining transformed coefficients. You need to use mi estimate's method for both linear and nonlinear combinations of coefficients. We are about to demonstrate how to use the method using the ratio of coefficients as an example, but what we are about to do would be equally necessary if we wanted to obtain the difference in two coefficients.

For the purpose of illustration, suppose that we want to estimate the ratio of the coefficients, say, age and `sqft` from example 2. We can do this by typing

```
. use http://www.stata-press.com/data/r12/mhouses1993s30
(Albuquerque Home Prices Feb15-Apr30, 1993)
. mi estimate (ratio: _b[age]/_b[sqft]):
> regress price tax sqft age nfeatures ne custom corner
```

Multiple-imputation estimates		Imputations	=	30
Linear regression		Number of obs	=	117
		Average RVI	=	0.0648
		Largest FMI	=	0.2533
		Complete DF	=	109
DF adjustment: Small sample		DF: min	=	69.12
		avg	=	94.02
		max	=	105.51
Model F test: Equal FMI		F(7, 106.5) =		67.18
Within VCE type: OLS		Prob > F	=	0.0000

price	Coef.	Std. Err.	t	P>|t|	[95% Conf. Interval]	
tax	.6768015	.1241568	5.45	0.000	.4301777	.9234253
sqft	.2118129	.069177	3.06	0.003	.0745091	.3491168
age	.2471445	1.653669	0.15	0.882	-3.051732	3.546021
nfeatures	9.288033	13.30469	0.70	0.487	-17.12017	35.69623
ne	2.518996	36.99365	0.07	0.946	-70.90416	75.94215
custom	134.2193	43.29755	3.10	0.002	48.35674	220.0818
corner	-68.58686	39.9488	-1.72	0.089	-147.7934	10.61972
_cons	123.9118	71.05816	1.74	0.085	-17.19932	265.0229

Transformations		Average RVI	=	0.2899
		Largest FMI	=	0.2316
		Complete DF	=	109
DF adjustment: Small sample		DF: min	=	72.51
		avg	=	72.51
Within VCE type: OLS		max	=	72.51
ratio: _b[age]/_b[sqft]				

price	Coef.	Std. Err.	t	P>|t|	[95% Conf. Interval]	
ratio	1.44401	8.217266	0.18	0.861	-14.93485	17.82287

We use the `nlcom` syntax to specify the transformation: `(ratio: _b[age]/_b[sqft])` defines the transformation and its name is `ratio`. All transformations must be specified following `mi estimate` and before the colon, and must be bound in parentheses.

A separate table containing the estimate of the ratio is displayed following the estimates of coefficients. If desired, we can suppress the table containing the estimates of coefficients by specifying the `nocoef` option. The header reports the average RVI due to nonresponse, the largest FMI, and the summaries of the degrees of freedom specific to the estimated transformations. Because we specified only one transformation, the minimum, average, and maximum degrees of freedom are the same. They correspond to the individual degrees of freedom for `ratio`.

See [MI] **mi test** for an example of linear transformation.

Example 6: Monte Carlo error estimates

Multiple imputation is a stochastic procedure. Each time we reimpute our data, we get different sets of imputations because of the randomness of the imputation step, and therefore we get different multiple-imputation estimates. However, we want to be able to reproduce MI results. Of course, we can always set the random-number seed to ensure reproducibility by obtaining the same imputed values. However, what if we use a different seed? Would we not want our results to be similar regardless of what seed we use? This leads us to a notion we call statistical reproducibility—we want results to be similar across repeated uses of the same imputation procedure; that is, we want to minimize the simulation error associated with our results.

To assess the level of simulation error, White, Royston, and Wood (2011) propose to use a Monte Carlo error of the MI results, defined as the standard deviation of the results across repeated runs of the same imputation procedure using the same data. The authors suggest evaluating Monte Carlo error estimates not only for parameter estimates but also for other statistics, including p-values and confidence intervals, as well as MI statistics including RVI and FMI.

Clearly, as the number of imputations increases, the simulation error decreases. Consider the total MI variance $T = \overline{U} + B + B/M$ of a single parameter, where $\overline{U}$ is the within-imputation variance and B is the between-imputation variance; see *Methods and formulas* for details. The term B/M reflects the increase in variance due to using a finite number of imputations, and its square root defines the Monte Carlo error associated with a single parameter. In general, Monte Carlo error estimates are obtained by applying a jackknife procedure to MI results. That is, an MCE estimate of an MI statistic is the standard error of the mean of the pseudovalues for that statistic, computed by omitting one imputation at a time; see [R] **jackknife** for technical details.

Consider our heart attack data analysis from example 1. Let's compute Monte Carlo error estimates of MI results. To obtain MCE estimates, we specify the `mcerror` option during estimation:

```
. use http://www.stata-press.com/data/r12/mheart1s20
(Fictional heart attack data; bmi missing)
. mi estimate, dots mcerror: logit attack smokes age bmi hsgrad female
Imputations (20):
  .........10........20 done
```

```
Multiple-imputation estimates              Imputations       =         20
Logistic regression                        Number of obs     =        154
                                           Average RVI       =     0.0312
                                           Largest FMI       =     0.1355
DF adjustment:    Large sample             DF:      min       =    1060.38
                                                    avg       =  223362.56
                                                    max       =  493335.88
Model F test:     Equal FMI                F(  5,71379.3)    =       3.59
Within VCE type:      OIM                  Prob > F          =     0.0030
```

attack	Coef.	Std. Err.	t	P>|t|	[95% Conf. Interval]	
smokes	1.198595	.3578195	3.35	0.001	.4972789	1.899911
	.0068541	.0008562	0.01	0.000	.0056572	.0082212
age	.0360159	.0154399	2.33	0.020	.0057541	.0662776
	.0002654	.0000351	0.01	0.001	.0002319	.0003108
bmi	.1039416	.0476136	2.18	0.029	.010514	.1973692
	.0038014	.0008904	0.09	0.006	.0039928	.0044049
hsgrad	.1578992	.4049257	0.39	0.697	-.6357464	.9515449
	.0091517	.0010209	0.02	0.016	.0086215	.0100602
female	-.1067433	.4164735	-0.26	0.798	-.9230191	.7095326
	.0077566	.0009279	0.02	0.015	.006985	.0088408
_cons	-5.478143	1.685075	-3.25	0.001	-8.782394	-2.173892
	.1079841	.0248274	0.07	0.000	.1310618	.1050817

```
Note: values displayed beneath estimates are Monte Carlo error estimates.
```

As the note describes, MCE estimates are displayed beneath parameter estimates. Following practical guidelines from White, Royston, and Wood (2011), MCE estimates of coefficients should be less than 10% of the standard errors of the coefficients; MCE estimates of test statistics should be approximately 0.1; and MCE estimates of p-values should be approximately 0.01 when the true p-value is 0.05 and 0.02 when the true p-value is 0.1. Our results based on 20 imputations satisfy these conditions, so we can be reasonably sure about the statistical reproducibility of our results.

We can also see Monte Carlo error estimates for other MI statistics reported by the `vartable` option. To redisplay Monte Carlo error estimates, we use the `mcerror` option upon replay. We also suppress the coefficient table by using the `nocitable` option.

```
. mi estimate, vartable mcerror nocitable
```

Multiple-imputation estimates   Imputations   =   20
Logistic regression

Variance information

| | Imputation variance | | | | | Relative |
	Within	Between	Total	RVI	FMI	efficiency
smokes	.127048	.00094	.128035	.007765	.007711	.999615
	.000559	.000211	.000613	.001744	.00172	.00009
age	.000237	1.4e-06	.000238	.006245	.00621	.99969
	8.6e-07	4.6e-07	1.1e-06	.002054	.002033	.000107
bmi	.001964	.000289	.002267	.154545	.135487	.993271
	.000026	.000077	.000085	.04134	.031986	.00166
hsgrad	.162206	.001675	.163965	.010843	.010739	.999463
	.000521	.000552	.000827	.003579	.003516	.000185
female	.172187	.001203	.17345	.007338	.00729	.999636
	.000614	.000297	.000773	.001811	.001788	.000094
_cons	2.5946	.233211	2.83948	.094377	.086953	.995671
	.029651	.070081	.083436	.028332	.024216	.001263

Note: values displayed beneath estimates are Monte Carlo error estimates.

MCE estimates of all statistics are small.

What if we want to see MCE estimates of odds ratios? We know that we can use the or option on replay to redisplay results as odds ratios. However, using this option in combination with mcerror upon replay will not display MCE estimates of odds ratios:

```
. mi estimate, or mcerror
```

Multiple-imputation estimates   Imputations   =   20
Logistic regression   Number of obs   =   154
                                           Average RVI   =   0.0312
                                           Largest FMI   =   0.1355
DF adjustment: Large sample   DF: min   =   1060.38
                                                   avg   = 223362.56
                                                   max   = 493335.88
Model F test: Equal FMI   F(5,71379.3) =   3.59
Within VCE type: OIM   Prob > F   =   0.0030

attack	Odds Ratio	Std. Err.	t	P>|t|	[95% Conf. Interval]	
smokes	3.315455	1.186334	3.35	0.001	1.644241	6.685298
age	1.036672	.0160061	2.33	0.020	1.005771	1.068523
bmi	1.109536	.052829	2.18	0.029	1.010569	1.218194
hsgrad	1.171048	.4741875	0.39	0.697	.5295401	2.589707
female	.8987564	.3743082	-0.26	0.798	.3973177	2.033041
_cons	.0041771	.0070387	-3.25	0.001	.0001534	.1137342

Note: Monte Carlo error estimates are not available for exponentiated
      coefficients.

The same applies to a combination of the level() and mcerror options specified on replay to try to display MCE estimates of confidence intervals for a confidence level other than the one used during estimation.

To compute MCE estimates for odds ratios in addition to coefficients, you need to specify the or option in combination with mcerror during estimation. Similarly, to compute MCE estimates for confidence intervals with a specific confidence level, you need to specify the level() option in combination with mcerror during estimation. Otherwise, MCE estimates of 95% confidence intervals are computed.

```
. mi estimate, mcerror or level(90): logit attack smokes age bmi hsgrad female
Multiple-imputation estimates             Imputations      =         20
Logistic regression                       Number of obs    =        154
                                          Average RVI      =     0.0312
                                          Largest FMI      =     0.1355
DF adjustment:   Large sample             DF:      min      =    1060.38
                                                   avg      =  223362.56
                                                   max      =  493335.88
Model F test:       Equal FMI             F(  5,71379.3) =         3.59
Within VCE type:         OIM              Prob > F         =     0.0030
```

attack	Odds Ratio	Std. Err.	t	P>|t|	[90% Conf.	Interval]
smokes	3.315455	1.186334	3.35	0.001	1.840491	5.97245
	.0227267	.0104806	0.01	0.000	.0107398	.0477351
age	1.036672	.0160061	2.33	0.020	1.010676	1.063337
	.0002752	.000039	0.01	0.001	.0002388	.0003221
bmi	1.109536	.052829	2.18	0.029	1.025885	1.200007
	.0042178	.001033	0.09	0.006	.0040064	.0051089
hsgrad	1.171048	.4741875	0.39	0.697	.6016087	2.279478
	.0107188	.0049031	0.02	0.016	.0052248	.02254
female	.8987564	.3743082	-0.26	0.798	.4530363	1.782998
	.0069686	.00341	0.02	0.015	.0032087	.0154128
_cons	.0041771	.0070387	-3.25	0.001	.000261	.0668412
	.0004519	.0007338	0.07	0.000	.0000336	.0068716

```
Note: values displayed beneath estimates are Monte Carlo error estimates.
```

Similarly to the MCE estimates for coefficients, the MCE estimates for odds ratios are within acceptable limits.

If you wish to obtain Monte Carlo error estimates of confidence intervals for a number of different confidence levels, a more computationally efficient way of doing so is to use mi estimate using (see [MI] **mi estimate using**).

First, use mi estimate to save individual estimation results from a model to an estimation file:

```
. mi estimate, saving(miest): ...
```

Then use mi estimate using to obtain MCE estimates for different confidence intervals,

```
. mi estimate using miest, mcerror level(90) ...
. mi estimate using miest, mcerror level(80) ...
```

or for odds ratios,

```
. mi estimate using miest, mcerror or ...
```

without refitting the model.

Potential problems that can arise when using mi estimate

There are two problems that can arise when using `mi estimate`:

1. The estimation sample varies across imputations.

2. Different covariates are omitted across the imputations.

`mi estimate` watches for and issues an error message if either of these problems occur. Below we explain how each can arise and what to do about it. If you see one of these messages, be glad that `mi estimate` mentioned the problem, because otherwise, it might have gone undetected. A varying-estimation sample may result in biased or inefficient estimates. Different covariates being omitted always results in the combined results being biased.

If the first problem arises, `mi estimate` issues the error message "estimation sample varies between $m = \#$ and $m = \#$". `mi estimate` expects that when it runs the estimation command on the first imputation, on the second, and so on, the estimation command will use the same observations in each imputation. `mi estimate` does not just count, it watches which observations are used.

Perhaps the difference is due to a past mistake, such as not having imputed all the missing values. Perhaps you even corrupted your `mi` data so that the imputed variable is missing in some imputations and not in others.

Another reason the error can arise is because you specified an `if` condition based on imputed or passive variables. `mi estimate` considers this a mistake but, if this is your intent, you can reissue the `mi estimate` command and include the `esampvaryok` option.

Finally, it is possible that the varying observations are merely a characteristic of the estimator when combined with the two different imputed datasets. In this case, just as in the previous one, you can reissue `mi estimate` with the `esampvaryok` option.

The easy way to diagnose why you got this error is to use `mi xeq` (see [MI] **mi xeq**) to run the estimation command separately on the two imputations mentioned in the error message. Alternatively, you can rerun the `mi estimate` command immediately with the `esampvaryok` option and with the `esample(`*varname*`)` option, which will create in new variable *varname* the `e(sample)` from each of the individual estimations. If you use the second approach, you must first `mi convert` your data to flong or flongsep if they are not recorded in that style already; see [MI] **mi convert** for details.

The second problem we mentioned concerns omitted variables as opposed to omitted observations. `mi estimate` reports that "omitted variables vary" and goes on to mention the two imputations between which the variation was detected.

This can be caused when you include factor variables but did not specify base categories. It was the base categories that differed in the two imputations. That could happen if you specified `i.group`. By default, Stata chooses to omit the most frequent category. If `group` were imputed or passive, then the most frequent category could vary between two imputations. The solution is to specify the base category for yourself by typing, for instance, `b2.group`; see [U] **11.4.3 Factor variables**.

There are other possible causes. Varying omitted variables 1) includes different variables being omitted in the two imputations and 2) includes no variables being omitted in one imputation and, in the other, one or more variables being omitted.

When different variables are being omitted, it is usually caused by collinearity, and one of the variables needs to be dropped from the model. Variables `x1` and `x2` are collinear; sometimes the estimation command is choosing to omit `x1` and other times, `x2`. The solution is that you choose which to omit by removing it from your model.

If no variables were omitted in one of the imputations, the problem is more difficult to explain. Say that you included `i.group` in your model, the base category remained the same for the two imputations, but in one of the imputations, no one is observed in group 3, and thus no coefficient for group 3 could be estimated. You choices are to accept that you cannot estimate a group 3 coefficient and combine group 3 with, say, group 4, or to drop all imputations in which there is no one in group 3. If you want to drop imputations 3, 9, and 12, you type `mi set m -= (3,9,12)`; see [MI] **mi set**.

❏ Technical note

As we already mentioned, `mi estimate` obtains MI estimates by using the combination rules to pool results from the specified command executed separately on each imputation. As such, certain concepts (for example, likelihood function) and most postestimation tools specific to the command may not be applicable to the MI estimates; see *Analysis of multiply imputed data* in [MI] **intro substantive**. MI estimates may not even have a valid variance–covariance matrix associated with them when the number of imputations is smaller than the number of estimated parameters. For these reasons, the system matrices `e(b)` and `e(V)` are not set by `mi estimate`. If desired, you can save the MI estimates and their variance–covariance estimates in `e(b)` and `e(V)` by specifying the `post` option. See [MI] **mi estimate postestimation** for postestimation tools available after `mi estimate`.

❏

Saved results

`mi estimate` saves the following in `e()`:

Scalars

`e(df_avg[_Q]_mi)`	average degrees of freedom
`e(df_c_mi)`	complete degrees of freedom (if originally saved by *estimation_command* in `e(df_r)`)
`e(df_max[_Q]_mi)`	maximum degrees of freedom
`e(df_min[_Q]_mi)`	minimum degrees of freedom
`e(df_m_mi)`	MI model test denominator (residual) degrees of freedom
`e(df_r_mi)`	MI model test numerator (model) degrees of freedom
`e(esampvary_mi)`	varying-estimation sample flag (0 or 1)
`e(F_mi)`	model test F statistic
`e(k_exp_mi)`	number of expressions (transformed coefficients)
`e(M_mi)`	number of imputations
`e(N_mi)`	number of observations (minimum, if varies)
`e(N_min_mi)`	minimum number of observations
`e(N_max_mi)`	maximum number of observations
`e(N_g_mi)`	number of groups
`e(g_min_mi)`	smallest group size
`e(g_avg_mi)`	average group size
`e(g_max_mi)`	largest group size
`e(p_mi)`	MI model test p-value
`e(cilevel_mi)`	confidence level used to compute Monte Carlo error estimates of confidence intervals
`e(fmi_max[_Q]_mi)`	largest FMI
`e(rvi_avg[_Q]_mi)`	average RVI
`e(rvi_avg_F_mi)`	average RVI associated with the residual degrees of freedom for model test
`e(ufmi_mi)`	1 if unrestricted FMI model test is performed, 0 if equal FMI model test is performed

Macros
e(mi)	mi
e(cmdline_mi)	command as typed
e(prefix_mi)	mi estimate
e(cmd_mi)	name of *estimation_command*
e(cmd)	mi estimate (equals e(cmd_mi) when post is used)
e(title_mi)	"Multiple-imputation estimates"
e(wvce_mi)	title used to label within-imputation variance in the table header
e(modeltest_mi)	title used to label the model test in the table header
e(dfadjust_mi)	title used to label the degrees-of-freedom adjustment in the table header
e(expnames_mi)	names of expressions specified in *spec*
e(exp#_mi)	expressions of the transformed coefficients specified in *spec*
e(rc_mi)	return codes for each imputation
e(m_mi)	specified imputation numbers
e(m_est_mi)	imputation numbers used in the computation
e(names_vvl_mi)	command-specific e() macro names that contents varied across imputations
e(names_vvm_mi)	command-specific e() matrix names that values varied across imputations (excluding b, V, and Cns)
e(names_vvs_mi)	command-specific e() scalar names that values varied across imputations

Matrices
e(b)	MI estimates of coefficients (equals e(b_mi), saved only if post is used)
e(V)	variance–covariance matrix (equals e(V_mi), saved only if post is used)
e(Cns)	constraint matrix (for constrained estimation only; equals e(Cns_mi), saved only if post is used)
e(N_g_mi)	group counts
e(g_min_mi)	group-size minimums
e(g_avg_mi)	group-size averages
e(g_max_mi)	group-size maximums
e(b[_Q]_mi)	MI estimates of coefficients (or transformed coefficients)
e(V[_Q]_mi)	variance–covariance matrix (total variance)
e(Cns_mi)	constraint matrix (for constrained estimation only)
e(W[_Q]_mi)	within-imputation variance matrix
e(B[_Q]_mi)	between-imputation variance matrix
e(re[_Q]_mi)	parameter-specific relative efficiencies
e(rvi[_Q]_mi)	parameter-specific RVIs
e(fmi[_Q]_mi)	parameter-specific FMIs
e(df[_Q]_mi)	parameter-specific degrees of freedom
e(pise[_Q]_mi)	parameter-specific percentages increase in standard errors
e(*vs_names*_vs_mi)	values of command-specific e() scalar *vs_names* that varied across imputations

vs_names include (but are not restricted to) df_r, N, N_strata, N_psu, N_pop, N_sub, N_postrata, N_stdize, N_subpop, N_over, and converged.

Results N_g_mi, g_min_mi, g_avg_mi, and g_max_mi are saved for panel-data models only. The results are saved as matrices for mixed-effects models and as scalars for other panel-data models.

If transformations are specified, the corresponding estimation results are saved with the _Q_mi suffix, as described above.

Command-specific e() results that remain constant across imputations are also saved. Command-specific results that vary from imputation to imputation are posted as missing, and their names are saved in the corresponding macros e(names_vvl_mi), e(names_vvm_mi), and e(names_vvs_mi). For some command-specific e() scalars (see *vs_names* above), their values from each imputation are saved in a corresponding matrix with the _vs_mi suffix.

Methods and formulas

Let $\mathbf{q}$ define a column vector of parameters of interest. For example, $\mathbf{q}$ may be a vector of coefficients (or functions of coefficients) from a regression model. Let $\{(\widehat{\mathbf{q}}_i, \widehat{\mathbf{U}}_i) : i = 1, 2, \ldots, M\}$

be the completed-data estimates of $\mathbf{q}$ and the respective variance–covariance estimates from M imputed datasets.

The MI estimate of $\mathbf{q}$ is

$$\overline{\mathbf{q}}_M = \frac{1}{M} \sum_{i=1}^{M} \widehat{\mathbf{q}}_i$$

The variance–covariance estimate (VCE) of $\overline{\mathbf{q}}_M$ (total variance) is

$$\mathbf{T} = \overline{\mathbf{U}} + (1 + \frac{1}{M})\mathbf{B}$$

where $\overline{\mathbf{U}} = \sum_{i=1}^{M} \widehat{\mathbf{U}}_i/M$ is the within-imputation variance–covariance matrix and $\mathbf{B} = \sum_{i=1}^{M}(\mathbf{q}_i - \overline{\mathbf{q}}_M)(\mathbf{q}_i - \overline{\mathbf{q}}_M)'/(M-1)$ is the between-imputation variance–covariance matrix.

Methods and formulas are presented under the following headings:

> *Univariate case*
> *Multivariate case*

Univariate case

Let Q, $\overline{Q}_M$, B, $\overline{U}$, and T correspond to the scalar analogues of the above formulas. Univariate inferences are based on the approximation

$$T^{-1/2}(Q - \overline{Q}_M)^2 \sim t_\nu \tag{1}$$

where t_ν denotes a Student's t distribution with ν degrees of freedom, which depends on the number of imputations, M, and the increase in variance of estimates due to missing data. Under the large-sample assumption with respect to complete data, the degrees of freedom is

$$\nu_{\mathrm{large}} = (M - 1)\left(1 + \frac{1}{r}\right)^2 \tag{2}$$

where

$$r = \frac{(1 + M^{-1})B}{\overline{U}} \tag{3}$$

is an RVI due to missing data. Under the small-sample assumption, the degrees of freedom is

$$\nu_{\mathrm{small}} = \left(\frac{1}{\nu_{\mathrm{large}}} + \frac{1}{\widehat{\nu}_{\mathrm{obs}}}\right)^{-1} \tag{4}$$

where $\widehat{\nu}_{\mathrm{obs}} = \nu_c(\nu_c + 1)(1 - \gamma)/(\nu_c + 3)$, $\gamma = (1 + 1/M)B/T$, and ν_c are the complete degrees of freedom, the degrees of freedom used for inference when data are complete (Barnard and Rubin 1999).

The small-sample adjustment (4) is applied to the degrees of freedom ν when the specified command saves the residual degrees of freedom in e(df_r). This number of degrees of freedom is used as the complete degrees of freedom, ν_c, in the computation. (If e(df_r) varies across imputations, the smallest is used in the computation, resulting in conservative inference.) If e(df_r) is not set by the specified command or if the nosmall option is specified, then (2) is used to compute the degrees of freedom, ν.

Parameter-specific significance levels, confidence intervals, and degrees of freedom as reported by `mi estimate` are computed using the formulas above.

The percentage of standard-error increase due to missing data, as reported by `mi estimate`, `dftable`, is computed as $\{(T/\overline{U})^{1/2} - 1\} \times 100\%$.

The FMIs due to missing data and relative efficiencies reported by `mi estimate`, `vartable` are computed as follows.

In the large-sample case, the fraction of information about Q missing due to nonresponse (Rubin 1987, 77) is

$$\lambda = \frac{r + 2/(\nu_{\text{large}} + 3)}{r + 1}$$

where the RVI, r, is defined in (3). In the small-sample case, the fraction of information about Q missing due to nonresponse (Barnard and Rubin 1999, 953) is

$$\lambda = 1 - \frac{\lambda(\nu_{\text{small}})}{\lambda(\nu_c)}\frac{\overline{U}}{T}$$

where $\lambda(u) = (u + 1)/(u + 3)$.

The relative (variance) efficiency of using M imputations versus the infinite number of imputations is $\text{RE} = (1 + \lambda/M)^{-1}$ (Rubin 1987, 114).

Also see Rubin (1987, 76–77) and Schafer (1997, 109–111) for details.

Multivariate case

The approximation (1) can be generalized to the multivariate case:

$$(\mathbf{q} - \overline{\mathbf{q}}_M)\mathbf{T}^{-1}(\mathbf{q} - \overline{\mathbf{q}}_M)'/k \sim F_{k,\nu} \tag{5}$$

where $F_{k,\nu}$ denotes an F distribution with $k = \text{rank}(T)$ numerator degrees of freedom and ν denominator degrees of freedom defined as in (2), where the RVI, r, is replaced with the average RVI, r_{ave}:

$$r_{\text{ave}} = (1 + 1/M)\text{tr}(\mathbf{B}\overline{\mathbf{U}}^{-1})/k$$

The approximation (5) is inadequate with a small number of imputations because the between-imputation variance, $\mathbf{B}$, cannot be estimated reliably based on small M. Moreover, when M is smaller than the number of estimated parameters, $\mathbf{B}$ does not have a full rank. As such, the total variance, $\mathbf{T}$, may not be a valid variance–covariance matrix for $\overline{\mathbf{q}}_M$.

One solution is to assume that the between-imputation and within-imputation matrices are proportional, that is $B = \overline{\lambda} \times \mathbf{U}$ (Rubin 1987, 78). This assumption implies that FMIs of all estimated parameters are equal. Under this assumption, approximation (5) becomes

$$(1 + r_{\text{ave}})^{-1}(\mathbf{q} - \overline{\mathbf{q}}_M)\overline{\mathbf{U}}^{-1}(\mathbf{q} - \overline{\mathbf{q}}_M)'/k \sim F_{k,\nu_{\star}} \tag{6}$$

where $k = \text{rank}(U)$ and $\nu_{\star}$ is computed as described in Li et al. (1991, 1067).

Also see Rubin (1987, 77–78) and Schafer (1997, 112–114) for details.

We refer to (6) as an equal FMI test and to (5) as the unrestricted FMI test. By default, `mi estimate` uses the approximation (6) for the model test. If the `ufmitest` option is specified, it uses the approximation (5) for the model test.

Similar to the univariate case, the degrees of freedom $\nu_\star$ and ν are adjusted for small samples when the command saves the completed-data residual degrees of freedom in e(df_r).

In the small-sample case, the degrees of freedom $\nu_\star$ is computed as described in Reiter (2007) (in the rare case, when $k(M-1) \leq 4$, $\nu_\star = (k+1)\nu_1/2$, where ν_1 is the degrees of freedom from Barnard and Rubin [1999]). In the small-sample case, the degrees of freedom ν is computed as described in Barnard and Rubin (1999) and Marchenko and Reiter (2009).

Acknowledgments

The mi estimate command was inspired by the user-written command mim by John Carlin and John Galati, both of the Murdoch Children's Research Institute and University of Melbourne, Patrick Royston of the MRC Clinical Trials Unit, and Ian White of the MRC Biostatistics Unit. We greatly appreciate the authors for their extensive body of work in Stata in the multiple-imputation area.

References

Barnard, J., and D. B. Rubin. 1999. Small-sample degrees of freedom with multiple imputation. *Biometrika* 86: 948–955.

Cleves, M. A., W. W. Gould, R. G. Gutierrez, and Y. V. Marchenko. 2010. *An Introduction to Survival Analysis Using Stata*. 3rd ed. College Station, TX: Stata Press.

Li, K.-H., X.-L. Meng, T. E. Raghunathan, and D. B. Rubin. 1991. Significance levels from repeated p-values with multiply-imputed data. *Statistica Sinica* 1: 65–92.

Marchenko, Y. V., and J. P. Reiter. 2009. Improved degrees of freedom for multivariate significance tests obtained from multiply imputed, small-sample data. *Stata Journal* 9: 388–397.

Mortimore, P., P. Sammons, L. Stoll, D. Lewis, and R. Ecob. 1988. *School Matters*. Berkeley, CA: University of California Press.

Reiter, J. P. 2007. Small-sample degrees of freedom for multi-component significance tests with multiple imputation for missing data. *Biometrika* 94: 502–508.

Royston, P., J. B. Carlin, and I. R. White. 2009. Multiple imputation of missing values: New features for mim. *Stata Journal* 9: 252–264.

Rubin, D. B. 1987. *Multiple Imputation for Nonresponse in Surveys*. New York: Wiley.

Schafer, J. L. 1997. *Analysis of Incomplete Multivariate Data*. Boca Raton, FL: Chapman & Hall/CRC.

White, I. R., and P. Royston. 2009. Imputing missing covariate values for the Cox model. *Statistics in Medicine* 28: 1982–1998.

White, I. R., P. Royston, and A. M. Wood. 2011. Multiple imputation using chained equations: Issues and guidance for practice. *Statistics in Medicine* 30: 377–399.

Also see

[MI] **mi estimate using** — Estimation using previously saved estimation results

[MI] **mi estimate postestimation** — Postestimation tools for mi estimate

[MI] **intro substantive** — Introduction to multiple-imputation analysis

[MI] **intro** — Introduction to mi

[MI] **Glossary**

Title

> **mi estimate using** — Estimation using previously saved estimation results

Syntax

Compute MI estimates of coefficients using previously saved estimation results

> mi est̲imate using *miestfile* [, *options*]

Compute MI estimates of transformed coefficients using previously saved estimation results

> mi est̲imate [*spec*] using *miestfile* [, *options*]

where *spec* may be one or more terms of the form ([*name:*] *exp*). *exp* is any function of the parameter estimates allowed by nlcom; see [R] **nlcom**.

miestfile.ster contains estimation results previously saved by mi estimate, saving(*miestfile*); see [MI] **mi estimate**.

options	Description
Options	
ni̲mputations(*#*)	specify number of imputations to use; default is to use all saved imputations
i̲mputations(*numlist*)	specify which imputations to use
est̲imations(*numlist*)	specify which estimation results to use
mcerr̲or	compute Monte Carlo error estimates
ufmitest	perform unrestricted FMI model test
nosmall	do not apply small-sample correction to degrees of freedom
Tables	
[no]citable	suppress/display standard estimation table containing parameter-specific confidence intervals; default is citable
df̲table	display degrees-of-freedom table; dftable implies nocitable
var̲table	display variance information about estimates; vartable implies citable
table_options	control table output
display_options	control column formats, row spacing, and display of omitted variables and base and empty cells
Reporting	
le̲vel(*#*)	set confidence level; default is level(95)
dots	display dots as estimations are performed
noi̲sily	display any output from nlcom if transformations are specified
trace	trace nlcom if transformations are specified; implies noisily
replay	replay command-specific results from each individual estimation in *miestfile*.ster; implies noisily
cmdlegend	display the command legend
nogroup	suppress summary about groups displayed for xt commands
xtme_options	control output from mixed-effects commands

Advanced

errorok	allow estimation even when nlcom errors out in some imputations; such imputations are discarded from the analysis
coeflegend	display legend instead of statistics
nowarning	suppress the warning about varying estimation samples
noerrnotes	suppress error notes associated with failed estimation results in *miestfile*.ster
showimputations	show imputations saved in *miestfile*.ster
eform_option	display coefficient table in exponentiated form
post	post estimated coefficients and VCE to e(b) and e(V)

coeflegend, nowarning, noerrnotes, showimputations, *eform_option*, and post do not appear in the dialog box.

table_options	Description
noheader	suppress table header(s)
notable	suppress table(s)
nocoef	suppress table output related to coefficients
nocmdlegend	suppress command legend that appears in the presence of transformed coefficients when nocoef is used
notrcoef	suppress table output related to transformed coefficients
nolegend	suppress table legend(s)
nocnsreport	do not display constraints

See [MI] **mi estimate postestimation** for features available after estimation. To replay results, type mi estimate without arguments.

Menu

Statistics > Multiple imputation

Description

mi estimate using *miestfile* is for use after mi estimate, saving(*miestfile*): .... It allows obtaining multiple-imputation (MI) estimates, including standard errors and confidence intervals, for transformed coefficients or the original coefficients, this time calculated on a subset of the imputations. The transformation can be linear or nonlinear.

Options

⌐ Options └

nimputations(#) specifies that the first # imputations be used; # must be $M_{\min} \leq \# \leq M$, where $M_{\min} = 3$ if mcerror is specified and $M_{\min} = 2$, otherwise. The default is to use all imputations, M. Only one of nimputations(), imputations(), or estimations() may be specified.

imputations(*numlist*) specifies which imputations to use. The default is to use all of them. *numlist* must contain at least two numbers corresponding to the imputations saved in *miestfile*.ster. If mcerror is specified, *numlist* must contain at least three numbers. You can use the show-imputations option to display imputations currently saved in *miestfile*.ster. Only one of nimputations(), imputations(), or estimations() may be specified.

estimations(*numlist*) does the same thing as imputations(*numlist*), but this time the imputations are numbered differently. Say that *miestfile*.ster was created by mi estimate and mi estimate was told to limit itself to imputations 1, 3, 5, and 9. With imputations(), the imputations are still numbered 1, 3, 5, and 9. With estimations(), they are numbered 1, 2, 3, and 4. Usually, one does not specify a subset of imputations when using mi estimate, and so usually, the imputations() and estimations() options are identical. The specified *numlist* must contain at least two numbers. If mcerror is specified, *numlist* must contain at least three numbers. Only one of nimputations(), imputations(), or estimations() may be specified.

mcerror specifies to compute Monte Carlo error (MCE) estimates for the results displayed in the estimation, degrees-of-freedom, and variance-information tables. MCE estimates reflect variability of MI results across repeated uses of the same imputation procedure and are useful for determining an adequate number of imputations to obtain stable MI results; see White, Royston, and Wood (2011) for details and guidelines.

MCE estimates are obtained by applying the jackknife procedure to multiple-imputation results. That is, the jackknife pseudovalues of MI results are obtained by omitting one imputation at a time; see [R] **jackknife** for details about the jackknife procedure. As such, the Monte Carlo error computation requires at least three imputations.

If level() is specified during estimation, MCE estimates are obtained for confidence intervals with the specified confidence level instead of using the default 95% confidence level. If any of the options described in [R] *eform_option* is specified during estimation, MCE estimates for the coefficients, standard errors, and confidence intervals in the exponentiated form are also computed. mcerror can also be used upon replay to display MCE estimates. Otherwise, MCE estimates are not reported upon replay even if they were previously computed.

ufmitest specifies that the unrestricted fraction missing information (FMI) model test be used. The default test performed assumes equal fractions of information missing due to nonresponse for all coefficients. This is equivalent to the assumption that the between-imputation and within-imputation variances are proportional. The unrestricted test may be preferable when this assumption is suspect provided the number of imputations is large relative to the number of estimated coefficients.

nosmall specifies that no small-sample correction be made to the degrees of freedom. By default, individual tests of coefficients (and transformed coefficients) use the small-sample correction of Barnard and Rubin (1999), and the overall model test uses the small-sample correction of Reiter (2007).

⌐ Tables ⌐

All table options below may be specified at estimation time or when redisplaying previously estimated results.

citable and nocitable specify whether the standard estimation table containing parameter-specific confidence intervals is displayed. The default is citable. nocitable can be used with vartable to suppress the confidence-interval table.

dftable displays a table containing parameter-specific degrees of freedom and percentages of increase in standard errors due to nonresponse. dftable implies nocitable.

vartable displays a table reporting variance information about MI estimates. The table contains estimates of within-imputation variances, between-imputation variances, total variances, relative increases in variance due to nonresponse, fractions of information about parameter estimates missing due to nonresponse, and relative efficiencies for using finite M rather than a hypothetically infinite number of imputations. vartable implies citable.

table_options control the appearance of all displayed table output:

noheader suppresses all header information from the output. The table output is still displayed.

notable suppresses all tables from the output. The header information is still displayed.

nocoef suppresses the display of tables containing coefficient estimates. This option affects the table output produced by citable, dftable, and vartable.

nocmdlegend suppresses the table legend showing the command line, used to produce results in *miestfile*.ster, from the output. This legend appears above the tables containing transformed coefficients (or above the variance-information table if vartable is used) when nocoef is specified.

notrcoef suppresses the display of tables containing estimates of transformed coefficients (if specified). This option affects the table output produced by citable, dftable, and vartable.

nolegend suppresses all table legends from the output.

nocnsreport; see [R] **estimation options**.

display_options: <u>noomit</u>ted, vsquish, noemptycells, <u>base</u>levels, <u>allbase</u>levels, cformat(%*fmt*), pformat(%*fmt*), and <u>sformat</u>(%*fmt*); see [R] **estimation options**.

⌐──────── ⌐ Reporting ⌐──

level(#); see [R] **estimation options**.

dots specifies that dots be displayed as estimations of transformed coefficients are successfully completed. An x is displayed if nlcom fails to estimate one of the transformed coefficients specified in *spec*. This option is relevant only if transformations are specified.

noisily specifies that any output from nlcom, used to obtain the estimates of transformed coefficients, be displayed. This option is relevant only if transformations are specified.

trace traces the execution of nlcom. trace implies noisily and is relevant only if transformations are specified.

replay replays estimation results from *miestfile*.ster, previously saved by mi estimate, saving(*miestfile*). This option implies noisily.

cmdlegend requests that the command line corresponding to the estimation command used to produce the estimation results saved in *miestfile*.ster be displayed. cmdlegend may be specified at run time or when redisplaying results.

nogroup suppresses the display of group summary information (number of groups, average group size, minimum, and maximum) as well as other command-specific information displayed for xt commands.

xtme_options: <u>var</u>iance, <u>noret</u>able, <u>nofet</u>able, and <u>estm</u>etric. These options are relevant only with the mixed-effects commands such as xtmixed (see [XT] **xtmixed**), xtmelogit (see [XT] **xtmelogit**), and xtmepoisson (see [XT] **xtmepoisson**). The estmetric option is implied when vartable or dftable is used.

⌐‾‾⌐ Advanced ⌐

errorok specifies that estimations of transformed coefficients that fail be skipped and the combined results be based on the successful estimation results. The default is that `mi estimate` stops if an individual estimation fails. If the *miestfile*.`ster` file contains failed estimation results, `mi estimate using` does not error out; it issues notes about which estimation results failed and discards these estimation results in the computation. You can use the `noerrnotes` option to suppress the display of the notes.

The following options are available with `mi estimate using` but are not shown in the dialog box:

coeflegend; see [R] **estimation options**. `coeflegend` implies `nocitable` and cannot be combined with `citable` or `dftable`.

nowarning suppresses the warning message at the bottom of table output that occurs if the estimation sample varies and `esampvaryok` is specified. See *Potential problems that can arise when using mi estimate* in [MI] **mi estimate** for details.

noerrnotes suppresses notes about failed estimation results. These notes appear when *miestfile*.`ster` contains estimation results, previously saved by `mi estimate, saving(`*miestfile*`)`, from imputations for which the estimation command used with `mi estimate` failed to estimate parameters.

showimputations displays imputation numbers corresponding to the estimation results saved in *miestfile*.`ster`. `showimputations` may be specified at run time or when redisplaying results.

eform_option; see [R] ***eform_option***. `mi estimate using` reports results in the coefficient metric under which the combination rules are applied. You may use the appropriate *eform_option* to redisplay results in exponentiated form, if desired. If `dftable` is also specified, the reported degrees of freedom and percentage increases in standard errors are not adjusted and correspond to the original coefficient metric.

post requests that MI estimates of coefficients and their respective VCEs be posted in the usual way. This allows the use of *estimation_command*-specific postestimation tools with MI estimates. There are issues; see *Using the command-specific postestimation tools* in [MI] **mi estimate postestimation**. `post` may be specified at estimation time or when redisplaying previously estimated results.

Remarks

`mi estimate using` is convenient when refitting models using `mi estimate` would be tedious or time consuming. In such cases, you can perform estimation once and save the uncombined, individual results by specifying `mi estimate`'s `saving(`*miestfile*`)` option. After that, you can repeatedly use `mi estimate using` *miestfile* to estimate linear and nonlinear transformations of coefficients or to obtain MI estimates using a subset of saved imputations.

`mi estimate using` performs the pooling step of the MI procedure; see [MI] **intro substantive**. That is, it combines completed-data estimates from the *miestfile*.`ster` file by applying Rubin's combination rules (Rubin 1987, 77).

▷ Example 1

Recall the analysis of house resale prices from *Example 2: Completed-data linear regression analysis* in [MI] **mi estimate**:

```
. use http://www.stata-press.com/data/r12/mhouses1993s30
(Albuquerque Home Prices Feb15-Apr30, 1993)
. mi estimate, saving(miest): regress price tax sqft age nfeatures ne custom corner
```

Multiple-imputation estimates			Imputations	=	30
Linear regression			Number of obs	=	117
			Average RVI	=	0.0648
			Largest FMI	=	0.2533
			Complete DF	=	109
DF adjustment:	Small sample		DF: min	=	69.12
			avg	=	94.02
			max	=	105.51
Model F test:	Equal FMI		F(7, 106.5)	=	67.18
Within VCE type:	OLS		Prob > F	=	0.0000

price	Coef.	Std. Err.	t	P>|t|	[95% Conf. Interval]	
tax	.6768015	.1241568	5.45	0.000	.4301777	.9234253
sqft	.2118129	.069177	3.06	0.003	.0745091	.3491168
age	.2471445	1.653669	0.15	0.882	-3.051732	3.546021
nfeatures	9.288033	13.30469	0.70	0.487	-17.12017	35.69623
ne	2.518996	36.99365	0.07	0.946	-70.90416	75.94215
custom	134.2193	43.29755	3.10	0.002	48.35674	220.0818
corner	-68.58686	39.9488	-1.72	0.089	-147.7934	10.61972
_cons	123.9118	71.05816	1.74	0.085	-17.19932	265.0229

In the above, we use the saving() option to save the individual completed-data estimates from a regression analysis in Stata estimation file miest.ster. We can now use mi estimate using to recombine the first 5 imputations, and ignore the remaining 25, without reestimation:

```
. mi estimate using miest, ni(5)
```

Multiple-imputation estimates			Imputations	=	5
Linear regression			Number of obs	=	117
			Average RVI	=	0.0685
			Largest FMI	=	0.2075
			Complete DF	=	109
DF adjustment:	Small sample		DF: min	=	48.59
			avg	=	85.22
			max	=	104.79
Model F test:	Equal FMI		F(7, 103.9)	=	67.50
Within VCE type:	OLS		Prob > F	=	0.0000

price	Coef.	Std. Err.	t	P>|t|	[95% Conf. Interval]	
tax	.6631356	.122443	5.42	0.000	.4195447	.9067265
sqft	.2185884	.0670182	3.26	0.002	.0856051	.3515718
age	-.0395402	1.613185	-0.02	0.981	-3.28205	3.202969
nfeatures	8.735622	13.42251	0.65	0.517	-18.01198	35.48323
ne	4.069381	36.94491	0.11	0.913	-69.4355	77.57426
custom	130.4925	42.93286	3.04	0.003	45.36257	215.6225
corner	-71.25406	40.06697	-1.78	0.078	-150.7152	8.207084
_cons	130.2002	70.38012	1.85	0.068	-9.624642	270.025

We obtain results identical to those shown in the example in [MI] **mi estimate**.

We can also obtain estimates of transformed coefficients without refitting the models to the imputed dataset. Recall the example from *Example 5: Estimating transformations* in [MI] **mi estimate**, where we estimated the ratio of the coefficients for age and sqft. We can obtain the same results by using the following:

```
. mi estimate (ratio: _b[age]/_b[sqft]) using miest
```

Multiple-imputation estimates			Imputations	=	30
Linear regression			Number of obs	=	117
			Average RVI	=	0.0648
			Largest FMI	=	0.2533
			Complete DF	=	109
DF adjustment:	Small sample		DF: min	=	69.12
			avg	=	94.02
			max	=	105.51
Model F test:	Equal FMI		F(7, 106.5)	=	67.18
Within VCE type:	OLS		Prob > F	=	0.0000

price	Coef.	Std. Err.	t	P>|t|	[95% Conf.	Interval]
tax	.6768015	.1241568	5.45	0.000	.4301777	.9234253
sqft	.2118129	.069177	3.06	0.003	.0745091	.3491168
age	.2471445	1.653669	0.15	0.882	-3.051732	3.546021
nfeatures	9.288033	13.30469	0.70	0.487	-17.12017	35.69623
ne	2.518996	36.99365	0.07	0.946	-70.90416	75.94215
custom	134.2193	43.29755	3.10	0.002	48.35674	220.0818
corner	-68.58686	39.9488	-1.72	0.089	-147.7934	10.61972
_cons	123.9118	71.05816	1.74	0.085	-17.19932	265.0229

Transformations			Average RVI	=	0.2899
			Largest FMI	=	0.2316
			Complete DF	=	109
DF adjustment:	Small sample		DF: min	=	72.51
			avg	=	72.51
Within VCE type:	OLS		max	=	72.51

ratio: _b[age]/_b[sqft]

price	Coef.	Std. Err.	t	P>|t|	[95% Conf.	Interval]
ratio	1.44401	8.217266	0.18	0.861	-14.93485	17.82287

The results are the same as in the example in [MI] **mi estimate**.

◁

For more examples, see [MI] **mi test**.

Saved results

See *Saved results* in [MI] **mi estimate**.

Methods and formulas

See *Methods and formulas* in [MI] **mi estimate**.

References

Barnard, J., and D. B. Rubin. 1999. Small-sample degrees of freedom with multiple imputation. *Biometrika* 86: 948–955.

Reiter, J. P. 2007. Small-sample degrees of freedom for multi-component significance tests with multiple imputation for missing data. *Biometrika* 94: 502–508.

Rubin, D. B. 1987. *Multiple Imputation for Nonresponse in Surveys.* New York: Wiley.

White, I. R., P. Royston, and A. M. Wood. 2011. Multiple imputation using chained equations: Issues and guidance for practice. *Statistics in Medicine* 30: 377–399.

Also see

[MI] **mi estimate** — Estimation using multiple imputations

[MI] **mi estimate postestimation** — Postestimation tools for mi estimate

[MI] **intro substantive** — Introduction to multiple-imputation analysis

[MI] **intro** — Introduction to mi

[MI] **Glossary**

Title

mi estimate postestimation — Postestimation tools for mi estimate

Description

The following postestimation commands are available after mi estimate and mi estimate using:

Command	Description
mi test	perform tests on coefficients
mi testtransform	perform tests on transformed coefficients
mi predict	obtain linear predictions
mi predictnl	obtain nonlinear predictions

See [MI] **mi test** and [MI] **mi predict**.

Remarks

After estimation by mi estimate: *estimation_command*, in general, you may not use the standard postestimation commands such as test, testnl, or predict; nor may you use *estimation_command*-specific postestimation commands such as estat. As we have mentioned often, mi estimate is its own estimation command, and the postestimation commands available after mi estimate (and mi estimate using) are listed in the table above.

Using the command-specific postestimation tools

After mi estimate: *estimation_command*, you may not use *estimation_command*'s postestimation features. More correctly, you may not use them unless you specify mi estimate's post option:

        . mi estimate, post: *estimation_command* ...

Specifying post causes many statistical issues, so do not be casual about specifying it.

First, the MI estimate of the VCE is poor unless the number of imputations, M, is sufficiently large. How large is uncertain, but you should not be thinking $M = 20$ rather than $M = 5$; you should be thinking of M in the hundreds. What is statistically true is that, asymptotically in M (and in the number of observations, N), the MI estimated coefficients approach normality and the VCE becomes well estimated.

Second, there are substantive issues about what is meant by *estimation_command*'s prediction after MI estimation that you are going to have to resolve for yourself. There is no one estimation sample. There are M of them, and as we have just argued, M is large. Do not expect postestimation commands that depend on predicted values such as margins, lroc, and the like, to produce correct results, if they produce results at all.

Which brings us to the third point. Even when you specify mi estimate's post option, mi estimate still does not post everything the estimation command expects to see. It does not post likelihood values, for instance, because there is no counterpart after MI estimation. Thus, you should be prepared to see unexpected and inelegant error messages if you use a postestimation command that depends on an unestimated and unposted result.

77

All of which is to say that if you specify the `post` option, you have a responsibility beyond the usual to ensure the validity of any statistical results.

Also see

[MI] **mi test** — Test hypotheses after mi estimate

[MI] **mi predict** — Obtain multiple-imputation predictions

[MI] **mi estimate** — Estimation using multiple imputations

[MI] **mi estimate using** — Estimation using previously saved estimation results

[MI] **intro substantive** — Introduction to multiple-imputation analysis

[MI] **intro** — Introduction to mi

[MI] **Glossary**

Title

mi expand — Expand mi data	

Syntax

mi expand $[=]exp$ $[if]$ $[$, *options* $]$

options	Description
<u>g</u>enerate(*newvar*)	create *newvar*; 0 = original, 1 = expanded
<u>noup</u>date	see [MI] **noupdate option**

Menu

Statistics > Multiple imputation

Description

mi expand is expand (see [D] **expand**) for mi data. The syntax is identical to expand except that in *range* is not allowed and the noupdate option is allowed.

mi expand replaces each observation in the dataset with n copies of the observation, where n is equal to the required expression rounded to the nearest integer. If the expression is less than 1 or equal to missing, it is interpreted as if it were 1, meaning that the observation is retained but not duplicated.

Options

generate(*newvar*) creates new variable *newvar* containing 0 if the observation originally appeared in the dataset and 1 if the observation is a duplication.

noupdate in some cases suppresses the automatic mi update this command might perform; see [MI] **noupdate option**.

Remarks

mi expand amounts to performing expand on $m = 0$, then duplicating the result on $m = 1$, $m = 2$, ..., $m = M$, and then combining the result back into mi format. Thus if the requested expansion specified by *exp* is a function of an imputed, passive, varying, or super-varying variable, then it is the values of the variable in $m = 0$ that will be used to produce the result for $m = 1$, $m = 2$, ..., $m = M$, too.

Also see

[MI] **intro** — Introduction to mi

[D] **expand** — Duplicate observations

Title

---
mi export — Export mi data

Syntax

```
mi export nhanes1 ...

mi export ice ...
```

See [MI] **mi export nhanes1** and [MI] **mi export ice**.

Description

Use `mi export nhanes1` to export data in the format used by the National Health and Nutrition Examination Survey.

Use `mi export ice` to export data in the format used by `ice` (Royston 2004, 2005a, 2005b, 2007, 2009).

If and when other standards develop for recording multiple-imputation data, other `mi export` subcommands will be added.

Remarks

If you wish to send data to other Stata users, ignore `mi export` and just send them your `mi` dataset(s).

To send data to users of other packages, however, you will have to negotiate the format you will use. The easiest way to send data to non–Stata users is probably to `mi convert` (see [MI] **mi convert**) your data to flongsep and then use `outfile` (see [D] **outfile**), `outsheet` (see [D] **outsheet**), or a transfer program such as Stat/Transfer. Also see [U] **21 Inputting and importing data**.

References

Royston, P. 2004. Multiple imputation of missing values. *Stata Journal* 4: 227–241.

——. 2005a. Multiple imputation of missing values: Update. *Stata Journal* 5: 188–201.

——. 2005b. Multiple imputation of missing values: Update of ice. *Stata Journal* 5: 527–536.

——. 2007. Multiple imputation of missing values: Further update of ice, with an emphasis on interval censoring. *Stata Journal* 7: 445–464.

——. 2009. Multiple imputation of missing values: Further update of ice, with an emphasis on categorical variables. *Stata Journal* 9: 466–477.

Also see

[MI] **intro** — Introduction to mi

[MI] **mi export nhanes1** — Export mi data to NHANES format

[MI] **mi export ice** — Export mi data to ice format

Title

> **mi export ice** — Export mi data to ice format

Syntax

> mi export ice [, clear]

Menu

Statistics > Multiple imputation

Description

> mi export ice converts the mi data in memory to ice format. See Royston (2004, 2005a, 2005b, 2007, 2009) for a description of ice.

Option

> clear specifies that it is okay to replace the data in memory even if they have changed since they were last saved to disk.

Remarks

> mi export ice is the inverse of mi import ice (see [MI] **mi import ice**). Below we use mi export ice to convert miproto.dta to ice format. miproto.dta happens to be in wide form, but that is irrelevant.

```
. use http://www.stata-press.com/data/r12/miproto
(mi prototype)
. mi describe
  Style:  wide
          last mi update 30mar2011 12:46:49, 1 day ago
  Obs.:   complete      1
          incomplete    1   (M = 2 imputations)
          _____
          total         2
  Vars.:  imputed:  1; b(1)
          passive:  1; c(1)
          regular:  1; a
          system:   1; _mi_miss
          (there are no unregistered variables)
. list
```

	a	b	c	_1_b	_2_b	_1_c	_2_c	_mi_miss
1.	1	2	3	2	2	3	3	0
2.	4	.	.	4.5	5.5	8.5	9.5	1

```
. mi export ice

. list, separator(2)
```

	a	b	c	_mj	_mi
1.	1	2	3	0	1
2.	4	.	.	0	2
3.	1	2	3	1	1
4.	4	4.5	8.5	1	2
5.	1	2	3	2	1
6.	4	5.5	9.5	2	2

References

Royston, P. 2004. Multiple imputation of missing values. *Stata Journal* 4: 227–241.

———. 2005a. Multiple imputation of missing values: Update. *Stata Journal* 5: 188–201.

———. 2005b. Multiple imputation of missing values: Update of ice. *Stata Journal* 5: 527–536.

———. 2007. Multiple imputation of missing values: Further update of ice, with an emphasis on interval censoring. *Stata Journal* 7: 445–464.

———. 2009. Multiple imputation of missing values: Further update of ice, with an emphasis on categorical variables. *Stata Journal* 9: 466–477.

Also see

[MI] **intro** — Introduction to mi

[MI] **mi export** — Export mi data

[MI] **mi import ice** — Import ice-format data into mi

Title

> **mi export nhanes1** — Export mi data to NHANES format

Syntax

mi export nhanes1 *filenamestub* $\left[\, , \, \textit{options odd\_options} \right]$

options	Description
replace	okay to replace existing files
uppercase	uppercase prefix and suffix
passiveok	include passive variables

odd_options	Description
nacode(#)	not applicable code; default is 0
obscode(#)	observed code; default is 1
impcode(#)	imputed code; default is 2
impprefix("*string*" "*string*")	variable prefix; default is "" ""
impsuffix("*string*" "*string*")	variable suffix; default is "if" "mi"

Note: The *odd_options* are not specified unless you want to create results that are nhanes1-like but not really nhanes1 format.

Menu

Statistics > Multiple imputation

Description

mi export nhanes1 writes the mi data in memory to disk files in nhanes1 format. The files will be named *filenamestub*.dta, *filenamestub*1.dta, *filenamestub*2.dta, and so on. In addition to the variables in the original mi data, new variable seqn will be added to record the sequence number. After using mi export nhanes1, you can use outfile (see [D] **outfile**) or outsheet (see [D] **outsheet**) or a transfer program such as Stat/Transfer to convert the resulting .dta files into a format suitable for sending to a non-Stata user. Also see [U] **21 Inputting and importing data**.

mi export nhanes1 leaves the data in memory unchanged.

Options

replace indicates that it is okay to overwrite existing files.

uppercase specifies that the new sequence variable SEQN and the variable suffixes IF and MI be in uppercase. The default is lowercase. (More correctly, when generalizing beyond nhanes1 format, the uppercase option specifies that SEQN be created in uppercase along with all prefixes and suffixes.)

passiveok specifies that passive variables are to be written as if they were imputed variables. The default is to issue an error if passive variables exist in the original data.

nacode(#), obscode(#), and impcode(#) are optional and are never specified when reading true nhanes1 data. The default nacode(0) obscode(1) impcode(2) corresponds to the nhanes1 definition. These options allow changing the codes for not applicable, observed, and imputed.

impprefix("*string*" "*string*") and impsuffix("*string*" "*string*") are optional and are never specified when reading true nhanes1 data. The default impprefix("" "") impsuffix("if" "mi") corresponds to the nhanes1 definition. These options allow setting different prefixes and suffixes.

Remarks

mi export nhanes1 is the inverse of mi import nhanes1; see [MI] **mi import nhanes1** for a description of the nhanes1 format.

Below we use mi export nhanes1 to convert miproto.dta to nhanes1 format. miproto.dta happens to be in wide form, but that is irrelevant.

```
. use http://www.stata-press.com/data/r12/miproto
(mi prototype)
. mi describe
  Style:  wide
          last mi update 30mar2011 12:46:49, 1 day ago
  Obs.:   complete        1
          incomplete      1  (M = 2 imputations)
          _____
          total           2
  Vars.:  imputed:  1; b(1)
          passive:  1; c(1)
          regular:  1; a
          system:   1; _mi_miss
          (there are no unregistered variables)
. list
```

	a	b	c	_1_b	_2_b	_1_c	_2_c	_mi_miss
1.	1	2	3	2	2	3	3	0
2.	4	.	.	4.5	5.5	8.5	9.5	1

```
. mi export nhanes1 mynh, passiveok replace
files mynh.dta mynh1.dta mynh2.dta created
```

```
. use mynh
(mi prototype)

. list
```

	seqn	a	bif	cif
1.	1	1	1	1
2.	2	4	2	2

```
. use mynh1
(mi prototype)

. list
```

	seqn	a	bmi	cmi
1.	1	1	2	3
2.	2	4	4.5	8.5

```
. use mynh2
(mi prototype)

. list
```

	seqn	a	bmi	cmi
1.	1	1	2	3
2.	2	4	5.5	9.5

Also see

[MI] **intro** — Introduction to mi

[MI] **mi export** — Export mi data

[MI] **mi import nhanes1** — Import NHANES-format data into mi

Title

> **mi extract** — Extract original or imputed data from mi data

Syntax

mi extract # $\left[\, , \; options \right]$

where $0 \le \# \le M$

options	Description
clear	okay to replace unsaved data in memory
esample(...)	rarely specified option
esample(*varname*)	... syntax when $\# > 0$
esample(*varname* $\#_e$)	... syntax when $\# = 0$; $1 \le \#_e \le M$

Menu

Statistics > Multiple imputation

Description

mi extract # replaces the data in memory with the data for $m = \#$. The data are not mi set.

Options

clear specifies that it is okay to replace the data in memory even if the current data have not been saved to disk.

esample(*varname* $\left[\#_e \right]$) is rarely specified. It is for use after mi estimate (see [MI] **mi estimate**) when the esample(*newvar*) option was specified to save in *newvar* the e(sample) for $m = 1$, $m = 2$, ..., $m = M$. It is now desired to extract the data for one m and for e(sample) set correspondingly.

mi extract #, esample(*varname*), $\# > 0$, is the usual case in this unlikely event. One extracts one of the imputation datasets and redefines e(sample) based on the e(sample) previously saved for $m = \#$.

The odd case is mi extract 0, esample(*varname* $\#_e$), where $\#_e > 0$. One extracts the original data but defines e(sample) based on the e(sample) previously saved for $m = \#_e$.

Specifying the esample() option changes the sort order of the data.

Remarks

If you wanted to give up on mi and just get your original data back, you could type

. mi extract 0

You might do this if you wanted to send your original data to a coworker or you wanted to try a different approach to dealing with the missing values in these data. Whatever the reason, the result is that the original data replace the data in memory. The data are not mi set. Your original mi data remain unchanged.

If you suspected there was something odd about the imputations in $m = 3$, you could type

. mi extract 3

You would then have a dataset in memory that looked just like your original, except the missing values of the imputed and passive variables would be replaced with the imputed and passive values from $m = 3$. The data are not mi set. Your original data remain unchanged.

Also see

[MI] **intro** — Introduction to mi

Title

┌───┐
│ **mi import** — Import data into mi   │
└───┘

Syntax

        mi import nhanes1 ...

        mi import ice ...

        mi import flong ...

        mi import flongsep ...

        mi import wide ...

See [MI] **mi import nhanes1**, [MI] **mi import ice**, [MI] **mi import flong**, [MI] **mi import flongsep**, and [MI] **mi import wide**.

Description

mi import imports into mi data that contain original data and imputed values.

Remarks

Remarks are presented under the following headings:

> *When to use which mi import command*
> *Import data into Stata before importing into mi*
> *Using mi import nhanes1, ice, flong, and flongsep*

When to use which mi import command

mi import nhanes1 imports data recorded in the format used by the National Health and Nutrition Examination Survey (NHANES) produced by the National Center for Health Statistics of the U.S. Centers for Disease Control and Prevention (CDC); see http://www.cdc.gov/nchs/nhanes.htm.

mi import ice imports data recorded in the format used by ice (Royston 2004, 2005a, 2005b, 2007, 2009).

mi import flong and mi import flongsep import data that are in flong- and flongsep-like format, which is to say, the data are repeated for $m = 0$, $m = 1$, ..., and $m = M$. mi import flong imports data in which the information is contained in one file. mi import flongsep imports data in which the information is recorded in a collection of files.

mi import wide imports data that are in wide-like format, where additional variables are used to record the imputed values.

Import data into Stata before importing into mi

With the exception of `mi import ice`, you must import the data into Stata before you can use `mi import` to import the data into mi. `mi import ice` is the exception only because the data are already in Stata format. That is, `mi import` requires that the data be stored in Stata-format `.dta` datasets. You perform the initial import into Stata by using any method described in [D] **import** or a transfer program such as Stat/Transfer.

Using mi import nhanes1, ice, flong, and flongsep

Import commands `mi import nhanes1` and `mi import flongsep` produce an flongsep result; `mi import ice` and `mi import flong` produce an flong result. You can use `mi convert` (see [MI] **mi convert**) afterward to convert the result to another style, and we usually recommend that. Before doing that, however, you need to examine the freshly imported data and verify that all imputed and passive variables are registered correctly. If they are not registered correctly, you risk losing imputed values.

To perform this verification, use the `mi describe` (see [MI] **mi describe**) and `mi varying` (see [MI] **mi varying**) commands immediately after `mi import`:

```
. mi import ...
. mi describe
. mi varying
```

`mi describe` will list the registration status of the variables. `mi varying` will report the varying and super-varying variables. Verify that all varying variables are registered as imputed or passive. If one or more is not, register them now:

```
. mi register imputed forgottenvar
. mi register passive another_forgottenvar
```

There is no statistical distinction between imputed and passive variables, so you may register variables about which you are unsure either way. If an unregistered variable is found to be varying and you are convinced that is an error, register the variable as regular:

```
. mi register regular variable_in_error
```

Next, if `mi varying` reports that your data contain any super-varying variables, determine whether the variables are due to errors in the source data or really are intended to be super varying. If they are errors, register the variables as imputed, passive, or regular, as appropriate. Leave any intended super-varying variables unregistered, however, and make a note to yourself: never convert these data to the wide or mlong styles. Data with super-varying variables can be stored only in the flong and flongsep styles.

Now run `mi describe` and `mi varying` again:

```
. mi describe
. mi varying
```

Ensure that you have registered variables correctly, and, if necessary, repeat the steps above to fix any remaining problems.

After that, you may use `mi convert` to switch the data to a more convenient style. We generally start with style wide:

```
. mi convert wide
```

Do not switch to wide, however, if you have any super-varying variables. Try flong instead:

. mi convert flong

Whichever style you choose, if you get an insufficient-memory error, you will have to either increase the amount of memory dedicated to Stata or use these data in the more inconvenient, but perfectly workable, flongsep style. Concerning increasing memory, see *Converting from flongsep* in [MI] **mi convert**. Concerning the workability of flongsep, see *Advice for using flongsep* in [MI] **styles**.

We said to perform the checks above before using mi convert. It is, however, safe to convert the just-imported flongsep data to flong, perform the checks, and then convert to the desired form. The checks will run more quickly if you convert to flong first.

You can vary how you perform the checks. The logic underlying our recommendations is as follows:

- It is possible that you did not specify all the imputed and passive variables when you imported the data, perhaps due to errors in the data's documentation. It is also possible that there are errors in the data that you imported. It is worth checking.

- As long as the imported data are recorded in the flongsep or flong style, unregistered variables will appear exactly as they appeared in the original source. It is only when the data are converted to the wide or mlong style that assumptions about the structure of the data are exploited to save memory. Thus you need to perform checks before converting the data to the more convenient wide or mlong style.

- If you find errors, you could go back and reimport the data correctly, but it is easier to use mi register after the fact. When you type mi register you are not only informing mi about how to deal with the variable but also asking mi register to examine the variable and fix any problems given its new registration status.

References

Royston, P. 2004. Multiple imputation of missing values. *Stata Journal* 4: 227–241.

———. 2005a. Multiple imputation of missing values: Update. *Stata Journal* 5: 188–201.

———. 2005b. Multiple imputation of missing values: Update of ice. *Stata Journal* 5: 527–536.

———. 2007. Multiple imputation of missing values: Further update of ice, with an emphasis on interval censoring. *Stata Journal* 7: 445–464.

———. 2009. Multiple imputation of missing values: Further update of ice, with an emphasis on categorical variables. *Stata Journal* 9: 466–477.

Also see

[MI] **intro** — Introduction to mi

[MI] **mi import nhanes1** — Import NHANES-format data into mi

[MI] **mi import ice** — Import ice-format data into mi

[MI] **mi import flong** — Import flong-like data into mi

[MI] **mi import flongsep** — Import flongsep-like data into mi

[MI] **mi import wide** — Import wide-like data into mi

[MI] **styles** — Dataset styles

Title

> **mi import flong** — Import flong-like data into mi

Syntax

mi import flong, *required_options* [*true_options*]

required_options	Description
m(*varname*)	name of variable containing m
id(*varlist*)	identifying variable(s)

true_options	Description
imputed(*varlist*)	imputed variables to be registered
passive(*varlist*)	passive variables to be registered
clear	okay to replace unsaved data in memory

Menu

Statistics > Multiple imputation

Description

mi import flong imports flong-like data, that is, data in which $m = 0$, $m = 1$, ..., $m = M$ are all recorded in one .dta dataset.

mi import flong converts the data to mi flong style. The data are mi set.

Options

m(*varname*) and id(*varlist*) are required. m(*varname*) specifies the variable that takes on values 0, 1, ..., M, the variable that identifies observations corresponding to $m = 0$, $m = 1$, ..., $m = M$. *varname* $= 0$ identifies the original data, *varname* $= 1$ identifies $m = 1$, and so on.

id(*varlist*) specifies the variable or variables that uniquely identify observations within m().

imputed(*varlist*) and passive(*varlist*) are truly optional options, although it would be unusual if imputed() were not specified.

imputed(*varlist*) specifies the names of the imputed variables.

passive(*varlist*) specifies the names of the passive variables, if any.

clear specifies that it is okay to replace the data in memory even if they have changed since they were saved to disk. Remember, mi import flong starts with flong-like data in memory and ends with mi flong data in memory.

Remarks

The procedure to convert flong-like data to mi flong is this:

1. use the unset data.

2. Issue the mi import flong command.

3. Perform the checks outlined in *Using mi import nhanes1, ice, flong, and flongsep* of [MI] **mi import**.

4. Use mi convert (see [MI] **mi convert**) to convert the data to a more convenient style, such as wide or mlong.

For instance, you have the following unset data:

```
. use http://www.stata-press.com/data/r12/ourunsetdata
(mi prototype)
. list, separator(2)
```

	m	subject	a	b	c
1.	0	101	1	2	3
2.	0	102	4	.	.
3.	1	101	1	2	3
4.	1	102	4	4.5	8.5
5.	2	101	1	2	3
6.	2	102	4	5.5	9.5

You are told that these data contain the original data ($m = 0$) and two imputations ($m = 1$ and $m = 2$), that variable b is imputed, and that variable c is passive and in fact equal to $a + b$. These are the same data discussed in [MI] **styles** but in unset form.

The fact that these data are nicely sorted is irrelevant. To import these data, type

```
. mi import flong, m(m) id(subject) imputed(b) passive(c)
```

These data are short enough that we can list the result:

```
. list, separator(2)
```

	m	subject	a	b	c	_mi_m	_mi_id	_mi_miss
1.	0	101	1	2	3	0	1	0
2.	0	102	4	.	.	0	2	1
3.	1	101	1	2	3	1	1	.
4.	1	102	4	4.5	8.5	1	2	.
5.	2	101	1	2	3	2	1	.
6.	2	102	4	5.5	9.5	2	2	.

We will now perform the checks outlined in *Using mi import nhanes1, ice, flong, and flongsep* of [MI] **mi import**, which are to run mi describe and mi varying to verify that variables are registered correctly:

```
. mi describe
  Style:  flong
          last mi update 30mar2011 11:54:58, 0 seconds ago
  Obs.:   complete          1
          incomplete        1  (M = 2 imputations)
          ─────────────────────
          total             2
  Vars.:  imputed:  1; b(1)
          passive:  1; c(1)
          regular:  0
          system:   3; _mi_m _mi_id _mi_miss
          (there are 3 unregistered variables; m subject a)
. mi varying
              Possible problem    variable names
         ─────────────────────────────────────────────────
             imputed nonvarying:  (none)
             passive nonvarying:  (none)
             unregistered varying:  (none)
        *unregistered super/varying:  (none)
         unregistered super varying:  m
         ─────────────────────────────────────────────────
```

> * super/varying means super varying but would be varying if registered as
>   imputed; variables vary only where equal to soft missing in *m*=0.

We discover that unregistered variable m is super varying (see [MI] **Glossary**). Here we no longer need m, so we will drop the variable and rerun mi varying. We will find that there are no remaining problems, so we will convert our data to our preferred wide style:

```
. drop m
. mi varying
              Possible problem    variable names
         ─────────────────────────────────────────────────
             imputed nonvarying:  (none)
             passive nonvarying:  (none)
             unregistered varying:  (none)
        *unregistered super/varying:  (none)
         unregistered super varying:  (none)
         ─────────────────────────────────────────────────
```

> * super/varying means super varying but would be varying if registered as
>   imputed; variables vary only where equal to soft missing in *m*=0.

```
. mi convert wide, clear
. list

      ┌──────────────────────────────────────────────────────────┐
      │ subject   a   b   c   _mi_miss   _1_b   _1_c   _2_b   _2_c │
      ├──────────────────────────────────────────────────────────┤
  1.  │    101    1   2   3        0        2      3      2      3 │
  2.  │    102    4   .   .        1      4.5    8.5    5.5    9.5 │
      └──────────────────────────────────────────────────────────┘
```

Also see

[MI] **intro** — Introduction to mi

[MI] **mi import** — Import data into mi

Title

---
mi import flongsep — Import flongsep-like data into mi

Syntax

mi import flongsep *name*, *required_options* [*true_options*]

where *name* is the name of the flongsep data to be created.

required_options	Description
using(*filenamelist*)	input filenames for $m = 1$, $m = 2$, ...
id(*varlist*)	identifying variable(s)

Note: use the input file for $m=0$ before issuing mi import flongsep.

true_options	Description
imputed(*varlist*)	imputed variables to be registered
passive(*varlist*)	passive variables to be registered
clear	okay to replace unsaved data in memory

Menu

Statistics > Multiple imputation

Description

mi import flongsep imports flongsep-like data, that is, data in which $m = 0$, $m = 1$, ..., $m = M$ are each recorded in separate .dta datasets.

mi import flongsep converts the data to mi flongsep and mi sets the data.

Options

using(*filenamelist*) is required; it specifies the names of the .dta datasets containing $m = 1$, $m = 2$, ..., $m = M$. The dataset corresponding to $m = 0$ is not specified; it is to be in memory at the time the mi import flongsep command is given.

The filenames might be specified as

    using(ds1 ds2 ds3 ds4 ds5)

which states that $m = 1$ is in file ds1.dta, $m = 2$ is in file ds2.dta, ..., and $m = 5$ is in file ds5.dta. Also, {#-#} is understood, so the above could just as well be specified as

    using(ds{1-5})

94

The braced numeric range may appear anywhere in the name, and thus

    using(ds{1-5}imp)

would mean that ds1imp.dta, ds2imp.dta, ..., ds5imp.dta contain $m = 1$, $m = 2$, ..., $m = 5$.

Alternatively, a comma-separated list can appear inside the braces. Filenames dsfirstm.dta, dssecondm.dta, ..., dsfifthm.dta can be specified as

    using(ds{first,second,third,fourth,fifth}m)

Filenames can be specified with or without the .dta suffix and may be enclosed in quotes if they contain special characters.

id(*varlist*) is required; it specifies the variable or variables that uniquely identify the observations in each dataset. The coding must be the same across datasets.

imputed(*varlist*) and passive(*varlist*) are truly optional options, although it would be unusual if imputed() were not specified.

imputed(*varlist*) specifies the names of the imputed variables.

passive(*varlist*) specifies the names of the passive variables.

clear specifies that it is okay to replace the data in memory even if they have changed since they were saved to disk.

Remarks

The procedure to convert flongsep-like data to mi flongsep is this:

1. use the dataset corresponding to $m = 0$.

2. Issue the mi import flongsep *name* command, where *name* is the name of the mi flongsep data to be created.

3. Perform the checks outlined in *Using mi import nhanes1, ice, flong, and flongsep* of [MI] **mi import**.

4. Use mi convert (see [MI] **mi convert**) to convert the data to a more convenient style such as wide, mlong, or flong.

For instance, you have been given the unset datasets imorig.dta, im1.dta, and im2.dta. You are told that these datasets contain the original data and two imputations, that variable b is imputed, and that variable c is passive and in fact equal to a + b. Here are the datasets:

    . use http://www.stata-press.com/data/r12/imorig
    . list

	subject	a	b	c
1.	101	1	2	3
2.	102	4	.	.

```
. use http://www.stata-press.com/data/r12/im1
. list
```

	subject	a	b	c
1.	101	1	2	3
2.	102	4	4.5	8.5

```
. use http://www.stata-press.com/data/r12/im2
. list
```

	subject	a	b	c
1.	101	1	2	3
2.	102	4	5.5	9.5

These are the same data discussed in [MI] **styles** but in unset form.

The fact that these datasets are nicely sorted is irrelevant. To import these datasets, you type

```
. use http://www.stata-press.com/data/r12/imorig
. mi import flongsep mymi, using(im1 im2) id(subject) imputed(b) passive(c)
```

We will now perform the checks outlined in *Using mi import nhanes1, ice, flong, and flongsep* of [MI] **mi import**, which are to run mi describe and mi varying to verify that variables are registered correctly:

```
. mi describe
  Style:  flongsep mymi
          last mi update 30mar2011 13:21:08, 0 seconds ago
  Obs.:   complete             1
          incomplete           1   (M = 2 imputations)
          ──────────────────────
          total                2
  Vars.:  imputed:  1; b(1)
          passive:  1; c(1)
          regular:  0
          system:   2; _mi_id _mi_miss
          (there are 2 unregistered variables; subject a)
. mi varying
              Possible problem    variable names
          ──────────────────────────────────────────────
                imputed nonvarying:   (none)
                passive nonvarying:   (none)
               unregistered varying:   (none)
          *unregistered super/varying:   (none)
           unregistered super varying:   (none)
          ──────────────────────────────────────────────
  * super/varying means super varying but would be varying if registered as
    imputed; variables vary only where equal to soft missing in m=0.
```

`mi varying` reported no problems. We finally convert to our preferred wide style:

```
. mi convert wide, clear
. list
```

	subject	a	b	c	_mi_miss	_1_b	_1_c	_2_b	_2_c
1.	101	1	2	3	0	2	3	2	3
2.	102	4	.	.	1	4.5	8.5	5.5	9.5

We are done with the converted data in flongsep format, so we will erase the files:

```
. mi erase mymi
(files mymi.dta _1_mymi.dta _2_mymi.dta erased)
```

Also see

[MI] **intro** — Introduction to mi

[MI] **mi import** — Import data into mi

Title

> **mi import ice** — Import ice-format data into mi

Syntax

mi import ice [, *options*]

options	Description
<u>auto</u>matic	register variables automatically
<u>imp</u>uted(*varlist*)	imputed variables to be registered
<u>pass</u>ive(*varlist*)	passive variables to be registered
clear	okay to replace unsaved data

Menu

Statistics > Multiple imputation

Description

mi import ice converts the data in memory to mi data, assuming the data in memory are in ice format. See Royston (2004, 2005a, 2005b, 2007, 2009) for a description of ice.

mi import ice converts the data to mi style flong. The data are mi set.

Options

automatic determines the identity of the imputed variables automatically. Use of this option is recommended.

imputed(*varlist*) specifies the names of the imputed variables. This option may be used with automatic, in which case automatic is taken to mean automatically determine the identity of imputed variables in addition to the imputed() variables specified. It is difficult to imagine why one would want to do this.

passive(*varlist*) specifies the names of the passive variables. This option may be used with automatic and usefully so. automatic cannot distinguish imputed variables from passive variables, so it assumes all variables that vary are imputed. passive() allows you to specify the subset of varying variables that are passive.

Concerning the above options: If none are specified, all variables are left unregistered in the result. You can then use mi varying to determine the varying variables and use mi register to register them appropriately; see [MI] **mi varying** and [MI] **mi set**. If you follow this approach, remember to register imputed variables before registering passive variables.

clear specifies that it is okay to replace the data in memory even if they have changed since they were last saved to disk. Remember, mi import ice starts with ice data in memory and ends with mi data in memory.

Remarks

The procedure to convert `ice` data to `mi` flong is

1. use the `ice` data.

2. Issue the `mi import ice` command, preferably with the `automatic` option and perhaps with the `passive()` option, too, although it really does not matter if passive variables are registered as imputed, so long as they are registered.

3. Perform the checks outlined in *Using mi import nhanes1, ice, flong, and flongsep* of [MI] **mi import**.

4. Use `mi convert` (see [MI] **mi convert**) to convert the data to a more convenient style such as wide or mlong.

For instance, you have the following ice data:

```
. use http://www.stata-press.com/data/r12/icedata
. list, separator(2)
```

	_mj	_mi	a	b	c
1.	0	1	1	2	3
2.	0	2	4	.	.
3.	1	1	1	2	3
4.	1	2	4	4.5	8.5
5.	2	1	1	2	3
6.	2	2	4	5.5	9.5

_mj and _mi are `ice` system variables. These data contain the original data and two imputations. Variable b is imputed, and variable c is passive and in fact equal to a + b. These are the same data discussed in [MI] **styles** but in `ice` format.

The fact that these data are nicely sorted is irrelevant. To import these data, you type

```
. mi import ice, automatic
(1 m=0 obs. now marked as incomplete)
```

although it would be even better if you typed

```
. mi import ice, automatic passive(c)
(1 m=0 obs. now marked as incomplete)
```

With the first command, both b and c will be registered as imputed. With the second, c will instead be registered as passive. Whether c is registered as imputed or passive makes no difference statistically.

These data are short enough that we can list the result:

. list, separator(2)

	a	b	c	_mi_m	_mi_id	_mi_miss
1.	1	2	3	0	1	0
2.	4	.	.	0	2	1
3.	1	2	3	1	1	.
4.	4	4.5	8.5	1	2	.
5.	1	2	3	2	1	.
6.	4	5.5	9.5	2	2	.

We will now perform the checks outlined in *Using mi import nhanes1, ice, flong, and flongsep* of [MI] **mi import**, which are to run mi describe and mi varying to verify that variables are registered correctly:

```
. mi describe
  Style:  flong
          last mi update 30mar2011 11:54:58, 0 seconds ago
   Obs.:  complete            1
          incomplete          1  (M = 2 imputations)
          ─────────────────────
          total               2
  Vars.:  imputed:   1; b(1)
          passive:   1; c(1)
          regular:   0
          system:    3; _mi_m _mi_id _mi_miss
          (there is one unregistered variable; a)
. mi varying
                 Possible problem    variable names
          ─────────────────────────────────────────
               imputed nonvarying:   (none)
               passive nonvarying:   (none)
            unregistered varying:    (none)
      *unregistered super/varying:   (none)
       unregistered super varying:   (none)
          ─────────────────────────────────────────

 * super/varying means super varying but would be varying if registered as
   imputed; variables vary only where equal to soft missing in m=0.
```

We find that there are no remaining problems, so we convert our data to our preferred wide style:

```
. mi convert wide, clear
. list
```

	a	b	c	_mi_miss	_1_b	_1_c	_2_b	_2_c
1.	1	2	3	0	2	3	2	3
2.	4	.	.	1	4.5	8.5	5.5	9.5

References

Royston, P. 2004. Multiple imputation of missing values. *Stata Journal* 4: 227–241.

———. 2005a. Multiple imputation of missing values: Update. *Stata Journal* 5: 188–201.

———. 2005b. Multiple imputation of missing values: Update of ice. *Stata Journal* 5: 527–536.

———. 2007. Multiple imputation of missing values: Further update of ice, with an emphasis on interval censoring. *Stata Journal* 7: 445–464.

———. 2009. Multiple imputation of missing values: Further update of ice, with an emphasis on categorical variables. *Stata Journal* 9: 466–477.

Also see

[MI] **intro** — Introduction to mi

[MI] **mi import** — Import data into mi

Title

> **mi import nhanes1** — Import NHANES-format data into mi

Syntax

 mi import nhanes1 *name*, *required_options* $\left[\ true\_options\ odd\_options\ \right]$

where *name* is the name of the flongsep data to be created.

required_options	Description
using(*filenamelist*)	input filenames for $m = 1$, $m = 2$, ...
id(*varlist*)	identifying variable(s)

Note: use the input file for $m=0$ before issuing mi import nhanes1.

true_options	Description
uppercase	prefix and suffix in uppercase
clear	okay to replace unsaved data in memory

odd_options	Description
nacode(*#*)	not applicable code; default is 0
obscode(*#*)	observed code; default is 1
impcode(*#*)	imputed code; default is 2
impprefix("*string*" "*string*")	variable prefix; default is "" ""
impsuffix("*string*" "*string*")	variable suffix; default is "if" "mi"

Note: The *odd_options* are not specified unless you need to import data that are nhanes1-like but not really nhanes1 format.

Menu

Statistics > Multiple imputation

Description

 mi import nhanes1 imports data recorded in the format used by the National Health and Nutrition Examination Survey (NHANES) produced by the National Center for Health Statistics (NCHS) of the U.S. Centers for Disease Control and Prevention (CDC); see http://www.cdc.gov/nchs/nhanes/nh3data.htm.

Options

 using(*filenamelist*) is required; it specifies the names of the .dta datasets containing $m = 1$, $m = 2$, ..., $m = M$. The dataset corresponding to $m = 0$ is not specified; it is to be in memory at the time the mi import nhanes1 command is given.

The filenames might be specified as

using(nh1 nh2 nh3 nh4 nh5)

which states that $m = 1$ is in file nh1.dta, $m = 2$ is in file nh2.dta, ..., and $m = 5$ is in file nh5.dta. Also, {#-#} is understood, so the files could just as well be specified as

using(nh{1-5})

The braced numeric range may appear anywhere in the name, and thus

using(nh{1-5}imp)

would mean that nh1imp.dta, nh2imp.dta, ..., nh5imp.dta contain $m = 1$, $m = 2$, ..., $m = 5$.

Alternatively, a comma-separated list can appear inside the braces. Filenames nhfirstm.dta, nhsecondm.dta, ..., nhfifthm.dta can be specified as

using(nh{first,second,third,fourth,fifth}m)

Filenames can be specified with or without the .dta suffix and must be enclosed in quotes if they contain special characters.

id(*varlist*) is required and is usually specified as id(seqn) or id(SEQN) depending on whether your variable names are in lowercase or uppercase. id() specifies the variable or variables that uniquely identify the observations in each dataset. Per the nhanes1 standard, the variable should be named seqn or SEQN.

uppercase is optional; it specifies that the variable suffixes IF and MI of the nhanes1 standard are in uppercase. The default is lowercase. (More correctly, when generalizing beyond nhanes1 format, the uppercase option specifies that all prefixes and suffixes are in uppercase.)

nacode(#), obscode(#), and impcode(#) are optional and are never specified when reading true nhanes1 data. The defaults nacode(0), obscode(1), and impcode(2) correspond to the nhanes1 definition. These options allow changing the codes for not applicable, observed, and imputed.

impprefix("*string*" "*string*") and impsuffix("*string*" "*string*") are optional and are never specified when reading true nhanes1 data. The defaults impprefix("" "") and impsuffix("if" "mi") correspond to the nhanes1 definition. These options allow setting different prefixes and suffixes.

clear specifies that it is okay to replace the data in memory even if they have changed since they were saved to disk. Remember, mi import nhanes1 starts with the first of the NHANES data in memory and ends with mi data in memory.

Remarks

Remarks are presented under the following headings:

Description of the nhanes1 format
Importing nhanes1 data

Description of the nhanes1 format

Nhanes1 is not really an official format; it is the format used for a particular dataset distributed by NCHS. Because there currently are no official or even informal standards for multiple-imputation data, perhaps the method used by the NCHS for NHANES will catch on, so we named it nhanes1. We included the 1 on the end of the name in case the format is modified.

Data in nhanes1 format consist of a collection of $M + 1$ separate files. The first file contains the original data. The remaining M files contain the imputed values for $m = 1$, $m = 2$, ..., $m = M$.

The first file contains a variable named seqn containing a sequence number. The file also contains other variables that comprise the nonimputed variables. Imputed variables, however, have their names suffixed with IF, standing for imputation flag, and those variables contain 1s, 2s, and 0s. 1 means that the value of the variable in that observation was observed, 2 means that the value was missing, and 0 means not applicable. Think of 0 as being equivalent to hard missing. The value is not observed for good reason and therefore was not imputed.

The remaining M files contain seqn and the imputed variables themselves. In these files, unobserved values are imputed. This time, imputed variable names are suffixed with MI.

Here is an example:

```
. use http://www.stata-press.com/data/r12/nhorig
. list
```

	seqn	a	bIF	cIF
1.	1	11	1	1
2.	2	14	2	2

The above is the first of the $M + 1$ datasets. The seqn variable is the sequence number. The a variable is a regular variable; we know that because the name does not end in IF. The b and c variables are imputed, and this dataset contains their imputation flags. Both variables are observed in the first observation and unobserved in the second.

Here is the corresponding dataset for $m = 1$:

```
. use http://www.stata-press.com/data/r12/nh1
. list
```

	seqn	bMI	cMI
1.	1	2	3
2.	2	4.5	8.5

This dataset states that in $m = 1$, b is equal to 2 and 4.5 and c is equal to 3 and 8.5.

We are about to show you the dataset for $m = 2$. Even before looking at it, however, we know that 1) it will have two observations; 2) it will have the seqn variable containing 1 and 2; 3) it will have two more variables named bMI and cMI; and 4) bMI will be equal to 2 and cMI will be equal to 3 in observations corresponding to seqn = 1. We know the last because in the first dataset, we learned that b and c were observed in seqn = 1.

```
. use http://www.stata-press.com/data/r12/nh2
. list
```

	seqn	a	bMI	cMI
1.	1	11	2	3
2.	2	14	5.5	9.5

Importing nhanes1 data

The procedure to import nhanes1 data is this:

1. use the dataset corresponding to $m = 0$; see [D] **use**.

2. Issue mi import nhanes1 *name* ..., where *name* is the name of the mi flongsep dataset to be created.

3. Perform the checks outlined in *Using mi import nhanes1, ice, flong, and flongsep* of [MI] **mi import**.

4. Use mi convert (see [MI] **mi convert**) to convert the data to a more convenient style such as wide, mlong, or flong.

To import the nhorig.dta, nh1.dta, and nh2.dta datasets described in the section above, we will specify mi import nhanes1's uppercase option because the suffixes were in uppercase. We type

```
. use http://www.stata-press.com/data/r12/nhorig
. mi import nhanes1 mymi, using(nh1 nh2) id(seqn) uppercase
```

The lack of any error message means that we have successfully converted nhanes1-format files nhorig.dta, nh1.dta, and nh2.dta to mi flongsep files mymi.dta, _1_mymi.dta, and _2_mymi.dta.

We will now perform the checks outlined in *Using mi import nhanes1, ice, flong, and flongsep* of [MI] **mi import**, which are to run mi describe and mi varying (see [MI] **mi describe** and [MI] **mi varying**) to verify that variables are registered correctly:

```
. mi describe
  Style:  flongsep mymi
          last mi update 30mar2011 12:58:46, 0 seconds ago
  Obs.:   complete           1
          incomplete         1   (M = 2 imputations)
          ─────────────────────
          total              2
  Vars.:  imputed:  2; b(1) c(1)
          passive:  0
          regular:  0
          system:   2; _mi_id _mi_miss
          (there are 2 unregistered variables; seqn a)
```

```
. mi varying
```

	Possible problem	variable names
imputed nonvarying:	(none)
passive nonvarying:	(none)
unregistered varying:	(none)
*unregistered super/varying:	(none)
unregistered super varying:	(none)

    * super/varying means super varying but would be varying if registered as
      imputed; variables vary only where equal to soft missing in $m=0$.

mi varying reported no problems.

We finally convert to style flong, although in real life we would choose styles mlong or wide. We are choosing flong because it is more readable:

```
. mi convert flong, clear
. list, separator(2)
```

	seqn	a	b	c	_mi_id	_mi_miss	_mi_m
1.	1	11	2	3	1	0	0
2.	2	14	.	.	2	1	0
3.	1	11	2	3	1	.	1
4.	2	14	4.5	8.5	2	.	1
5.	1	11	2	3	1	.	2
6.	2	14	5.5	9.5	2	.	2

The flong data are in memory. We are done with the converted data in flongsep format, so we erase the files:

```
. mi erase mymi
(files mymi.dta _1_mymi.dta _2_mymi.dta erased)
```

Also see

[MI] **intro** — Introduction to mi

[MI] **mi import** — Import data into mi

Title

> **mi import wide** — Import wide-like data into mi

Syntax

mi import wide [, *options*]

options	Description
imputed(*mvlist*)	imputed variables
passive(*mvlist*)	passive variables
dupsok	allow variable to be posted repeatedly
drop	drop imputed and passive after posting
clear	okay to replace unsaved data in memory

See description of options below for definition of *mvlist*.

Menu

Statistics > Multiple imputation

Description

mi import wide imports wide-like data, that is, data in which $m = 0$, $m = 1$, ..., $m = M$ values of imputed and passive variables are recorded in separate variables.

mi import wide converts the data to mi wide style and mi sets the data.

Options

imputed(*mvlist*) and passive(*mvlist*) specify the imputed and passive variables.

For instance, if the data had two imputed variables, x and y; x and y contained the $m = 0$ values; the corresponding $m = 1$, $m = 2$, and $m = 3$ values of x were in x1, x2, and x3; and the corresponding values of y were in y1, y2, and y3, then the imputed() option would be specified as

    imputed(x=x1 x2 x3 y=y1 y2 y3)

If variable y2 were missing from the data, you would specify

    imputed(x=x1 x2 x3 y=y1 . y3)

The same number of imputations must be specified for each variable.

dupsok specifies that it is okay if you specify the same variable name for two different imputations. This would be an odd thing to do, but if you specify dupsok, then you can specify

    imputed(x=x1 x1 x3 y=y1 y2 y3)

Without the dupsok option, the above would be treated as an error.

drop specifies that the original variables containing values for $m = 1$, $m = 2$, ..., $m = M$ are to be dropped from the data once mi import wide has recorded the values. This option is recommended.

clear specifies that it is okay to replace the data in memory even if they have changed since they were last saved to disk.

Remarks

The procedure to convert wide-like data to mi wide style is this:

1. use the unset data; see [D] **use**.

2. Issue the mi import wide command.

3. Use mi describe (see [MI] **mi describe**) and mi varying (see [MI] **mi varying**) to verify that the result is as you anticipated.

4. Optionally, use mi convert (see [MI] **mi convert**) to convert the data to what you consider a more convenient style.

For instance, you have been given unset dataset wi.dta and have been told that it contains variables a, b, and c; that variable b is imputed and contains $m = 0$ values; that variables b1 and b2 contain the $m = 1$ and $m = 2$ values; that variable c is passive (equal to $a + b$) and contains $m = 0$ values; and that variables c1 and c2 contain the corresponding $m = 1$ and $m = 2$ values. Here are the data:

```
. use http://www.stata-press.com/data/r12/wi
(mi prototype)
. list
```

	a	b	c	b1	b2	c1	c2
1.	1	2	3	2	2	3	3
2.	4	.	.	4.5	5.5	8.5	9.5

These are the same data discussed in [MI] **styles**. To import these data, type

```
. mi import wide, imputed(b=b1 b2  c=c1 c2) drop
```

These data are short enough that we can list the result:

```
. list
```

	a	b	c	_mi_miss	_1_b	_2_b	_1_c	_2_c
1.	1	2	3	0	2	2	3	3
2.	4	.	.	1	4.5	5.5	8.5	9.5

Returning to the procedure, we run mi describe and mi varying on the result:

```
. mi describe
  Style:  wide
          last mi update 30mar2011 08:51:09, 0 seconds ago
  Obs.:   complete                 1
          incomplete               1   (M = 2 imputations)
          ─────────────────────────
          total                    2
  Vars.:  imputed:  2; b(1) c(1)
          passive:  0
          regular:  0
          system:   1; _mi_miss
          (there is one unregistered variable; a)
. mi varying
                Possible problem   variable names
          ──────────────────────────────────────────────────────────
                imputed nonvarying:  (none)
                passive nonvarying:  (none)
          ──────────────────────────────────────────────────────────
```

Perhaps you would prefer seeing these data in flong style:

```
. mi convert flong, clear
. list, separator(2)
```

	a	b	c	_mi_miss	_mi_m	_mi_id
1.	1	2	3	0	0	1
2.	4	.	.	1	0	2
3.	1	2	3	.	1	1
4.	4	4.5	8.5	.	1	2
5.	1	2	3	.	2	1
6.	4	5.5	9.5	.	2	2

Also see

[MI] **intro** — Introduction to mi

[MI] **mi import** — Import data into mi

Title

> **mi impute** — Impute missing values

Syntax

mi <u>impu</u>te *method* ... [, *impute_options* ...]

method	Description
Univariate	
<u>regress</u>	linear regression for a continuous variable
pmm	predictive mean matching for a continuous variable
<u>truncreg</u>	truncated regression for a continuous variable with a restricted range
<u>intreg</u>	interval regression for a continuous partially observed (censored) variable
<u>logit</u>	logistic regression for a binary variable
<u>ologit</u>	ordered logistic regression for an ordinal variable
<u>mlogit</u>	multinomial logistic regression for a nominal variable
poisson	Poisson regression for a count variable
nbreg	negative binomial regression for an overdispersed count variable
Multivariate	
<u>monot</u>one	sequential imputation using a monotone-missing pattern
<u>chain</u>ed	sequential imputation using chained equations
mvn	multivariate normal regression

impute_options	Description
Main	
* add(#)	specify number of imputations to add; required when no imputations exist
* replace	replace imputed values in existing imputations
rseed(#)	specify random-number seed
double	save imputed values in double precision; the default is to save them as float
by(*varlist*[, *byopts*])	impute separately on each group formed by *varlist*
Reporting	
dots	display dots as imputations are performed
<u>noisi</u>ly	display intermediate output
<u>noleg</u>end	suppress all table legends
Advanced	
force	proceed with imputation, even when missing imputed values are encountered
<u>noupd</u>ate	do not perform mi update; see [MI] **noupdate option**

*add(#) is required when no imputations exist; add(#) or replace is required if imputations exist.
noupdate does not appear in the dialog box.
You must mi set your data before using mi impute; see [MI] **mi set**.

Menu

Description

mi impute fills in missing values (.) of a single variable or of multiple variables using the specified method. The available methods (by variable type and missing-data pattern) are summarized in the tables below.

Single imputation variable (univariate imputation)

Pattern	Type	Imputation method
	continuous	regress, pmm, truncreg, intreg
always monotone	binary	logit
	categorical	ologit, mlogit
	count	poisson, nbreg

Multiple imputation variables (multivariate imputation)

Pattern	Type	Imputation method
monotone missing	mixture	monotone
arbitrary missing	mixture	chained
arbitrary missing	continuous	mvn

The suggested reading order of mi impute's subentries is

[MI] **mi impute regress**

[MI] **mi impute pmm**

[MI] **mi impute truncreg**

[MI] **mi impute intreg**

[MI] **mi impute logit**

[MI] **mi impute ologit**

[MI] **mi impute mlogit**

[MI] **mi impute poisson**

[MI] **mi impute nbreg**

[MI] **mi impute monotone**

[MI] **mi impute chained**

[MI] **mi impute mvn**

Options

add(#) specifies the number of imputations to add to the mi data. This option is required if there are no imputations in the data. If imputations exist, then add() is optional. The total number of imputations cannot exceed 1,000.

replace specifies to replace existing imputed values with new ones. One of replace or add() must be specified when mi data already have imputations.

rseed(#) sets the random-number seed. This option can be used to reproduce results. rseed(#) is equivalent to typing set seed # prior to calling mi impute; see [R] **set seed**.

double specifies that the imputed values be stored as doubles. By default, they are stored as floats. mi impute makes this distinction only when necessary. For example, if the logit method is used, the imputed values are stored as bytes.

by(*varlist*[, *byopts*]) specifies that imputation be performed separately for each by-group. By-groups are identified by equal values of the variables in *varlist* in the original data ($m = 0$). Missing categories in *varlist* are omitted, unless the missing suboption is specified within by(). Imputed and passive variables may not be specified within by().

   byopts are missing, noreport, nolegend, and nostop.

   missing specifies that missing categories in *varlist* are not omitted.

   noreport suppresses reporting of intermediate information about each group.

   nolegend suppresses the display of group legends that appear before the imputation table when long group descriptions are encountered.

   nostop specifies to proceed with imputation when imputation fails in some groups. By default, mi impute terminates with error when this happens.

dots specifies to display dots as imputations are successfully completed. An x is displayed if any of the specified imputation variables still have missing values.

noisily specifies that intermediate output from mi impute be displayed.

nolegend suppresses the display of all legends that appear before the imputation table.

force specifies to proceed with imputation even when missing imputed values are encountered. By default, mi impute terminates with error if missing imputed values are encountered.

The following option is available with mi impute but is not shown in the dialog box:

noupdate in some cases suppresses the automatic mi update this command might perform; see [MI] **noupdate option**. This option is rarely used.

Remarks

Remarks are presented under the following headings:

Imputation methods

mi impute supports both univariate and multivariate imputation under the missing at random assumption (see *Assumptions about missing data* under *Remarks* in [MI] **intro substantive**).

Univariate imputation is used to impute a single variable. It can be used repeatedly to impute multiple variables only when the variables are independent and will be used in separate analyses. To impute a single variable, you can choose from the following methods: regress, pmm, truncreg, intreg, logit, ologit, mlogit, poisson, and nbreg; see [MI] **mi impute regress**, [MI] **mi impute pmm**, [MI] **mi impute truncreg**, [MI] **mi impute intreg**, [MI] **mi impute logit**, [MI] **mi impute ologit**, [MI] **mi impute mlogit**, [MI] **mi impute poisson**, and [MI] **mi impute nbreg**.

For a continuous variable, either regress or pmm can be used (for example, Rubin [1987] and Schenker and Taylor [1996]). For a continuous variable with a restricted range, a truncated variable, either pmm or truncreg (Raghunathan et al. 2001) can be used. For a continuous partially observed or censored variable, intreg can be used (Royston 2007). For a binary variable, logit can be used (Rubin 1987). For a categorical variable, ologit can be used to impute missing categories if they are ordered, and mlogit can be used to impute missing categories if they are unordered (Raghunathan et al. 2001). For a count variable, either poisson (Raghunathan et al. 2001) or nbreg (Royston 2009), in the presence of overdispersion, is often suggested. Also see van Buuren (2007) for a detailed list of univariate imputation methods.

Theory dictates that multiple variables usually must be imputed simultaneously, and that requires using a multivariate imputation method. The choice of an imputation method in this case also depends on the pattern of missing values.

If variables follow a monotone-missing pattern (see *Patterns of missing data* under *Remarks* in [MI] **intro substantive**), they can be imputed sequentially using univariate conditional distributions, which is implemented in the monotone method (see [MI] **mi impute monotone**). A separate univariate imputation model can be specified for each imputation variable, which allows simultaneous imputation of variables of different types (Rubin 1987).

When a pattern of missing values is arbitrary, iterative methods are used to fill in missing values. The mvn method (see [MI] **mi impute mvn**) uses multivariate normal data augmentation to impute missing values of continuous imputation variables (Schafer 1997). Allison (2001), for example, also discusses how to use this method to impute binary and categorical variables.

Another multivariate imputation method that accommodates arbitrary missing-value patterns is imputation using chained equations (ICE), also known as imputation using fully conditional specifications (van Buuren, Boshuizen, and Knook 1999) and as sequential regression multivariate imputation (Raghunathan et al. 2001) in the literature. The ICE method is implemented in the chained method (see [MI] **mi impute chained**) and uses a Gibbs-like algorithm to impute multiple variables sequentially using univariate fully conditional specifications. Despite a lack of rigorous theoretical justification, the flexibility of ICE has made it one of the most popular choices used in practice.

For a recent comparison of ICE and multivariate normal imputation, see Lee and Carlin (2010).

Imputation modeling

As discussed in [MI] **intro substantive**, imputation modeling is important to obtain proper imputations. Imputation modeling is not confined to the specification of an imputation method and an imputation model. It also requires careful consideration of how to handle complex data structures, such as survey or longitudinal data, and how to preserve existing relationships in the data during the imputation step. Rubin (1987), Meng (1994), Schafer (1997), Allison (2001), Royston (2007), Graham (2009), White, Royston, and Wood (2011), and others provide guidelines about imputation modeling. We summarize some of them below.

As with any statistical procedure, choosing an appropriate imputation approach is an art, and the choice should ultimately be determined by your data and research objectives. Regardless of which imputation approach you decide to pursue, it is good practice to check that your imputations are sensible before performing primary data analysis (see *Imputation diagnostics*) and to perform sensitivity analysis (for example, Kenward and Carpenter [2007]).

Model building

Perhaps the most important component of imputation modeling is the construction of an imputation model that preserves all the main characteristics of the observed data. This includes the following:

1. Use as many predictors as possible in the model to avoid making incorrect assumptions about the relationships between the variables. Omitting key predictors from the imputation model may lead to biased estimates for these predictors in the analysis. On the other hand, including insignificant predictors will result in less efficient yet still statistically valid results.

2. Include design variables representing the structure of the data in your imputation model. For example, sampling weights, strata and cluster identifiers of survey data, repeated-measures identifiers of longitudinal data must be included in the imputation model.

3. Specify the correct functional form of an imputation model. For example, include interactions of variables (or impute missing values separately using different subsamples; see *Imputing on subsamples*) to preserve higher-order dependencies.

The imputation model must be compatible with any model that can be used for the analysis. If variable X is to be included in the analysis model, it should also be used in the imputation model. If the analysis model estimates a correlation of X_1 and X_2, then both variables should be present in the imputation model. Accordingly, the outcome variable should always be present in the imputation model. Also, in addition to all the variables that may be used in the analysis model, you should include any auxiliary variables that may contain information about missing data. This will make the MAR assumption more plausible and will improve the quality of the imputed values. For more information about congeniality between the imputation and complete-data models, see Meng (1994).

As we mentioned above, it is important to specify the correct functional form of an imputation model to obtain proper imputations. The failure to accommodate such model features as interactions and nonlinearities during imputation may lead to severely biased results. There is no definitive recommendation for the best way to incorporate various functional forms into the imputation model. Currently, two main approaches are the joint modeling of all functional terms and modeling using passive variables (variables derived from imputation variables) also known as *passive imputation*. The joint modeling approach simply treats all functional terms as separate variables and imputes them together with the underlying imputation variables using a multivariate model, often a multivariate normal model. On the other hand, passive imputation—available within the ICE framework—fills in only the underlying imputation variables and computes the respective functional terms from the imputed variables, maintaining functional dependencies between the imputed and derived variables. The joint modeling approach imposes a rather stringent assumption of multivariate normality for possibly highly nonlinear terms and does not recognize functional dependencies between the imputed and derived variables. The naïve application of passive imputation, however, may omit certain functional relationships and thus lead to biased results. So, careful consideration for the specification of each conditional model is important. See White, Royston, and Wood (2011) for more details and some guidelines.

Outcome variables

Imputing outcome variables receive special attention in the literature because of the controversy about whether they should be imputed. As we already mentioned, it is important to include the outcome variable in the imputation model to obtain valid results. But what if the outcome variable itself has missing values? Should it be imputed? Should missing values be discarded from the analysis? There is no definitive answer to this question. The answer ultimately comes down to whether the specified imputation model describes the missing data adequately. When the percentage of missing values is low, using an incorrect imputation model may have little effect on the resulting repeated-imputation inference. With a large fraction of missing observations, a misspecified imputation model may distort the observed relationship between the outcome and predictor variables. In general, with large fractions of missing observations on any variable, the imputed values have more influence on the results, and thus more careful consideration of the imputation probability model is needed.

Transformations

Although the choice of an imputation method may not have significant impact on the results with low fractions of missing data, it may with larger fractions. A number of different imputation methods are available to model various types of imputation variables: continuous, categorical, count, and so on. However, in practice, these methods in no way cover all possible distributions that imputation variables may have. Often, the imputation variables can be transformed to the scale appropriate for an imputation method. For example, a log transformation (or, more generally, a Box–Cox transformation) can be used for highly skewed continuous variables to make them suitable for imputation using the linear regression method. If desired, the imputed values can be transformed back after the imputation. Transformations are useful when a variable has a restricted range. For instance, a preimputation logit transformation and a postimputation inverse logit transformation can be used to ensure that the imputed values are between 0 and 1.

It is important to remember that although the choice of a transformation is often determined based on the variable of interest alone, it is the conditional distribution of that variable given other predictors that is being modeled, and so the transformation must be suitable for it.

Categorical variables

To impute one categorical variable, you can use one of the categorical imputation methods: logistic, ordered logistic, or multinomial logistic regressions (see [MI] **mi impute logit**, [MI] **mi impute ologit**, or [MI] **mi impute mlogit**). These methods can also be used to impute multiple categorical variables with a monotone missing-data pattern using monotone imputation (see [MI] **mi impute monotone**) and with an arbitrary missing-data pattern using ICE (see [MI] **mi impute chained**). Also, for multiple categorical variables with only two categories (binary or dummy variables), a multivariate normal approach (see [MI] **mi impute mvn**) can be used to impute missing values and then, if needed, the imputed values can be rounded to 0 if the value is smaller than 0.5, or 1 otherwise. For categorical variables with more than two categories, Allison (2001) describes how to use the normal model to impute missing values.

The issue of perfect prediction during imputation of categorical data

Perfect prediction (or separation—for example, see Albert and Anderson [1984]) occurs often in the analysis of categorical data. The issue of perfect prediction is inherent to the discrete nature of categorical data and arises in the presence of covariate patterns for which outcomes of a categorical variable can be predicted almost perfectly. Perfect prediction usually leads to infinite coefficients with infinite standard errors and often causes numerical instability during estimation. This issue is often resolved by discarding the observations corresponding to offending covariate patterns as well as the independent variables perfectly predicting outcomes during estimation; see, for example, *Model identification* in [R] **logit**.

Perfect prediction is even more likely to arise during imputation because imputation models, per imputation modeling guidelines, tend to include many variables and thus may include many categorical variables. Perfect prediction may arise when variables are imputed using one of these imputation methods: logit, ologit, or mlogit.

Let's discuss how perfect prediction affects imputation. Recall that to obtain proper imputations (*Proper imputation methods* in [MI] **intro substantive**), imputed values must be simulated from the posterior predictive distribution of missing data given observed data. The categorical imputation methods achieve this by first drawing a new set of regression coefficients from a normal distribution (a large-sample approximation to their posterior distribution) with mean and variance determined by the maximum likelihood estimates of the coefficients from the observed data and their variance–covariance matrix. The imputed values are then obtained using the new set of coefficients; see *Methods and formulas* in the method-specific manual entries for details.

In the presence of perfect prediction, very large estimates of coefficients and their standard errors arise during estimation. As a result, new coefficients, drawn from the corresponding asymptotic normal distribution, will either be large and positive or large and negative. As such, missing values—say, of a binary imputation variable—may all be imputed as ones in some imputations and may all be imputed as zeros in other imputations. This will clearly bias the multiple-imputation estimate of the proportion of ones (or zeros) in the sample of perfectly predicted cases.

To eliminate the issue of perfect prediction during imputation, we cannot, unfortunately, drop observations and variables when estimating model parameters as is normally done during estimation using, for example, the logit command. Doing so would violate one of the main requirements of imputation modeling: all variables and cases that may be used during primary, completed-data analysis must be included in the imputation model. So, what can you do?

When perfect prediction is detected, `mi impute` issues an error message:

```
. mi impute logit x1 z1 z2 ..., ...
mi impute logit: perfect predictor(s) detected
    Variables that perfectly predict an outcome were detected when logit
    executed on the observed data. First, specify mi impute's option noisily
    to identify the problem covariates.  Then either remove perfect predictors
    from the model or specify mi impute logit's option augment to perform
    augmented regression; see The issue of perfect prediction during imputation
    of categorical data under Remarks in [MI] mi impute for details.
r(498);
```

You have two alternatives at this point.

You can fit the specified imputation model to the observed data using the corresponding command (in our example, `logit`) to identify the observations and variables causing perfect prediction in your data. Depending on the research objective and specifics of the data collection process, it may be reasonable to omit the offending covariate patterns and perfect predictors from your analysis. If you do so, you must carefully document which observations and variables were removed and adjust your inferential conclusions accordingly. Once offending instances are removed, you can proceed with imputation followed by your primary data analysis. Make sure that the instances you removed from the imputation model are not used in your further analysis.

The above approach may be difficult to pursue when imputing a large number of variables, among which are many categorical variables. Another option is to handle perfect prediction directly during imputation via the `augment` option, which is available for all categorical imputation methods: `logit`, `ologit`, and `mlogit`.

`mi impute ..., augment ...` implements an augmented-regression approach, an ad hoc but computationally convenient approach suggested by White, Daniel, and Royston (2010). According to this approach, a few extra observations with small weights are added to the data during estimation of model parameters in a way that prevents perfect prediction. See White, Daniel, and Royston (2010) for simulation results and computational details.

Convergence of iterative methods

When the missing-value pattern is arbitrary, iterative Markov chain Monte Carlo (MCMC-like) imputation methods are used to simulate imputed values from the posterior predictive distribution of the missing data given the observed data; also see *Multivariate imputation*. In this case, the resulting sequences (chains) of simulated parameters or imputed values should be examined to verify the convergence of the algorithm. The modeling task may be influenced by the convergence process of the algorithm given the data. For example, a different prior distribution for the model parameters may be needed with `mi impute mvn` when some aspects of the model cannot be estimated because of the sparseness of the missing data.

Markov chain simulation is often done in one of two ways: subsampling a single chain or running multiple independent chains. Subsampling a chain involves running a single chain for a prespecified number of iterations T, discarding the first b iterations until the chain reaches stationarity (the burn-in period), and sampling the chain each kth iteration to produce a final sequence of independent draws $\{\mathbf{X}^{(b)}, \mathbf{X}^{(b+k)}, \mathbf{X}^{(b+2k)}, \ldots\}$ from the target distribution. The number of between iterations k is chosen such that draws $\mathbf{X}^{(t)}$ and $\mathbf{X}^{(t+k)}$ are approximately independent. Alternatively, one can obtain independent draws by running multiple independent chains using different starting values $\{\mathbf{X}^{(i,t)} : t = 0, 1, \ldots\}$, $i = 1, 2, \ldots$, and discarding the first b iterations of each to obtain a final sample $\{\mathbf{X}^{(1,b)}, \mathbf{X}^{(2,b)}, \mathbf{X}^{(3,b)}, \ldots\}$ from the target distribution.

mi impute mvn subsamples the chain, whereas mi impute chained runs multiple independent chains; see [MI] **mi impute mvn** and [MI] **mi impute chained** for details on how to monitor convergence of each method.

Imputation diagnostics

After imputation, it is important to examine the sensibility of the obtained imputed values. If any abnormalities are detected, the imputation model must be revised. Diagnostics for imputations is still an ongoing research topic, but two general recommendations are to check model fit of the specified imputation model to the observed data and to compare distributions of the imputed and observed data. To check model fit of an imputation model to the observed data, you can use any standard postestimation tools usually used with that type of model. Also see, for example, [R] **mfp** to help determine an appropriate functional form of the imputation model. The differences (if any) between the distributions of the observed and of the imputed data should be plausible within the context of your study. For more information, see for example, Gelman et al. (2005), Abayomi, Gelman, and Levy (2008), and Marchenko and Eddings (2011) for how to perform multiple-imputation diagnostics in Stata.

Using mi impute

To use mi impute, you first mi set your data; see [MI] **mi set**. Next you register all variables whose missing values are to be imputed; see mi register in [MI] **mi set**.

mi impute has two main options: add() and replace. If you do not have imputations, use add() to create them. If you already have imputations, you have three choices:

1. Add new imputations to the existing ones by specifying the add() option.

2. Add new imputations and also replace the existing ones by specifying both the add() and the replace options.

3. Replace existing imputed values by specifying the replace option.

add() is required if no imputations exist in the mi data, and either add() or replace must be specified if imputations exist. See *Univariate imputation* for examples. Note that with replace, only imputed values of the specified imputation variables within the specified subsample will be updated.

For reproducibility, use the rseed() option to set the random-number seed, or equivalently, set the seed by using set seed immediately before calling mi impute. If you forget and still have mi impute's saved results in memory, you can retrieve the seed from the saved result r(rseed); see *Saved results* below.

By default, mi impute saves the imputed values using float precision. If you need more accuracy, you can specify the double option. Depending on the mi data style, the type of the imputed variable may change in the original data, $m = 0$. For example, if your data are in the mlong (or flong) style and you are imputing a binary variable using the regression method, the type of the variable will become float. If you are using the logistic method, the type of the variable may become byte even if originally your variable was declared as float or int. mi impute will never demote a variable if that would result in loss of precision.

Use the by(*varlist*) option to perform imputation separately on each group formed by *varlist*. Specifying by() is equivalent to the repeated use of an if condition with mi impute to restrict the imputation sample to each of the categories formed by *varlist*. Use the missing option within by() to prevent mi impute from omitting missing categories in *varlist*. By default, mi impute terminates with error if imputation fails in any of the groups; use by()'s nostop option to proceed with imputation. You may not specify imputation and passive variables within by().

mi impute terminates with error if the imputation procedure results in missing imputed values. This may happen if you include variables containing missing values as predictors in your imputation model. If desired, you can override this behavior with the `force` option.

mi impute may change the sort order of the data.

Univariate imputation

Univariate imputation by itself has limited application in practice. The situations in which only one variable needs to be imputed or in which multiple incomplete variables can be imputed independently are rare in real-data applications. Univariate imputation is most useful when it is used as a building block of sequential multivariate imputation methods; see *Multivariate imputation*. It is thus beneficial to first become familiar with univariate imputation.

Consider the heart attack data in which `bmi` contains missing values, as described in *A brief introduction to MI using Stata* of [MI] **intro substantive**. Here we use the already `mi set` version of the data with a subset of covariates of interest:

```
. use http://www.stata-press.com/data/r12/mheart1s0
(Fictional heart attack data; bmi missing)

. mi describe
  Style:  mlong
          last mi update 30mar2011 12:46:48, 1 day ago
  Obs.:   complete        132
          incomplete       22  (M = 0 imputations)
          ----------------------
          total           154
  Vars.:  imputed:  1; bmi(22)

          passive:  0

          regular:  5; attack smokes age female hsgrad

          system:   3; _mi_m _mi_id _mi_miss

          (there are no unregistered variables)
```

According to `mi describe`, the `mi` data style is mlong, and the dataset contains no imputations and 22 incomplete observations. The only registered imputed variable is `bmi` containing the 22 missing values. The other variables are registered as regular. See [MI] **mi describe** for details.

In the example in [MI] **intro substantive**, we used `mi impute regress` to impute missing values of `bmi`. Let's concentrate on the imputation step in more detail here:

```
. mi impute regress bmi attack smokes age female hsgrad, add(20)
Univariate imputation                     Imputations =     20
Linear regression                               added =     20
Imputed: m=1 through m=20                     updated =      0
```

	Observations per *m*			
Variable	Complete	Incomplete	Imputed	Total
bmi	132	22	22	154

```
(complete + incomplete = total; imputed is the minimum across m
 of the number of filled-in observations.)
```

The above output is common to all imputation methods of `mi impute`. In the left column, `mi impute` reports information about which imputation method was used and which imputations were created or updated. The right column contains the total number of imputations, and how many of them are new and how many are updated. The table contains the number of complete, incomplete, and imputed observations, and the total number of observations in the imputation sample, per imputation for each variable (see *Imputation and estimation samples* below). As indicated by the note, complete and incomplete observations sum to the total number of observations. The `imputed` column reports how many incomplete observations were actually imputed. This number represents the minimum across all imputations used ($m = 1$ through $m = 20$ in our example).

In the above example, we used `add(20)` to create 20 new imputations. Suppose that we decided that 20 is not enough and we want to add 30 more:

```
. mi impute regress bmi attack smokes age female hsgrad, add(30)
Univariate imputation                    Imputations =        50
Linear regression                              added =        30
Imputed: m=21 through m=50                   updated =         0
```

	Observations per *m*			
Variable	Complete	Incomplete	Imputed	Total
bmi	132	22	22	154

```
(complete + incomplete = total; imputed is the minimum across m
 of the number of filled-in observations.)
```

The table output is unchanged, but the header reports that total number of imputations is now 50. Thirty new imputations (from $m = 21$ to $m = 50$) were added, and the existing 20 imputations were left unchanged.

Suppose that we decide we want to impute `bmi` using the predictive mean matching (PMM) imputation method instead of the regression method. We use `mi impute pmm` and specify the `replace` option to update all existing imputations with new ones:

```
. mi impute pmm bmi attack smokes age female hsgrad, replace
Univariate imputation                    Imputations =        50
Predictive mean matching                       added =         0
Imputed: m=1 through m=50                     updated =        50
                                     Nearest neighbors =         1
```

	Observations per *m*			
Variable	Complete	Incomplete	Imputed	Total
bmi	132	22	22	154

```
(complete + incomplete = total; imputed is the minimum across m
 of the number of filled-in observations.)
```

The header reports that all 50 existing imputations, from $m = 1$ to $m = 50$, are replaced with new ones.

Later we decide to use more nearest neighbors than the default with `mi impute pmm` and also add 15 more imputations. We can do the latter by combining `replace` and `add()`. We specify `replace` to update the existing imputations with imputations from PMM with 3 nearest neighbors (`knn(3)`) and use `add(15)` to add 15 more imputations.

```
. mi impute pmm bmi attack smokes age female hsgrad, add(15) replace knn(3) dots
Imputing m=1 through m=65:
........10........20........30........40........50........60..... done
Univariate imputation              Imputations =       65
Predictive mean matching                  added =       15
Imputed: m=1 through m=65                updated =       50
                                 Nearest neighbors =        3
```

	Observations per m			
Variable	Complete	Incomplete	Imputed	Total
bmi	132	22	22	154

(complete + incomplete = total; imputed is the minimum across m
of the number of filled-in observations.)

The header reports a total of 65 imputations, among which 15 are new and 50 are updated. In this example, we also used the dots option to see the imputation progress. This option is useful with larger datasets to monitor the imputation process.

See *Imputing on subsamples* for other usage of add() and replace.

Multivariate imputation

When imputing multiple variables, their missing-data pattern must first be considered. As we briefly mentioned in *Patterns of missing data* in [MI] **intro substantive**, when a missing-data pattern is monotone distinct, multiple variables can be imputed sequentially without iteration using univariate conditional models (or monotone imputation). That is, a complicated multivariate imputation task can be replaced with a sequence of simpler univariate imputation tasks; see [MI] **mi impute monotone**.

Monotone missing-data patterns rarely arise naturally in practice. As such, it is important to be able to handle arbitrary missing-data patterns during imputation. Before we describe imputation methods accommodating arbitrary missing-data patterns, we will first discuss the difficulties arising with such patterns during imputation.

Monotone imputation is possible because variables can be ordered such that the complete observations of a variable being imputed are also complete in all prior imputed variables used to predict it. This means that the estimates of the parameters, which are obtained from complete data, do not depend on any previously imputed values (see Rubin [1987] for details). With an arbitrary pattern of missing data, such an ordering may not be possible because some variables may contain incomplete values in observations for which other variables are complete (and vice versa), resulting in estimated parameters being dependent on imputed values. The simultaneous imputation of multiple variables becomes more challenging when missingness is nonmonotone.

Consider the following example. Variable X_1 is complete in observation 1 and missing in observation 2, and variable X_2 is missing in observation 1 and complete in observation 2. We need to impute the two variables simultaneously. Suppose that we impute variable X_2 using previously imputed variable X_1. Observation 1, which contains an imputed value of X_1, is used to estimate the model parameters for X_2. As a result, the model parameters are obtained by treating the imputed value of X_1 as if it were true, thus ignoring the imputation variability in X_1. To account for the uncertainty in the imputed values during estimation, we need to iterate between the estimation step and the imputation step until the estimates of the model parameters depend only on the observed data.

Two main approaches for multivariate imputation with arbitrary missing-data patterns are joint modeling (JM) and fully conditional specification (FCS).

The JM approach assumes a genuine multivariate distribution for all imputation variables and imputes missing values as draws from the resulting posterior predictive distribution of the missing data given the observed data. The predictive distribution is often difficult to draw from directly, so the imputed values are often obtained by approximating this distribution using one of the MCMC methods. One such JM approach for continuous data is based on the multivariate normal distribution, the MVN method (Schafer 1997). The MVN method is implemented in [MI] **mi impute mvn** and uses the data augmentation MCMC method.

The FCS approach does not assume an explicit multivariate distribution for all imputation variables. Instead, it provides a set of chained equations, that is, univariate conditional distributions of each variable with fully conditional specifications of prediction equations. This approach is also known as ICE (van Buuren, Boshuizen, and Knook 1999) or sequential regression multivariate imputation (SRMI; Raghunathan et al. 2001). We will be using the terms ICE, FCS, and SRMI interchangeably throughout the documentation. ICE is similar in spirit to the Gibbs sampler, a popular MCMC method for simulating data from complicated multivariate distributions. Unlike the Gibbs sampler, however, conditional specifications within the ICE method are not guaranteed to correspond to a genuine multivariate distribution because ICE does not start from an explicit multivariate density. Regardless, ICE remains one of the popular imputation methods in practice. The ICE method is implemented in [MI] **mi impute chained**.

Currently, there is no definitive recommendation in the literature as to which approach, JM or FCS, is preferable. The JM approach ensures that imputed values are drawn from a genuine multivariate distribution, and it thus may be more attractive from a theoretical standpoint. However, except for simpler cases such as a multivariate normal model for continuous data, it may not be feasible to formulate a joint model for general data structures. In this regard, the FCS approach is more appealing because it not only can accommodate mixtures of different types of variables, but also can preserve some important characteristics often observed in real data, such as restrictions to subpopulations for certain variables and range restrictions. The tradeoff for such flexibility is a current lack of theoretical justification. See Lee and Carlin (2010) and references therein for more discussion about the two approaches.

Consider the heart attack data in which both bmi and age contain missing values. Again we will use data that have already been mi set.

```
. use http://www.stata-press.com/data/r12/mheart5s0, clear
(Fictional heart attack data; bmi and age missing)

. mi describe
  Style:  mlong
          last mi update 30mar2011 12:46:48, 1 day ago
  Obs.:   complete          126
          incomplete         28    (M = 0 imputations)
          _____
          total             154
  Vars.:  imputed:   2; bmi(28) age(12)
          passive:   0
          regular:   4; attack smokes female hsgrad
          system:    3; _mi_m _mi_id _mi_miss
          (there are no unregistered variables)
```

There are 28 incomplete observations in the dataset. The bmi variable contains 28 missing values and the age variable contains 12 missing values. Both bmi and age are registered as imputed. If we assume that age and BMI are independent, we can impute each of them separately by using the previously described univariate imputation methods. It is likely, however, that these variables are related, and so we use multivariate imputation.

First, we examine missing-value patterns of the data.

```
. mi misstable patterns
    Missing-value patterns
      (1 means complete)

                     |    Pattern
       Percent       |    1   2
    -----------------+------------
          82%        |    1   1
                     |
          10         |    1   0
           8         |    0   0
    -----------------+------------
         100%        |

    Variables are   (1) age   (2) bmi
```

From the output, 82% of observations are complete, 10% of observations contain missing values for `bmi`, and 8% of observations have both `bmi` and `age` missing. We can see that the dataset has a monotone-missing pattern (see [MI] **intro substantive**), that is, missing values of `age` are nested within missing values of `bmi`. Another way to see if the pattern of missingness is monotone is to use `mi misstable nested` ([MI] **mi misstable**):

```
. mi misstable nested
    1.  age(12) -> bmi(28)
```

Because the missing-data pattern is monotone, we can use `mi impute monotone` to impute missing values of `bmi` and `age` simultaneously:

```
. mi impute monotone (regress) age bmi = attack smokes hsgrad female, add(10)

Conditional models:
                age: regress age attack smokes hsgrad female
                bmi: regress bmi age attack smokes hsgrad female

Multivariate imputation                    Imputations =       10
Monotone method                                  added =       10
Imputed: m=1 through m=10                      updated =        0

                age: linear regression
                bmi: linear regression
```

	Observations per m			
Variable	Complete	Incomplete	Imputed	Total
age	142	12	12	154
bmi	126	28	28	154

```
(complete + incomplete = total; imputed is the minimum across m
 of the number of filled-in observations.)
```

Without going into detail, `mi impute monotone` imputes missing values of multiple variables by performing a sequence of independent univariate conditional imputations. In the above example, the regression method is used to impute missing values of both variables. `age` is imputed first from the observed variables `attack`, `smokes`, `hsgrad`, and `female`. Then `bmi` is imputed using the imputed `age` variable in addition to other observed variables. The output is consistent with that of the univariate imputation methods described earlier, with some additional information. See [MI] **mi impute monotone** for details.

We can also impute missing values of bmi and age simultaneously using either mi impute mvn

```
. mi impute mvn age bmi = attack smokes hsgrad female, replace nolog
```

Multivariate imputation   Imputations =   10
Multivariate normal regression   added =   0
Imputed: $m=1$ through $m=10$ updated =   10

Prior: uniform   Iterations =   1000
                          burn-in =   100
                        between =   100

	Observations per m			
Variable	Complete	Incomplete	Imputed	Total
age	142	12	12	154
bmi	126	28	28	154

(complete + incomplete = total; imputed is the minimum across m
of the number of filled-in observations.)

or mi impute chained

```
. mi impute chained (regress) age bmi = attack smokes hsgrad female, replace
```
note: missing-value pattern is monotone; no iteration performed
Conditional models (monotone):
          age: regress age attack smokes hsgrad female
          bmi: regress bmi age attack smokes hsgrad female

Performing chained iterations ...

Multivariate imputation   Imputations =   10
Chained equations   added =   0
Imputed: $m=1$ through $m=10$ updated =   10

Initialization: monotone   Iterations =   0
                           burn-in =   0

          age: linear regression
          bmi: linear regression

	Observations per m			
Variable	Complete	Incomplete	Imputed	Total
age	142	12	12	154
bmi	126	28	28	154

(complete + incomplete = total; imputed is the minimum across m
of the number of filled-in observations.)

Neither mi impute mvn nor mi impute chained requires the missing-data pattern to be monotone. mi impute mvn iterates to produce imputations. When the data are monotone missing, however, no iteration is required, and because mi impute monotone executes more quickly, it is preferred. mi impute chained also iterates to produce imputations, unless the missing-data pattern is monotone. However, mi impute monotone is still faster because it performs estimation only once on the original data, whereas mi impute chained performs estimation on each imputation. Use mi impute mvn and mi impute chained when there is an arbitrary missing-data pattern. See [MI] **mi impute mvn** and [MI] **mi impute chained** for details.

Imputing on subsamples

Consider the earlier example of the univariate imputation of `bmi`. Suppose that we want to perform imputation separately for females and males. Imputation on subsamples is useful when the imputation model must accommodate the interaction effects (see, for example, Allison [2001]). For example, if we want the effect of `bmi` on `attack` to vary by gender, we can perform imputation of `bmi` separately for females and males.

We first show how to do it manually using `if` and the `add()` and `replace` options:

```
. use http://www.stata-press.com/data/r12/mheart1s0, clear
(Fictional heart attack data; bmi missing)

. mi impute regress bmi attack smokes age hsgrad if female==1, add(20)
Univariate imputation              Imputations =        20
Linear regression                       added =        20
Imputed: m=1 through m=20             updated =         0
```

	Observations per m			
Variable	Complete	Incomplete	Imputed	Total
bmi	33	5	5	38

```
(complete + incomplete = total; imputed is the minimum across m
 of the number of filled-in observations.)

. mi impute regress bmi attack smokes age hsgrad if female==0, replace
Univariate imputation              Imputations =        20
Linear regression                       added =         0
Imputed: m=1 through m=20             updated =        20
```

	Observations per m			
Variable	Complete	Incomplete	Imputed	Total
bmi	99	17	17	116

```
(complete + incomplete = total; imputed is the minimum across m
 of the number of filled-in observations.)
```

First, we created 20 imputations and filled in the missing values of `bmi` for females by using the corresponding subset of observations. Then we filled in the remaining missing values of `bmi` for males in the existing imputations by using the subset of male observations. We will now be able to include the interaction between `bmi` and `female` in our logistic model.

A much easier way to do the above is to use `by()`:

```
. use http://www.stata-press.com/data/r12/mheart1s0
(Fictional heart attack data; bmi missing)

. mi impute regress bmi attack smokes age hsgrad, add(20) by(female)
```

Performing setup for each by() group:

```
-> female = 0

-> female = 1
```

```
Univariate imputation                      Imputations =         20
Linear regression                                added =         20
Imputed: m=1 through m=20                      updated =          0
```

by()	Observations per m			
Variable	Complete	Incomplete	Imputed	Total
female = 0				
bmi	99	17	17	116
female = 1				
bmi	33	5	5	38
Overall				
bmi	132	22	22	154

```
(complete + incomplete = total; imputed is the minimum across m
 of the number of filled-in observations.)
```

Conditional imputation

Often in practice, some variables are defined only within what we call a conditional sample, a subset of observations satisfying certain restrictions (Raghunathan et al. 2001, Royston 2009). For example, the number of cigarettes smoked is relevant to smokers only, the number of pregnancies is relevant to females only, etc. Outside the conditional sample, such variables are assumed to contain soft missing values and a nonmissing constant value, further referred to as a conditional constant, which represents a known value or an inadmissible value. We will refer to conditional imputation as imputation of such variables. So, the task of conditional imputation is to impute missing values of a variable within a conditional sample using only observations from that sample and to replace missing values outside the conditional sample with a conditional constant.

In the previous section, we learned that we can specify an `if` condition with `mi impute` to restrict imputation of variables to a subset of observations. Is this sufficient to accommodate conditional imputation? To answer this question, let's consider several examples.

We use our heart attack data as an example. Suppose that our only variable containing missing values is `hightar`, the indicator for smoking high-tar cigarettes. We want to impute missing values in `hightar` and use it among other predictors in the logistic analysis of heart attacks. Because `hightar` is relevant to smokers only, we want to impute `hightar` using the subset of observations with `smokes==1`.

Thus to impute `hightar`, we restrict our imputation sample to smokers:

```
. mi impute logit hightar attack age bmi ... if smokes==1, ...
```

Are we now ready to proceed with our primary logistic analysis of heart attacks? Not quite. Recall that we wish to use all observations of `hightar` in our analysis. If `hightar` contains missing values only in the conditional sample, `smokes==1`, we are finished. Otherwise, we need to replace all remaining missing values outside the conditional sample, for `smokes==0`, with the conditional constant, the nonmissing value of `hightar` in observations with `smokes==0`. In our example, this value is zero, so our final step is

```
. mi xeq: replace hightar = 0 if smokes==0
```

What if we have several imputation variables? Suppose now that `age` and `bmi` also contain missing values. Without making any assumptions about a missing-data pattern, we use `mi impute chained` to impute variables of different types: `age`, `bmi`, and `hightar`. We need to impute `hightar` for `smokes==1` but use the unrestricted sample to impute `age` and `bmi`. Can we still accomplish this by specifying an `if` condition? The answer is yes, but we need to replace missing values of `hightar` for `smokes==0` before imputation to ensure that `age` and `bmi` are imputed properly, using all observations, when `hightar` is used in their prediction equations:

```
. mi xeq: replace hightar = 0 if smokes==0
. mi impute chained (regress) bmi age (logit if smokes==1) hightar = ..., ...
```

It seems that we can get away with using `if` to perform conditional imputation. What is the catch? So far, we assumed that `smokes` does not contain any missing values. Let's see what happens if it does.

Because `hightar` depends on `smokes`, we must first impute missing values of `smokes` before we can impute missing values of `hightar`. As such, the set of observations for which `smokes==1` will vary from imputation to imputation and, in the case of `mi impute chained`, from iteration to iteration. The replacement of missing values of `hightar` outside the conditional sample should be performed each time a new set of imputed values is obtained for `smokes`, and thus must be directly incorporated into the imputation procedure.

The answer to our earlier question about using an `if` condition to perform conditional imputation is no, in general. To perform conditional imputation, use the `conditional()` option:

```
. mi imp chained (reg) bmi age (logit) smokes (logit, conditional(if smokes==1))
> hightar ...
```

Every univariate imputation method supports option `conditional()`. This option is most useful within specifications of univariate methods when multiple variables are being imputed using `mi impute monotone` or `mi impute chained`, as we showed above. Although in some cases, as we saw earlier, specifying an `if` condition in combination with manual replacement of missing values outside the conditional sample may produce equivalent results, such use should generally be avoided and `conditional()` should be used instead.

When you specify option `conditional()`, `mi impute` performs checks necessary for proper conditional imputation. For example, the imputed variable is verified to be constant outside the conditional sample and an error message is issued if it is not:

```
. mi impute logit hightar age bmi ..., conditional(if smokes==1)
conditional(): imputation variable not constant outside conditional sample;
    hightar is not constant outside the subset identified by (smokes==1)
    within the imputation sample
r(459);
```

mi impute also requires that missing values of all variables involved in conditional specifications (restrictions)—that is, conditioning variables—be nested within missing values of the conditional variable being imputed. If this does not hold true, mi impute issues an error message:

```
. mi impute logit hightar age bmi ..., conditional(if smokes==1)
conditional(): conditioning variables not nested;
    conditioning variable smokes is not nested within hightar
r(459);
```

Because missing values of all conditioning variables are assumed to be nested within missing values of a conditional variable, that conditional variable is not included in the prediction equations of the corresponding conditioning variables.

As an example, let's continue with our heart attack data, in which variables hightar and smokes contain missing values, as do age and bmi:

```
. use http://www.stata-press.com/data/r12/mheart7s0
(Fictional heart attack data; bmi, age, hightar, and smokes missing)
. mi describe
  Style:  mlong
          last mi update 25mar2011 11:00:38, 3 days ago
  Obs.:   complete        124
          incomplete       30   (M = 0 imputations)
          ───────────────────
          total           154
  Vars.:  imputed:  4; bmi(24) age(30) hightar(8) smokes(5)
          passive:  0
          regular:  3; attack female hsgrad
          system:   3; _mi_m _mi_id _mi_miss
          (there are no unregistered variables)
. mi misstable nested
    1.  smokes(5) -> hightar(8) -> bmi(24) -> age(30)
```

Our data are already mi set, so we proceed with imputation. According to mi misstable nested, all imputation variables are monotone missing, so we use mi impute monotone for imputation. For the purpose of illustration, we create only two imputations:

```
. mi impute monotone (regress) bmi age
>                    (logit, conditional(if smokes==1)) hightar
>                    (logit) smokes
>                                = attack hsgrad female, add(2)
Conditional models:
           smokes: logit smokes attack hsgrad female
          hightar: logit hightar i.smokes attack hsgrad female ,
                   conditional(if smokes==1)
              bmi: regress bmi i.hightar i.smokes attack hsgrad female
              age: regress age bmi i.hightar i.smokes attack hsgrad female
note: 1.smokes omitted because of collinearity

Multivariate imputation              Imputations =         2
Monotone method                            added =         2
Imputed: m=1 through m=2                  updated =         0
Conditional imputation:
  hightar: incomplete out-of-sample obs. replaced with value 0
              bmi: linear regression
              age: linear regression
          hightar: logistic regression
           smokes: logistic regression
```

	Observations per m			
Variable	Complete	Incomplete	Imputed	Total
bmi	130	24	24	154
age	124	30	30	154
hightar	146	8	8	154
smokes	149	5	5	154

```
(complete + incomplete = total; imputed is the minimum across m
of the number of filled-in observations.)
```

For each variable that was imputed conditionally, mi impute reports the conditional value used to replace all missing observations outside the conditional sample in a legend about conditional imputation. In our example, all missing values of hightar outside smokes==1 are replaced with zero. The reported numbers of complete, incomplete, and imputed observations for hightar correspond to the entire imputation sample (see *Imputation and estimation samples*) and not only to the conditional sample. For example, there are 146 complete and 8 incomplete observations of hightar in the combined sample of smokers and nonsmokers. The minimum number of imputed values across imputations is 8, so all incomplete observations of hightar were filled in—either imputed directly or replaced with a conditional value—in both imputations. Because smokes is being imputed, the numbers of incomplete and imputed observations of hightar for smokers and nonsmokers will vary across imputations.

You can accommodate more complicated restrictions or skip patterns, which often arise with questionnaire data, by specifying more elaborate restrictions within conditional() or by specifying the conditional() option with other variables. For example, suppose that the information about cigarette tar level (hightar) was collected only for heavy smokers, identified by an indicator variable heavysmoker. The heavysmoker variable contains missing values and needs to be imputed before hightar can be imputed. To impute heavysmoker, we need to restrict our sample to smokers only. Then to impute hightar, we need to use only heavy smokers among all smokers. We can do so as follows:

```
. mi impute chained (logit) smokes                                    ///
              (logit, conditional(if smokes==1)) heavysmoker           ///
              (logit, conditional(if smokes==1 & heavysmoker==1)) hightar ...
```

Imputation and estimation samples

Rubin (1987, 160–166) describes the imputation process as three tasks: modeling, estimation, and imputation. We concentrate on the latter two tasks here. The posterior distribution of the model parameters is estimated during the estimation task. This posterior distribution is used in the imputation task to simulate the parameters of the posterior predictive distribution of the missing data from which an imputed value is drawn. Accordingly, mi impute distinguishes between two main samples: imputation and estimation.

The imputation sample is determined by the imputation variables used in the imputation task. It is comprised of all observations for which the imputation variables contain no hard missing values (or no extended missing values). In other words, the imputation sample consists of the complete and incomplete observations as identified by the specified imputation variables. The estimation sample is comprised of all observations used by the model fit to the observed data during the estimation task.

For example,

```
. use http://www.stata-press.com/data/r12/mheart1s0, clear
(Fictional heart attack data; bmi missing)

. mi impute regress bmi attack smokes age hsgrad female, add(1) noisily

Running regress on observed data:
```

Source	SS	df	MS		Number of obs =	132
					F(5, 126) =	1.24
Model	99.5998228	5	19.9199646		Prob > F =	0.2946
Residual	2024.93667	126	16.070926		R-squared =	0.0469
					Adj R-squared =	0.0091
Total	2124.5365	131	16.2178358		Root MSE =	4.0089

bmi	Coef.	Std. Err.	t	P>|t|	[95% Conf. Interval]	
attack	1.71356	.7515229	2.28	0.024	.2263179	3.200801
smokes	-.5153181	.761685	-0.68	0.500	-2.02267	.9920341
age	-.033553	.0305745	-1.10	0.275	-.0940591	.026953
hsgrad	-.4674308	.8112327	-0.58	0.566	-2.072836	1.137975
female	-.3072767	.8074763	-0.38	0.704	-1.905249	1.290695
_cons	26.96559	1.884309	14.31	0.000	23.2366	30.69458

```
Univariate imputation                    Imputations =        1
Linear regression                            added =        1
Imputed: m=1                               updated =        0
```

		Observations per m			
Variable		Complete	Incomplete	Imputed	Total
bmi		132	22	22	154

```
(complete + incomplete = total; imputed is the minimum across m
 of the number of filled-in observations.)
```

The imputation sample contains 154 observations and the estimation sample contains 132 observations (from the regression output). The estimation task of mi impute regress consists of fitting a linear regression of bmi on other variables to the observed data. We specified the noisily option to see results from the estimation task. Usually, the number of complete observations in the imputation sample (132 in this example) will be equal to the number of observations used in the estimation. Sometimes, however, observations may be dropped from the estimation—for example, if independent variables contain missing values. In this case, the number of complete observations in the imputation sample and the number of observations used in the estimation will be different, and the following note will appear following the table output:

```
Note: right-hand-side variables (or weights) have missing values;
      model parameters estimated using listwise deletion
```

You should evaluate such cases to verify that results are as expected.

In general, missing values in independent variables (or in a weighting variable) do not affect the imputation sample but they may lead to missing imputed values. In the above example, if age contained missing values in incomplete observations of bmi, the linear prediction for those observations would have been missing and thus the resulting imputed values would have been missing, too.

Imputing on subsamples, or in other words, using an if condition with mi impute, restricts both imputation and estimation samples to include only observations satisfying the if condition. Conditional imputation (the conditional() option), on the other hand, affects only the estimation sample. All values, within and outside of a conditional sample, except extended missing values, are included in the imputation sample. With conditional imputation, the reported number of complete observations will almost always be different from the number of observations in the estimation sample, unless the conditional sample coincides with the imputation sample. In the case of observations being dropped from a conditional sample during estimation, a note as shown above will appear following the table output.

Imputing transformations of incomplete variables

Continuing with the univariate example above, say that we discover that the distribution of bmi is skewed to the right, and thus we decide to impute bmi on the logarithmic scale instead of the original one. We can do this by creating a new variable, lnbmi, and imputing it instead of bmi.

What we will do is create lnbmi, register it as imputed, impute it, and then create bmi as a passive variable based on the formula bmi = exp(lnbmi).

We need to be careful when we create lnbmi to get its missing values right. mi respects two kinds of missing values, called soft and hard missing. Soft missing values are missing values eligible for imputation. Hard missing values are missing values that are to remain missing even in the imputed data. Soft missing are recorded as ordinary missing (.), and hard missing are recorded as any of extended missing (.a–.z).

The issue here is that missing values could arise because of our application of the transform lnbmi = ln(bmi). In the case of the ln() transform, missing values will be created whenever bmi $\leq$ 0. (In general, transformations leading to undefined values should be avoided so that all available observed data are used during imputation.) Body mass index does not contain such values, but let's pretend it did. Here is what we would do:

1. Create lnbmi = ln(bmi).

2. Replace lnbmi to contain .z in observations for which lnbmi contains missing but bmi does not.

3. Register lnbmi as an imputed variable and impute it.

4. Create passive variable newbmi = exp(lnbmi).

5. Replace newbmi equal to bmi in observations for which newbmi is missing and bmi is not.

Alternatively, to avoid creating hard missing values in step 2, we could consider a different transformation; see, for example, [R] lnskew0.

As we said, for lnbmi $= \ln(\text{bmi})$ we need not perform all the steps above because bmi > 0. In the bmi case, all we need to do is

1. Create lnbmi $= \ln(\text{bmi})$.

2. Register lnbmi as an imputed variable and impute it.

3. Create passive variable newbmi $= \exp(\text{lnbmi})$.

If all we wanted to do was impute lnbmi $= \ln(\text{bmi})$ and, from that point on, just work with lnbmi, we would perform only the first two steps of the three-step procedure.

All that said, we are going to perform the five-step procedure because it will always work. We will continue from where we left off in the last example, so we will discard our previous imputation efforts by typing mi set M = 0. (Instead of typing mi set M = 0, we could just as easily begin by typing use http://www.stata-press.com/data/r12/mheart1s0.)

```
. mi set M = 0                                      // start again
. mi unregister bmi                                 // we do not impute bmi
. generate lnbmi = ln(bmi)                          // create lnbmi
. replace lnbmi = .z if lnbmi==. & bmi!=.
. mi register imputed lnbmi
. mi impute regress lnbmi attack smokes age hsgrad female, add(5)
. mi passive: gen newbmi = exp(lbmi)
. mi passive: replace newbmi = bmi if bmi!=.
```

The important thing about the above is the mechanical definition of an imputed variable. An imputed variable is a variable we actually impute, not a variable we desire to impute. In this case, we imputed lnbmi and derived bmi from it. Thus the variable we desired to impute became, mechanically, a passive variable.

Saved results

mi impute saves the following in r():

Scalars

r(M)	total number of imputations
r(M_add)	number of added imputations
r(M_update)	number of updated imputations
r(k_ivars)	number of imputed variables
r(N_g)	number of imputed groups (1 if by() is not specified)

Macros

r(method)	name of imputation method
r(ivars)	names of imputation variables
r(rseed)	random-number seed
r(by)	names of variables specified within by()

Matrices

r(N)	number of observations in imputation sample in each group (per variable)
r(N_complete)	number of complete observations in imputation sample in each group (per variable)
r(N_incomplete)	number of incomplete observations in imputation sample in each group (per variable)
r(N_imputed)	number of imputed observations in imputation sample in each group (per variable)

Also see *Saved results* in the method-specific manual entries for additional saved results.

Methods and formulas

All imputation methods (except predictive mean matching) are based on simulating from a Bayesian (approximate) posterior predictive distribution of missing data. Univariate imputation methods and the sequential monotone method use noniterative techniques for simulating from the posterior predictive distribution of missing data. The imputation method based on multivariate normal regression uses an iterative MCMC technique to simulate from the posterior predictive distribution of missing data. The ICE method uses a Gibbs-like algorithm to obtain imputed values.

See *Methods and formulas* in the method-specific manual entries for details.

Herman Otto Hartley (1912–1980) was born in Germany as Herman Otto Hirschfeld and immigrated to England in 1934 after completing his PhD in mathematics at Berlin University. He completed a second PhD in mathematical statistics under John Wishart at Cambridge in 1940 and went on to hold positions at Harper Adams Agricultural College, Scientific Computing Services (London), University College (London), Iowa State College, Texas A&M University, and Duke University. Among other awards he received and distinguished titles he held, Professor Hartley served as the president of the American Statistical Association in 1979. Known affectionately as HOH by almost all who knew him, he founded the Institute of Statistics, later to become the Department of Statistics, at Texas A&M University. His contributions to statistical computing are particularly notable considering the available equipment at the time. Professor Hartley is best known for his two-volume *Biometrika Tables for Statisticians* (jointly written with Egon Pearson) and for his fundamental contributions to sampling theory, missing-data methodology, variance-component estimation, and computational statistics.

References

Abayomi, K., A. Gelman, and M. Levy. 2008. Diagnostics for multivariate imputations. *Journal of the Royal Statistical Society, Series C* 57: 273–291.

Albert, A., and J. A. Anderson. 1984. On the existence of maximum likelihood estimates in logistic regression models. *Biometrika* 71: 1–10.

Allison, P. D. 2001. *Missing Data*. Thousand Oaks, CA: Sage.

Gelman, A., I. Van Mechelen, G. Verbeke, D. F. Heitjan, and M. Meulders. 2005. Multiple imputation for model checking: Completed-data plots with missing and latent data. *Biometrics* 61: 74–85.

Graham, J. W. 2009. Missing data analysis: Making it work in the real world. *Annual Review of Psychology* 60: 549–576.

Kenward, M. G., and J. R. Carpenter. 2007. Multiple imputation: Current perspectives. *Statistical Methods in Medical Research* 16: 199–218.

Lee, K. J., and J. B. Carlin. 2010. Multiple imputation for missing data: Fully conditional specification versus multivariate normal imputation. *American Journal of Epidemiology* 171: 624–632.

Marchenko, Y. V., and W. D. Eddings. 2011. A note on how to perform multiple-imputation diagnostics in Stata. http://www.stata.com/users/ymarchenko/midiagnote.pdf.

Meng, X.-L. 1994. Multiple-imputation inferences with uncongenial sources of input (with discussion). *Statistical Science* 9: 538–573.

Raghunathan, T. E., J. M. Lepkowski, J. Van Hoewyk, and P. Solenberger. 2001. A multivariate technique for multiply imputing missing values using a sequence of regression models. *Survey Methodology* 27: 85–95.

Royston, P. 2007. Multiple imputation of missing values: Further update of ice, with an emphasis on interval censoring. *Stata Journal* 7: 445–464.

———. 2009. Multiple imputation of missing values: Further update of ice, with an emphasis on categorical variables. *Stata Journal* 9: 466–477.

Rubin, D. B. 1987. *Multiple Imputation for Nonresponse in Surveys.* New York: Wiley.

Schafer, J. L. 1997. *Analysis of Incomplete Multivariate Data.* Boca Raton, FL: Chapman & Hall/CRC.

Schenker, N., and J. M. G. Taylor. 1996. Partially parametric techniques for multiple imputation. *Computational Statistics & Data Analysis* 22: 425–446.

van Buuren, S. 2007. Multiple imputation of discrete and continuous data by fully conditional specification. *Statistical Methods in Medical Research* 16: 219–242.

van Buuren, S., H. C. Boshuizen, and D. L. Knook. 1999. Multiple imputation of missing blood pressure covariates in survival analysis. *Statistics in Medicine* 18: 681–694.

White, I. R., R. Daniel, and P. Royston. 2010. Avoiding bias due to perfect prediction in multiple imputation of incomplete categorical data. *Computational Statistics & Data Analysis* 54: 2267–2275.

White, I. R., P. Royston, and A. M. Wood. 2011. Multiple imputation using chained equations: Issues and guidance for practice. *Statistics in Medicine* 30: 377–399.

Also see

[MI] **mi estimate** — Estimation using multiple imputations

[MI] **intro substantive** — Introduction to multiple-imputation analysis

[MI] **intro** — Introduction to mi

[MI] **Glossary**

Title

> **mi impute chained** — Impute missing values using chained equations

Syntax

Default specification of prediction equations, basic syntax

> mi <u>imp</u>ute <u>chain</u>ed (*uvmethod*) *ivars* [= *indepvars*] [*if*] [*weight*] [, *impute_options options*]

Default specification of prediction equations, full syntax

> mi <u>imp</u>ute <u>chain</u>ed *lhs* [= *indepvars*] [*if*] [*weight*] [, *impute_options options*]

Custom specification of prediction equations

> mi <u>imp</u>ute <u>chain</u>ed *lhsc* [= *indepvars*] [*if*] [*weight*] [, *impute_options options*]

where *lhs* is *lhs_spec* [*lhs_spec* [...]] and *lhs_spec* is

(*uvmethod* [*if*] [, *uvspec_options*]) *ivars*

lhsc is *lhsc_spec* [*lhsc_spec* [...]] and *lhsc_spec* is

(*uvmethod* [*if*] [, <u>incl</u>ude(*xspec*) omit(*varlist*) <u>noimp</u>uted *uvspec_options*]) *ivars*

ivars (or *newivar* if *uvmethod* is intreg) are the names of the imputation variables.

uvspec_options are <u>asc</u>ontinuous, <u>nois</u>ily, and the method-specific *options* as described in the manual entry for each univariate imputation method.

The include(), omit(), and noimputed options allow you to customize the default prediction equations.

uvmethod	Description
<u>regress</u>	linear regression for a continuous variable; [MI] **mi impute regress**
pmm	predictive mean matching for a continuous variable; [MI] **mi impute pmm**
truncreg	truncated regression for a continuous variable with a restricted range; [MI] **mi impute truncreg**
intreg	interval regression for a continuous censored variable; [MI] **mi impute intreg**
<u>logit</u>	logistic regression for a binary variable; [MI] **mi impute logit**
<u>ologit</u>	ordered logistic regression for an ordinal variable; [MI] **mi impute ologit**
<u>mlogit</u>	multinomial logistic regression for a nominal variable; [MI] **mi impute mlogit**
poisson	Poisson regression for a count variable; [MI] **mi impute poisson**
nbreg	negative binomial regression for an overdispersed count variable; [MI] **mi impute nbreg**

135

impute_options	Description
Main	
*add(#)	specify number of imputations to add; required when no imputations exist
*replace	replace imputed values in existing imputations
rseed(#)	specify random-number seed
double	save imputed values in double precision; the default is to save them as float
by(*varlist*[, *byopts*])	impute separately on each group formed by *varlist*
Reporting	
dots	display dots as imputations are performed
noisily	display intermediate output
nolegend	suppress all table legends
Advanced	
force	proceed with imputation, even when missing imputed values are encountered
noupdate	do not perform mi update; see [MI] **noupdate option**

*add(#) is required when no imputations exist; add(#) or replace is required if imputations exist.
noupdate does not appear in the dialog box.

options	Description
ICE options	
burnin(#)	specify number of iterations for the burn-in period; default is burnin(10)
chainonly	perform chained iterations for the length of the burn-in period without creating imputations in the data
augment	perform augmented regression in the presence of perfect prediction for all categorical imputation variables
noimputed	do not include imputation variables in any prediction equation
bootstrap	estimate model parameters using sampling with replacement
savetrace(...)	save summaries of imputed values from each iteration in *filename*.dta
Reporting	
dryrun	show conditional specifications without imputing data
report	show report about each conditional specification
chaindots	display dots as chained iterations are performed
showevery(#)	display intermediate results from every #th iteration
showiter(*numlist*)	display intermediate results from every iteration in *numlist*
Advanced	
orderasis	impute variables in the specified order
nomonotone	impute using chained equations even when variables follow a monotone-missing pattern; default is to use monotone method
nomonotonechk	do not check whether variables follow a monotone-missing pattern

You must mi set your data before using mi impute chained; see [MI] **mi set**.

You must mi register *ivars* as imputed before using mi impute chained; see [MI] **mi set**.

indepvars may contain factor variables; see [U] **11.4.3 Factor variables**.

fweights, aweights (regress, pmm, truncreg, and intreg only), iweights, and pweights are allowed; see [U] **11.1.6 weight**.

Menu

Statistics > Multiple imputation

Description

mi impute chained fills in missing values in multiple variables iteratively by using chained equations, a sequence of univariate imputation methods with fully conditional specification (FCS) of prediction equations. It accommodates arbitrary missing-value patterns. You can perform separate imputations on different subsets of the data by specifying the by() option. You can also account for frequency, analytic (with continuous variables only), importance, and sampling weights.

Options

```
                 ┌─ Main ┐
```

add(), replace, rseed(), double, by(); see [MI] **mi impute**.

The following options appear on a Specification dialog that appears when you click on the **Create ...** button on the **Main** tab. The include(), omit(), and noimputed options allow you to customize the default prediction equations.

include(*xspec*) specifies that *xspec* be included in prediction equations of all imputation variables corresponding to the current left-hand-side specification *lhsc_spec*. *xspec* includes complete variables and expressions of imputation variables bound in parentheses. If the noimputed option is specified within *lhsc_spec* or with mi impute chained, then *xspec* may also include imputation variables. *xspec* may contain factor variables; see [U] **11.4.3 Factor variables**.

omit(*varlist*) specifies that *varlist* be omitted from the prediction equations of all imputation variables corresponding to the current left-hand-side specification *lhsc_spec*. *varlist* may include complete variables or imputation variables. *varlist* may contain factor variables; see [U] **11.4.3 Factor variables**. In omit(), you should list variables to be omitted exactly as they appear in the prediction equation (abbreviations are allowed). For example, if variable x1 is listed as a factor variable, use omit(i.x1) to omit it from the prediction equation.

noimputed specifies that no imputation variables automatically be included in prediction equations of imputation variables corresponding to the current *uvmethod*.

uvspec_options are options specified within each univariate imputation method, *uvmethod*. *uvspec_options* include ascontinuous, noisily, and the method-specific *options* as described in the manual entry for each univariate imputation method.

ascontinuous specifies that categorical imputation variables corresponding to the current *uvmethod* be included as continuous in all prediction equations. This option is only allowed when *uvmethod* is logit, ologit, or mlogit.

noisily specifies that the output from the current univariate model fit to the observed data be displayed. This option is useful in combination with the showevery(#) or showiter(*numlist*) option to display results from a particular univariate imputation model for specific iterations.

burnin(#) specifies the number of iterations for the burn-in period for each chain (one chain per imputation). The default is burnin(10). This option specifies the number of iterations necessary for a chain to reach approximate stationarity or, equivalently, to converge to a stationary distribution. The required length of the burn-in period will depend on the starting values used and the missing-data patterns observed in the data. It is important to examine the chain for convergence to determine an adequate length of the burn-in period prior to obtaining imputations; see *Convergence of ICE*. The provided default is what current literature recommends. However, you are responsible for determining that sufficient iterations are performed.

chainonly specifies that mi impute chained perform chained iterations for the length of the burn-in period and then stop. This option is useful in combination with savetrace() to examine the convergence of the method prior to imputation. No imputations are created when chainonly is specified, so add() or replace is not required with mi impute chained, chainonly and they are ignored if specified.

augment specifies that augmented regression be performed if perfect prediction is detected. By default, an error is issued when perfect prediction is detected. The idea behind the augmented-regression approach is to add a few observations with small weights to the data during estimation to avoid perfect prediction. See *The issue of perfect prediction during imputation of categorical data* under *Remarks* in [MI] **mi impute** for more information. augment is not allowed with importance weights. This option is equivalent to specifying augment within univariate specifications of all categorical imputation methods: logit, ologit, and mlogit.

noimputed specifies that no imputation variables automatically be included in any of the prediction equations. This option is seldom used. This option is convenient if you wish to use different sets of imputation variables in all prediction equations. It is equivalent to specifying noimputed within all univariate specifications.

bootstrap specifies that posterior estimates of model parameters be obtained using sampling with replacement; that is, posterior estimates are estimated from a bootstrap sample. The default is to sample the estimates from the posterior distribution of model parameters or from the large-sample normal approximation of the posterior distribution. This option is useful when asymptotic normality of parameter estimates is suspect. This option is equivalent to specifying bootstrap within all univariate specifications.

savetrace(*filename*[, *traceopts*]) specifies to save the means and standard deviations of imputed values from each iteration to a Stata dataset called *filename*.dta. If the file already exists, the replace suboption specifies to overwrite the existing file. savetrace() is useful for monitoring convergence of the chained algorithm.

traceopts are replace, double, and detail.

replace indicates that *filename*.dta be overwritten if it exists.

double specifies that the variables be stored as doubles, meaning 8-byte reals. By default, they are stored as floats, meaning 4-byte reals. See [D] **data types**.

detail specifies that additional summaries of imputed values including the smallest and the largest values and the 25th, 50th, and 75th percentiles are saved in *filename*.dta.

dots, noisily, nolegend; see [MI] **mi impute**. noisily specifies that the output from all univariate conditional models fit to the observed data be displayed. nolegend suppresses all imputation table legends that include a legend with the titles of the univariate imputation methods used, a legend about conditional imputation when `conditional()` is used within univariate specifications, and group legends when `by()` is specified.

`dryrun` specifies to show the conditional specifications that would be used to impute each variable without actually imputing data. This option is recommended for checking specifications of conditional models prior to imputation.

`report` specifies to show a report about each univariate conditional specification. This option, in combination with `dryrun`, is recommended for checking specifications of conditional models prior to imputation.

`chaindots` specifies that all chained iterations be displayed as dots. An `x` is displayed for every failed iteration.

`showevery(#)` specifies that intermediate regression output be displayed for every #th iteration. This option requires `noisily`. If `noisily` is specified with `mi impute chained`, then the output from the specified iterations is displayed for all univariate conditional models. If `noisily` is used within a univariate specification, then the output from the corresponding univariate model from the specified iterations is displayed.

`showiter(numlist)` specifies that intermediate regression output be displayed for each iteration in *numlist*. This option requires `noisily`. If `noisily` is specified with `mi impute chained`, then the output from the specified iterations is displayed for all univariate conditional models. If `noisily` is used within a univariate specification, then the output from the corresponding univariate model from the specified iterations is displayed.

⌐ Advanced ⌐

`force`; see [MI] **mi impute**.

`orderasis` requests that the variables be imputed in the specified order. By default, variables are imputed in order from the most observed to the least observed.

`nomonotone`, a rarely used option, specifies not to use monotone imputation and to proceed with chained iterations even when imputation variables follow a monotone-missing pattern. `mi impute chained` checks whether imputation variables have a monotone missing-data pattern and, if they do, imputes them using the monotone method (without iteration). If `nomonotone` is used, `mi impute chained` imputes variables iteratively even if variables are monotone-missing.

`nomonotonechk` specifies not to check whether imputation variables follow a monotone-missing pattern. By default, `mi impute chained` checks whether imputation variables have a monotone missing-data pattern and, if they do, imputes them using the monotone method (without iteration). If `nomonotonechk` is used, `mi impute chained` does not check the missing-data pattern and imputes variables iteratively even if variables are monotone-missing. Once imputation variables are established to have an arbitrary missing-data pattern, this option may be used to avoid potentially time-consuming checks; the monotonicity check may be time consuming when a large number of variables is being imputed.

The following option is available with `mi impute` but is not shown in the dialog box:

`noupdate`; see [MI] **noupdate option**.

Remarks

Remarks are presented under the following headings:

Multivariate imputation using chained equations
Compatibility of conditionals
Convergence of ICE
First use
Using mi impute chained
Default prediction equations
Custom prediction equations
Link between mi impute chained and mi impute monotone
Examples

See [MI] **mi impute** for a general description and details about options common to all imputation methods, *impute_options*. Also see [MI] **workflow** for general advice on working with mi.

Multivariate imputation using chained equations

When a missing-data structure is monotone distinct, multiple variables can be imputed sequentially without iteration by using univariate conditional models (see [MI] **mi impute monotone**). Such monotone imputation is impossible with arbitrary missing-data patterns, and simultaneous imputation of multiple variables in such cases requires iteration. We described the impact of an arbitrary missing-data pattern on multivariate imputation and two common imputation approaches used in such cases, the multivariate normal method and imputation using chained equations (ICE), in *Multivariate imputation* in [MI] **mi impute**. In this entry, we describe ICE, also known as imputation using FCS (van Buuren, Boshuizen, and Knook 1999) or sequential regression multivariate imputation (SRMI; Raghunathan et al. 2001), in more detail. We use the terms ICE, FCS, and SRMI interchangeably throughout the documentation.

ICE is similar to monotone imputation in the sense that it is also based on a series of univariate imputation models. Unlike monotone imputation, ICE uses FCSs of prediction equations (chained equations) and requires iteration. Iteration is needed to account for possible dependence of the estimated model parameters on the imputed data when a missing-data structure is not monotone distinct.

The general idea behind ICE is to impute multiple variables iteratively via a sequence of univariate imputation models, one for each imputation variable, with fully conditional specifications of prediction equations: all variables except the one being imputed are included in a prediction equation. Formally, for imputation variables $X_1, X_2, \ldots, X_p$ and complete predictors (independent variables) $\mathbf{Z}$, this procedure can be described as follows. Imputed values are drawn from

$$
\begin{aligned}
X_1^{(t+1)} &\sim g_1(X_1 | X_2^{(t)}, \ldots, X_p^{(t)}, \mathbf{Z}, \phi_1) \\
X_2^{(t+1)} &\sim g_2(X_2 | X_1^{(t+1)}, X_3^{(t)}, \ldots, X_p^{(t)}, \mathbf{Z}, \phi_2) \\
&\cdots \\
X_p^{(t+1)} &\sim g_p(X_p | X_1^{(t+1)}, X_2^{(t+1)}, \ldots, X_{p-1}^{(t+1)}, \mathbf{Z}, \phi_p)
\end{aligned}
\tag{1}
$$

for iterations $t = 0, 1, \ldots, T$ until convergence at $t = T$, where ϕ_j are the corresponding model parameters with a uniform prior. The univariate imputation models, $g_j(\cdot)$, can each be of a different type (normal, logistic, etc.), as is appropriate for imputing X_j.

Fully conditional specifications (1) are similar to the Gibbs sampling algorithm (Geman and Geman 1984; Gelfand and Smith 1990), one of the MCMC methods for simulating from complicated multivariate distributions. In fact, in certain cases these specifications do correspond to a genuine Gibbs sampler. For example, when all X_js are continuous and all $g_j(\cdot)$s are normal linear regressions with constant variances, then (1) corresponds to a Gibbs sampler based on a multivariate normal distribution with a uniform prior for model parameters. Such correspondence does not hold in general because unlike the Gibbs sampler, the conditional densities $\{g_j(\cdot), \ j = 1, 2, \ldots, p\}$ may not correspond to any multivariate joint conditional distribution of $X_1, X_2, \ldots, X_p$ given $\mathbf{Z}$ (Arnold, Castillo, and Sarabia 2001). This issue is known as incompatibility of conditionals (for example, Arnold, Castillo, and Sarabia [1999]). When conditionals are not compatible, the ICE procedure may not converge to any stationary distribution, which can raise concerns about its validity as a principled statistical method; see *Compatibility of conditionals* and *Convergence of ICE* for more details.

Despite the lack of a general theoretical justification, ICE is very popular in practice. Its popularity is mainly due to the tremendous flexibility it offers for imputing various types of data arising in observational studies. Similarly to monotone imputation, the variable-by-variable specification of ICE allows practitioners to simultaneously impute variables of different types by choosing from several univariate imputation methods appropriate for each variable. Being able to specify a separate model for each variable provides an imputer with great flexibility in incorporating certain characteristics specific to each variable. For example, we can use predictive mean matching ([MI] **mi impute pmm**) or truncated regression ([MI] **mi impute truncreg**) to impute a variable with a restricted range. We can impute variables defined on a subsample using only observations in that subsample while using the entire sample to impute other variables; see *Conditional imputation* in [MI] **mi impute** for details. For more information about imputation using chained equations, see van Buuren, Boshuizen, and Knook (1999); Raghunathan et al. (2001); van Buuren et al. (2006); van Buuren (2007); White, Royston, and Wood (2011); and Royston (2004, 2005a, 2005b, 2007, 2009), among others.

The specification of a conditional imputation model $g_j(\cdot)$ includes an imputation method and a prediction equation relating an imputation variable to other explanatory variables. In what follows, we distinguish between the default specification (of prediction equations) in which the identities of the complete explanatory variables are the same across all prediction equations, and the custom specification in which the identities are allowed to differ.

Under the default specification, prediction equations of each imputation variable include all complete independent variables and all imputation variables except the one being imputed. Under the custom specification, each prediction equation may include a subset of the predictors that would be used under the default specification. The custom specification also allows expressions of imputation variables in prediction equations.

Model (1) corresponds to the default specification. For example, consider imputation variables X_1, X_2, and X_3 and complete predictors Z_1 and Z_2. Under the default specification, the individual prediction equations are determined as follows. The most observed variable—say, X_1—is predicted from X_2, X_3, Z_1, and Z_2. The next most observed variable—say, X_2—is predicted from X_3, Z_1, Z_2, and previously imputed X_1. The least observed variable, X_3, is predicted from Z_1, Z_2, and previously imputed X_1 and X_2. (A constant is included in all prediction equations, by default.) We use the following notation to refer to the above sequence of prediction equations (imputation sequence): $X_1|\mathbf{X}_{-1}, Z_1, Z_2 \to X_2|\mathbf{X}_{-2}, Z_1, Z_2 \to X_3|\mathbf{X}_{-3}, Z_1, Z_2$, where $\mathbf{X}_{-j}$ denotes all imputed or to-be-imputed variables except X_j.

A sequence such as $X_1|\mathbf{X}_{-1}, Z_1 \to X_2|\mathbf{X}_{-2}, Z_1, Z_2 \to X_3|\mathbf{X}_{-3}, Z_2$ would correspond to a custom specification. Here X_1 is assumed to be conditionally independent of Z_2, and X_3 is assumed to be conditionally independent of Z_1.

Compatibility of conditionals

A concern with ICE is its lack of a formal theoretical justification. Its theoretical weakness is possible incompatibility of fully conditional specifications (1). As we briefly mentioned earlier, it is possible to specify a set of full conditionals with ICE for which no multivariate distribution exists (for example, van Buuren et al. [2006] and van Buuren [2007]). In such a case, the validity of ICE as a statistical procedure is questionable.

The impact of incompatibility of conditional specifications in practice is still under investigation. For example, van Buuren et al. (2006) performed several simulations to investigate the consequences of strongly incompatible specifications on multiple-imputation (MI) results in a simple setting and found very little impact of it on estimated parameters. The effect of incompatible conditionals on the quality of imputations and final MI inference in general is not yet known. Of course, if a joint model is of main scientific interest, then incompatibility of conditionals poses a problem. In the discussion of Arnold, Castillo, and Sarabia (2001), Andrew Gelman and Trivellore Raghunathan mention that the existence of an underlying joint distribution may be less important within the imputation context than the ability to incorporate the unique features of the data.

For more information about the compatibility of conditional specifications, see Arnold, Castillo, and Sarabia (2001); van Buuren (2007); and Arnold, Castillo, and Sarabia (1999) and references therein.

Convergence of ICE

ICE is an iterative method and is similar in spirit to the Gibbs sampler, an MCMC method. Similarly to MCMC methods, ICE builds a sequence of draws $\left\{\mathbf{X}_m^{(t)} : t = 1, 2, \ldots\right\}$, a chain, and iterates until this chain reaches a stationary distribution. So as with any MCMC method, monitoring convergence is important with ICE.

ICE performs simulation by running multiple independent chains (see *Convergence of iterative methods* in [MI] **mi impute**). To assess convergence of multiple chains, we need to examine the stationarity of each chain by the end of the specified burn-in period b. In practice, convergence of ICE is often examined visually. Trace plots—plots of summaries of the distribution (means, standard deviations, quantiles, etc.) of imputed values against iteration numbers—are used to examine stationarity of the chain. Long-term trends in trace plots are indicative of slow convergence to stationarity. A suitable value for the burn-in period b can be inferred from a trace plot as the earliest iteration after which each chain does not exhibit a visible trend and the fluctuations in values become more regular. When the initial values are close to the mode of the target posterior distribution (when one exists), b will generally be small. When the initial values are far off in the tails of the posterior distribution, the initial number of iterations b will generally be larger.

The number of iterations necessary for ICE to converge depends on, among other things, the fractions of missing information and initial values. The higher the fractions of missing information and the farther the initial values are from the mode of the posterior predictive distribution of missing data, the slower the convergence, and thus the larger the number of iterations required. Current literature suggests that in many practical applications a low number of burn-in iterations, somewhere between 5 and 20 iterations, is usually sufficient for convergence (for example, van Buuren [2007]). In any case, examination of the data and missing-data patterns is highly recommended when investigating convergence of ICE.

The convergence of ICE may not be achieved when specified conditional models are incompatible, as described in *Compatibility of conditionals*. The simulation draws will depend on the order in which variables are imputed and on the chosen length of the burn-in period. It is important to evaluate the quality of imputations (see *Imputation diagnostics* in [MI] **mi impute**) to determine the impact of incompatibility on MI analysis.

First use

Before we describe various uses of mi impute chained, let's look at a simple example first.

Consider the heart attack data example examining the relationship between heart attacks and smoking from *Multivariate imputation* of [MI] **mi impute**, where the age and bmi variables contain missing values. In another version of the dataset, bmi and age have a nonmonotone missing-data pattern, and thus monotone imputation is not possible:

```
. use http://www.stata-press.com/data/r12/mheart8s0
(Fictional heart attack data; bmi and age missing; arbitrary pattern)
. mi misstable patterns, frequency
   Missing-value patterns
    (1 means complete)
                |  Pattern
     Frequency  | 1  2
    ------------+--------
          118   | 1  1
                |
           24   | 1  0
            8   | 0  1
            4   | 0  0
    ------------+--------
          154   |
  Variables are  (1) age  (2) bmi
```

mi impute chained does not require missing data to be monotone, so we can use it to impute missing values of age and bmi in this dataset. We use the same model specification as before:

```
. mi impute chained (regress) bmi age = attack smokes hsgrad female, add(10)
Conditional models:
          age: regress age bmi attack smokes hsgrad female
          bmi: regress bmi age attack smokes hsgrad female

Performing chained iterations ...

Multivariate imputation                  Imputations =       10
Chained equations                              added =       10
Imputed: m=1 through m=10                     updated =        0

Initialization: monotone                  Iterations =      100
                                             burn-in =       10

             bmi: linear regression
             age: linear regression
```

	Observations per *m*			
Variable	Complete	Incomplete	Imputed	Total
bmi	126	28	28	154
age	142	12	12	154

```
(complete + incomplete = total; imputed is the minimum across m
 of the number of filled-in observations.)
```

As before, 10 imputations are created (the add(10) option). The linear regression imputation method (regress) is used to impute both continuous variables. The attack, smokes, hsgrad, and female variables are used as complete predictors (independent variables).

`mi impute chained` reports the conditional specifications used to impute each variable and the order in which they were imputed. By default, `mi impute chained` imputes variables in order from the most observed to the least observed. In our example, `age` has the least number of missing values and so is imputed first, even though we listed `bmi` before `age` in the command specification.

With the default specification, `mi impute chained` builds appropriate FCSs automatically using the supplied imputation variables and complete predictors, specified as right-hand-side variables. The default prediction equation for `age` includes `bmi` and all the complete predictors, and the default prediction equation for `bmi` includes `age` and all the complete predictors.

The main header and table output were described in detail in [MI] **mi impute**. The information specific to `mi impute chained` includes the type of initialization, the burn-in period, and the number of iterations. By default, `mi impute chained` uses 10 burn-in iterations (also referred to as cycles in the literature) before drawing imputed values. The total number of iterations performed by `mi impute chained` to obtain 10 imputations is 100. Also, similarly to `mi impute monotone`, the additional information above the table includes the legend describing what univariate imputation method was used to impute each variable. (If desired, this legend may be suppressed by specifying the `nolegend` option.)

Using mi impute chained

Below we summarize general capabilities of `mi impute chained`.

1. `mi impute chained` offers two main syntaxes—one using the default prediction equations and the other allowing customization of prediction equations. We will refer to the two syntaxes as default and custom, respectively. We describe the two syntaxes in detail in the next two sections.

2. `mi impute chained` allows specification of a global (outer) `if` condition,

   `. mi impute chained ... if` *exp* `...`

   and equation-specific (inner) `if` conditions,

   `. mi impute chained ... (... if` *exp* `...) ...`

   A global `if` is applied to all equations. You may combine global and equation-specific `if` conditions:

   `. mi impute chained ... (... if` *exp* `...) ... if` *exp* `...`

3. `mi impute chained` allows specification of global weights, which are applied to all equations:

   `. mi impute chained ... [`*weight*`] ...`

4. `mi impute chained` uses fully specified prediction equations by default. Customize prediction equations by including or omitting desired terms:

   `. mi imp chain (... , include(z3) ...) (..., omit(z1) ...) ...`

5. `mi impute chained` automatically includes appropriate imputation variables in prediction equations. Use a global `noimputed` option to prevent inclusion of imputation variables in all prediction equations:

   `. mi impute chained ..., noimputed ...`

   Or use an equation-specific `noimputed` option to prevent inclusion of imputation variables in only some prediction equations:

   `. mi impute chained ... (..., noimputed ...) ...`

As we mentioned earlier, `mi impute chained` is an iterative imputation method. By default, it performs 10 burn-in iterations for each imputation before drawing the final set of imputed values. The number of iterations is determined by the length of the burn-in period after which a random sequence (chain) is assumed to converge to its stationary distribution. The provided default may not be applicable to all situations, so you can use the `burnin()` option to modify it.

Use the `chainonly` and `savetrace()` options to determine the appropriate burn-in period. For example,

        . mi impute chained ..., burnin(100) chainonly savetrace(impstats) ...

saves summaries of imputed values from 100 iterations for each of the imputation variables to `impstats.dta` without proceeding to impute data. You can apply techniques from *Convergence of ICE* to the data in `impstats.dta` to determine an adequate burn-in period.

Use a combination of the `dryrun` and `report` options to check the specification of each univariate imputation model prior to imputing data.

In the next two sections, we describe the use of `mi impute chained` first using hypothetical situations and then using real examples.

Default prediction equations

We showed in *First use* an example of `mi impute chained` with default prediction equations using the heart attack data. Here we provide more details about this default specification.

By default, `mi impute chained` imputes missing values by using the default prediction equations. It builds the corresponding univariate imputation models based on the supplied information: *uvmethod*, the imputation method; *ivars*, the imputation variables; and *indepvars*, the complete predictors or independent variables.

Suppose that continuous variables `x1`, `x2`, and `x3` contain missing values and are ordered from the most observed to the least observed. We want to impute these variables, and we decide to use the same univariate imputation method, say, linear regression, for all. We can do this by typing

        . mi impute chained (regress) x1 x2 x3 ...

The above command corresponds to the first syntax diagram of `mi impute chained`: *uvmethod* is `regress` and *ivars* is `x1 x2 x3`. Relating the above to the model notation used in (1), g_1, g_2, g_3 represent linear regression imputation models and the prediction sequence is $X_1|X_2, X_3 \rightarrow X_2|X_1, X_3 \rightarrow X_3|X_1, X_2$.

By default, `mi impute chained` imputes variables in order from the most observed to the least observed, regardless of the order in which variables were specified. For example, we can list imputation variables in the reverse order,

        . mi impute chained (regress) x3 x2 x1 ...

and `mi impute chained` will still impute `x1` first, `x2` second, and `x3` last. You can use the `orderasis` option to instruct `mi impute chained` to perform imputation of variables in the specified order.

If we have additional covariates containing no missing values (say, `z1` and `z2`) that we want to include in the imputation model, we can do so by typing

        . mi impute chained (regress) x1 x2 x3 = z1 z2 ...

Now *indepvars* is `z1 z2` and the prediction sequence is $X_1|X_2, X_3, Z_1, Z_2 \rightarrow X_2|X_1, X_3, Z_1, Z_2 \rightarrow X_3|X_1, X_2, Z_1, Z_2$. Independent variables are included in the prediction equations of all univariate models.

Suppose that we want to use a different imputation method for one of the variables—we want to impute x3 using predictive mean matching. We can do this by typing

```
. mi impute chained (regress) x1 x2  (pmm) x3  = z1 z2 ...
```

The above corresponds to the second syntax diagram of mi impute chained, a generalization of the first that accommodates differing imputation methods. The right-hand side of the equation is unchanged. z1 and z2 are included in all three prediction equations. The left-hand side now has two specifications: (regress) x1 x2 and (pmm) x3. In previous examples, we had only one left-hand-side specification, *lhs_spec*—(regress) x1 x2 x3. (The number of left-hand-side specifications does not necessarily correspond to the number of univariate models; the latter is determined by the number of imputation variables.) In this example, x1 and x2 are imputed using linear regression, and x3 is imputed using predictive mean matching.

Now, instead of using the default one nearest neighbor with pmm, say that we want to use three, which requires pmm's knn(3) option. All method-specific options must be specified within the parentheses surrounding the method:

```
. mi impute chained (regress) x1 x2  (pmm, knn(3)) x3  = z1 z2 ...
```

Suppose now we want to restrict the imputation sample for x2 to observations where z1 is one; also see *Imputing on subsamples* of [MI] **mi impute**. (We also omit pmm's knn() option here.) The corresponding syntax is

```
. mi impute chained (regress) x1  (regress if z1==1) x2  (pmm) x3  = z1 z2 ...
```

If, in addition to the above, we want to impute all variables using an overall subsample where z3 is one, we can specify the global if z3==1 condition:

```
. mi impute chained (regress) x1  (regress if z1==1) x2  (pmm) x3  = z1 z2
> if z3==1 ...
```

In the above, restrictions included only complete variables. When restrictions include imputation variables, you should use the conditional() option instead of an if condition; see *Conditional imputation* in [MI] **mi impute**. Suppose that we need to impute x2 using only observations for which x1 is positive, provided that missing values of x1 are nested within missing values of x2. We can do this by typing

```
. mi impute chained (regress) x1  (regress, cond(if x1>0)) x2  (pmm) x3  = z1 z2 ...
```

When any imputation variable is imputed using a categorical method, mi impute chained automatically includes it as a factor variable in the prediction equations of other imputation variables. Suppose that x1 is a categorical variable and is imputed using the multinomial logistic method:

```
. mi impute chained (mlogit) x1  (regress) x2 x3  ...
```

The above will result in the prediction sequence $X_1|X_2, X_3 \rightarrow X_2|\text{i}.X_1, X_3 \rightarrow X_3|\text{i}.X_1, X_2$ where $\text{i}.X_1$ denotes the factors of X_1.

If you wish to include a factor variable as continuous in prediction equations, you can use the ascontinuous option within the specification of the univariate imputation method for that variable:

```
. mi impute chained (mlogit, ascontinuous) x1  (regress) x2 x3  ...
```

As we discussed in *The issue of perfect prediction during imputation of categorical data* of [MI] **mi impute**, perfect prediction often occurs during imputation of categorical variables. One way of dealing with it is to use the augmented-regression approach (White, Daniel, and Royston 2010), available through the augment option. For example, if perfect prediction occurs during imputation of x1 in the above, you can specify augment within the method specification of x1 to perform augmented regression:

```
. mi impute chained (mlogit, augment) x1  (regress) x2 x3  ...
```

Alternatively, you can use the `augment` option with `mi impute chained` to perform augmented regression for all categorical variables for which the issue of perfect prediction arises:

```
. mi impute chained (mlogit) x1  (logit) x2  (regress) x3  ..., augment ...
```

The above is equivalent to specifying `augment` within each specification of a univariate categorical imputation method:

```
. mi impute chained (mlogit, augment) x1  (logit, augment) x2  (regress) x3  ...
```

Custom prediction equations

In the previous section, we considered various uses of `mi impute chained` with default prediction equations. Often, however, you may want to use different prediction equations for some or even all imputation variables. We can easily modify the above specifications to accommodate this.

Let's consider situations in which we want to use different sets of complete variables for some imputation variables first. Recall our following hypothetical example:

```
. mi impute chained (regress) x1 x2 x3 = z1 z2 ...                    (M1)
```

Suppose that we want to omit `z2` from the prediction equation for `x3`. To accommodate this, we need to include two separate specifications: one for `x1` and `x2` and one for `x3`:

```
. mi impute chained (regress) x1 x2  (regress, omit(z2)) x3  = z1 z2 ...
```

The above corresponds to the custom specification, the third syntax diagram, of `mi impute chained`. As before, we list all the complete variables *indepvars* to be included in all prediction equations to the right of the = sign. So, *indepvars* is still `z1 z2`. The prediction equation for `x3`, however, omits variable `z2`, specified within the `omit()` option. The prediction sequence for the above specification is $X_1|X_2, X_3, Z_1, Z_2 \rightarrow X_2|X_1, X_3, Z_1, Z_2 \rightarrow X_3|X_1, X_2, Z_1$.

Alternatively, we could have achieved the above by including variable `z1` in all prediction equations, as a right-hand-side specification *indepvars*, and using the `include()` option to add variable `z2` to the prediction equations of `x1` and `x2`:

```
. mi impute chained (regress, include(z2)) x1 x2  (regress) x3  = z1 ...
```

You may also want to modify the sets of imputation variables to be included in prediction equations. By default, `mi impute chained` automatically includes the appropriate fully conditional specifications of imputation variables in all prediction equations.

Suppose that in addition to different sets of complete predictors, we assume that X_1 and X_2 are conditionally independent given X_3, which implies that prediction equations for `x1` and `x2` include only `x3` and not each other. We can accommodate this with the command

```
. mi impute chained (regress, include(x3 z2) noimputed) x1 x2  (regress) x3  = z1 ...
```

which corresponds to the prediction sequence $X_1|X_3, Z_1, Z_2 \rightarrow X_2|X_3, Z_1, Z_2 \rightarrow X_3|X_1, X_2, Z_1$.

The above is also equivalent to the command

```
. mi impute chained (regress, omit(x1 x2)) x1 x2  (regress, omit(z2)) x3  = z1 z2 ...
```

There are other equivalent ways of achieving the above custom specifications by using various combinations of `include()`, `omit()`, and `noimputed`. The most convenient specification will depend on your particular structure of the prediction equations. You can also combine these options within the same univariate specification.

It is important to realize that equivalent syntaxes may produce different (yet equivalent with stable imputation models) sequences of imputed values when they have different ordering of variables in prediction equations. `mi impute chained` builds prediction equations as follows. Appropriate imputation variables are included first, unless the `noimputed` option is specified. By default, imputation variables are included in order from the most observed to the least observed. If the `orderasis` option is used, the variables are included in the specified order. Next, terms specified in the `include()` option are included in the listed order. Then right-hand-side variables (*indepvars*) are included in the listed order. Finally, variables listed in the `omit()` option are removed from the prediction equation. When you specify `omit()`, it is important to specify variables as they are included in the prediction equation; if `x1` is included as a factor variable, `omit(i.x1)` should be used.

You can also include functions of imputation variables in prediction equations with the custom specification of `mi impute chained`. As we discussed in *Model building* in [MI] **mi impute**, there are two ways to do that. You can include functions of imputation variables as separate imputation variables directly in your imputation model or you can impute them passively using `mi impute chained`.

For example, using model (M1), suppose that we would like to include the interaction between `x1` and `x2` in the conditional model for `x3`:

```
. mi impute chained (regress) x1 x2                    ///
                    (regress, include((x1*x2))) x3     ///
                                         = z1 z2 ...
```

The expression `x1*x2`, specified in the `include()` option, is enclosed in parentheses.

We also could have typed

```
. mi impute chained (regress, include((x1*x2))) x1 x2 x3 = z1 z2 ...
```

and `mi impute chained` would appropriately include the interaction term $X_1 X_2$ only in the prediction equation of X_3.

You can include any other expressions of imputation variables in `include()` within any of the left-hand-side specifications. Just remember to enclose such expressions in parentheses.

All the examples we considered in *Default prediction equations* are also applicable to `mi impute chained` with custom prediction equations. For example, to restrict imputation of `x2` to observations where `z1==1` in one of our earlier examples, we can type

```
. mi impute chained (reg) x1 (reg if z1==1) x2 (reg, omit(z2)) x3 = z1 z2 ...
```

Link between mi impute chained and mi impute monotone

Similarly to `mi impute monotone` (see [MI] **mi impute monotone**), `mi impute chained` uses a sequence of univariate imputation models to impute variables. So the use of `mi impute chained` is very similar to that of `mi impute monotone` except:

1. `mi impute chained` does not require that the specified imputation variables follow a monotone-missing pattern.

2. `mi impute chained` requires iteration to accommodate arbitrary missing-data patterns.

3. `mi impute chained`, by default, uses FCSs of the prediction equations where all specified complete variables and all imputation variables except the one being imputed are included in prediction equations.

4. `mi impute chained` provides an alternative way of specifying custom prediction equations to accommodate FCS of imputation variables.

When a missing-value pattern is monotone, `mi impute chained` defaults to the monotone method (unless `nomonotone` is specified) and produces the same results as `mi impute monotone`. However, using `mi impute monotone` in this case is faster because it performs the estimation step only once, on the original data, whereas `mi impute chained` performs estimation on every chained iteration.

The best approach to follow is

1. Check the missing-data pattern using `misstable nested` (or `mi misstable nested` if the data are already `mi set`; see [R] **misstable** or [MI] **mi misstable**) first.

2. If the missing-data pattern is monotone, use `mi impute monotone` to impute variables. If the missing-data pattern is not monotone, use `mi impute chained` to impute variables.

It is worth mentioning the difference between the documented custom syntaxes of `mi impute chained` and `mi impute monotone`.

With monotone imputation, variables are imputed in a particular, monotone-missing order and prediction equations are built in a particular way: previously imputed variables are added sequentially to the prediction equations of other imputation variables. So when building custom prediction equations, it is easier to construct one equation at a time in the order of the monotone missing pattern. As such, the custom syntax of `mi impute monotone`, as documented in [MI] **mi impute monotone**, requires full specification of a separate conditional model for each imputation variable in the monotone-missing order.

Imputation using chained equations does not require specific ordering in which variables must be imputed, although imputing variables in order from the most observed to the least observed usually leads to faster convergence. Also, because all imputation variables except the one being imputed are included in prediction equations, it does not matter in what order prediction equations are specified. The custom syntax of `mi impute chained` reflects this.

Examples

For the purpose of illustration, we use five imputations in our examples.

▷ Example 1: Different imputation methods

Recall the heart attack example from *First use*. If we wanted to impute `bmi` using predictive mean matching with, say, three nearest neighbors instead of linear regression, we could type

```
. use http://www.stata-press.com/data/r12/mheart8s0
(Fictional heart attack data; bmi and age missing; arbitrary pattern)
. mi impute chained (pmm, knn(3)) bmi (reg) age = attack smokes hsgrad female,
> add(5)

Conditional models:
             age: regress age bmi attack smokes hsgrad female
             bmi: pmm bmi age attack smokes hsgrad female , knn(3)

Performing chained iterations ...

Multivariate imputation                     Imputations =          5
Chained equations                                 added =          5
Imputed: m=1 through m=5                         updated =          0

Initialization: monotone                     Iterations =         50
                                                burn-in =         10

             bmi: predictive mean matching
             age: linear regression
```

		Observations per m		
Variable	Complete	Incomplete	Imputed	Total
bmi	126	28	28	154
age	142	12	12	154

```
(complete + incomplete = total; imputed is the minimum across m
 of the number of filled-in observations.)
```

As shown previously, `mi impute chained` imputed `age` first and `bmi` second, because `age` is the variable with the fewest missing values.

◁

▷ Example 2: Convergence of ICE

In *Convergence of ICE*, we described ways to assess convergence of the ICE algorithm. Continuing our previous example, let's investigate the trends in the summaries of imputed values of `age` and `bmi` over iterations.

Following the recommendation from *Using mi impute chained*, we use a combination of `chainonly` and `savetrace()` to perform chained iterations without creating imputations in the data and save summaries of imputed values to the new dataset `impstats.dta`. We perform 100 iterations and specify a random-number seed for reproducibility:

```
. use http://www.stata-press.com/data/r12/mheart8s0
(Fictional heart attack data; bmi and age missing; arbitrary pattern)
. mi impute chained (pmm, knn(3)) bmi (reg) age = attack smokes hsgrad female,
> chainonly burnin(100) savetrace(impstats) rseed(1359)

Conditional models:
             age: regress age bmi attack smokes hsgrad female
             bmi: pmm bmi age attack smokes hsgrad female , knn(3)

Performing chained iterations ...

Note: no imputation performed.
```

By default, means and standard deviations of imputed values for each imputation variable are saved along with iteration and imputation numbers (imputation number is always 0 when `chainonly` is used):

```
. use impstats
(Summaries of imputed values from -mi impute chained-)

. describe

Contains data from impstats.dta
  obs:             101                        Summaries of imputed values
                                                from -mi impute chained-
  vars:              6                        2 Apr 2011 11:02
  size:          1,818

              storage   display    value
variable name   type    format     label    variable label

iter           byte     %12.0g              Iteration numbers
m              byte     %12.0g              Imputation numbers
age_mean       float    %9.0g               Mean of age
age_sd         float    %9.0g               Std. Dev. of age
bmi_mean       float    %9.0g               Mean of bmi
bmi_sd         float    %9.0g               Std. Dev. of bmi

Sorted by:
```

We use the time-series command tsline (see [TS] **tsline**) to plot summaries of imputed values with respect to the iteration number. We first use tsset to set iter as the "time" variable and then use tsline to obtain trace plots. We create trace plots for all variables and combine them in one graph using graph combine:

```
. tsset iter
        time variable:  iter, 0 to 100
                delta:  1 unit

. tsline bmi_mean, name(gr1) nodraw

. tsline bmi_sd, name(gr2) nodraw

. tsline age_mean, name(gr3) nodraw

. tsline age_sd, name(gr4) nodraw

. graph combine gr1 gr2 gr3 gr4, title(Trace plots of summaries of imputed values)
> rows(2)
```

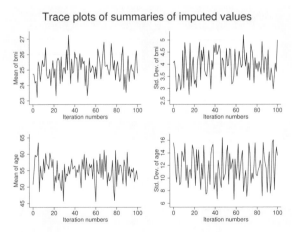

The trace plots show no apparent trends in the summaries of the imputed values, so the default number of burn-in iterations, 10, seems adequate. Although a low number of burn-in iterations may be sufficient in some applications, there are situations when larger numbers are required (for example, van Buuren [2007]).

It is also useful to look at several chains, each obtained using a different set of initial values, to check convergence and stability of the algorithm.

Let's look at three separate chains. The easiest way to do this is to use the add() option instead of chainonly to create three imputations. Remember that mi impute chained starts a new chain for each imputation, so a different set of initial values is used for each imputation. When savetrace() is specified, mi impute chained stores summaries of imputed values for each imputation.

```
. use http://www.stata-press.com/data/r12/mheart8s0
(Fictional heart attack data; bmi and age missing; arbitrary pattern)

. qui mi impute chained (pmm, knn(3)) bmi (reg) age = attack smokes hsgrad female,
> add(3) burnin(100) savetrace(impstats, replace) rseed(1359)
```

The results are saved in a long form. If we want to overlay separate chains in one graph, we need to convert our data to a wide form first—one variable per chain. We use the reshape command for this (see [D] **reshape**):

```
. use impstats, clear
(Summaries of imputed values from -mi impute chained-)

. reshape wide *mean *sd, i(iter) j(m)
(note: j = 1 2 3)
```

Data	long	->	wide
Number of obs.	303	->	101
Number of variables	6	->	13
j variable (3 values)	m	->	(dropped)
xij variables:			
	age_mean	->	age_mean1 age_mean2 age_mean3
	bmi_mean	->	bmi_mean1 bmi_mean2 bmi_mean3
	age_sd	->	age_sd1 age_sd2 age_sd3
	bmi_sd	->	bmi_sd1 bmi_sd2 bmi_sd3

We can now plot the three chains for, say, the mean of bmi using tsline:

```
. tsset iter
        time variable:  iter, 0 to 100
                delta:  1 unit

. tsline bmi_mean1 bmi_mean2 bmi_mean3, ytitle(Mean of bmi) yline(25.24)
> legend(rows(1) label(1 "Chain 1") label(2 "Chain 2") label(3 "Chain 3"))
```

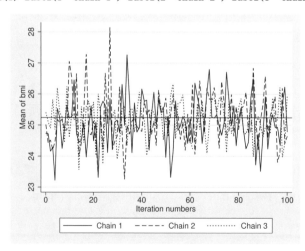

There are no apparent trends in any of the chains. All three chains seem to oscillate around the observed mean estimate of bmi of 25.24, providing some evidence of convergence of the algorithm.

◁

▷ Example 3: Custom prediction equations

Continuing example 1, we believe that there is no association between bmi and hsgrad conditional on other predictors, so we want to use hsgrad to model only age and omit it from the model for bmi:

```
. use http://www.stata-press.com/data/r12/mheart8s0
(Fictional heart attack data; bmi and age missing; arbitrary pattern)

. mi impute chained
>     (pmm, knn(3) omit(hsgrad)) bmi
>     (regress)                  age
>                                      = attack smokes hsgrad female, add(5)

Conditional models:
                age: regress age bmi attack smokes hsgrad female
                bmi: pmm bmi age attack smokes female , knn(3)

Performing chained iterations ...

Multivariate imputation              Imputations =          5
Chained equations                          added =          5
Imputed: m=1 through m=5                  updated =          0

Initialization: monotone               Iterations =         50
                                          burn-in =         10

                bmi: predictive mean matching
                age: linear regression
```

	Observations per m			
Variable	Complete	Incomplete	Imputed	Total
bmi	126	28	28	154
age	142	12	12	154

```
(complete + incomplete = total; imputed is the minimum across m
 of the number of filled-in observations.)
```

All right-hand-side complete predictors (attack, smokes, and female) are used in both prediction equations. The prediction equation for age additionally includes the hsgrad variable.

◁

▷ Example 4: Imputing variables of different types

We now consider an mi set version of the heart attack data containing an indicator for smoking high-tar cigarettes (variable hightar):

```
. use http://www.stata-press.com/data/r12/mheart9s0
(Fictional heart attack data; bmi, age, and hightar missing; arbitrary pattern)
. mi describe
  Style:  mlong
          last mi update 25mar2011 11:00:38, 3 days ago
  Obs.:   complete          98
          incomplete        56  (M = 0 imputations)
          ─────────────────────
          total            154
  Vars.:  imputed:  3; bmi(24) age(30) hightar(12)
          passive:  0
          regular:  4; attack smokes female hsgrad
          system:   3; _mi_m _mi_id _mi_miss
          (there are no unregistered variables)
. mi misstable nested
      1.  hightar(12)
      2.  bmi(24)
      3.  age(30)
```

According to mi describe, there are no imputations, three registered imputed variables (age, bmi, and hightar), and four registered regular variables. mi misstable nested reports that missing values of the three imputation variables are not nested.

The hightar variable is a binary variable, so we choose the logistic method to impute its values (see [MI] **mi impute logit**). Because hightar records whether a subject smokes high-tar cigarettes, we use only those who smoke to impute its missing values. As such, including smokes as a predictor of hightar is redundant, so we omit this variable from the prediction equation for hightar:

```
. mi impute chained
>    (pmm, knn(3) omit(hsgrad))          bmi
>    (regress)                           age
>    (logit if smokes==1, omit(smokes))  hightar
>                                        = attack smokes hsgrad female, add(5)
Conditional models:
          hightar: logit hightar bmi age attack hsgrad female if smokes==1
             bmi: pmm bmi i.hightar age attack smokes female , knn(3)
             age: regress age i.hightar bmi attack smokes hsgrad female

Performing chained iterations ...

Multivariate imputation          Imputations =        5
Chained equations                      added =        5
Imputed: m=1 through m=5             updated =        0

Initialization: monotone           Iterations =       50
                                      burn-in =       10

          bmi: predictive mean matching
          age: linear regression
      hightar: logistic regression
```

	Observations per m			
Variable	Complete	Incomplete	Imputed	Total
bmi	130	24	24	154
age	124	30	30	154
hightar	52	12	12	64

```
(complete + incomplete = total; imputed is the minimum across m
 of the number of filled-in observations.)
```

From the output, we see that all incomplete values of each of the variables are imputed in all imputations. Because we restricted the imputation sample of hightar to smokers, the total number of observations reported for hightar is 64 and not 154. mi impute chained also automatically included the binary variable hightar as a factor variable in prediction equations for age and bmi because we used logit to impute it.

As we described in *Conditional imputation*, you should be careful when using an if statement for imputing variables conditionally on other variables. It was safe to use if here, because smokes did not contain missing values and there were no missing values of hightar for the subjects who do not smoke.

◁

▷ Example 5: Conditional imputation

Continuing example 4, suppose now that the smokes variable also contains missing values:

```
. use http://www.stata-press.com/data/r12/mheart10s0
(Fict. heart attack data; bmi, age, hightar, & smokes missing; arbitrary pattern)
. mi describe
  Style:  mlong
          last mi update 25mar2011 11:00:38, 3 days ago
  Obs.:   complete        92
          incomplete      62   (M = 0 imputations)
          ──────────────────
          total          154
  Vars.:  imputed:  4; bmi(24) age(30) hightar(19) smokes(14)
          passive:  0
          regular:  3; attack female hsgrad
          system:   3; _mi_m _mi_id _mi_miss
          (there are no unregistered variables)
. mi misstable nested
     1.  smokes(14) -> hightar(19)
     2.  bmi(24)
     3.  age(30)
```

The smokes variable is now registered as imputed and the three regular variables are now attack, female, and hsgrad. mi misstable nested reports that although the missing-data pattern with respect to all four imputation variables is not monotone, the missing-data pattern with respect to smokes and hightar is monotone. Recall from *Conditional imputation* that one of the requirements of conditional imputation is that missing values of all conditioning variables (smokes) are nested within missing values of the conditional variable (hightar). So this requirement is satisfied in our data.

Because smokes contains missing values, we cannot use an if condition to restrict the imputation sample of hightar to those who smoke. We must use the conditional() option. We use the logistic method (see [MI] **mi impute logit**) to fill in missing values of smokes.

```
. mi impute chained
>  (pmm, knn(3) omit(hsgrad))            bmi
>  (regress)                             age
>  (logit, cond(if smokes==1) omit(i.smokes))  hightar
>  (logit)                               smokes
>                                        = attack hsgrad female, add(5)

Conditional models:
          smokes: logit smokes bmi age attack hsgrad female
         hightar: logit hightar bmi age attack hsgrad female , cond(if smokes==1)
             bmi: pmm bmi i.smokes i.hightar age attack female , knn(3)
             age: regress age i.smokes i.hightar bmi attack hsgrad female

Performing chained iterations ...

Multivariate imputation            Imputations =          5
Chained equations                        added =          5
Imputed: m=1 through m=5               updated =          0

Initialization: monotone             Iterations =         50
                                        burn-in =         10

Conditional imputation:
  hightar: incomplete out-of-sample obs. replaced with value 0

             bmi: predictive mean matching
             age: linear regression
         hightar: logistic regression
          smokes: logistic regression
```

	Observations per *m*			
Variable	Complete	Incomplete	Imputed	Total
bmi	130	24	24	154
age	124	30	30	154
hightar	135	19	19	154
smokes	140	14	14	154

```
(complete + incomplete = total; imputed is the minimum across m
 of the number of filled-in observations.)
```

With conditional imputation, a legend appears before the imputation table, reporting the conditional constant, the value that was used to replace all incomplete values of an imputation variable outside the conditional sample. The missing values of hightar in that sample were replaced with 0.

The smokes variable is imputed using logit and thus is included in prediction equations as a factor variable, i.smokes. As such, we specified omit(i.smokes) to omit smokes from the prediction equation for hightar.

Also notice that compared with imputation on a restricted subsample using an if condition, the reported total number of observations in the imputation sample for hightar is still 154. All incomplete observations, within and outside the conditional sample, are included in the imputation sample during conditional imputation. So the reported numbers of complete, incomplete, and imputed observations correspond with observations within and outside the conditional sample.

◁

▷ Example 6: Including expressions of imputation variables

In *Model building* of [MI] **mi impute**, we described two ways of accommodating functional relationships during imputation. Here we demonstrate a passive imputation approach that includes expressions of imputation variables directly into the imputation model.

Continuing example 5, suppose we assume that the conditional distribution of bmi exhibits some curvature with respect to age. We want to include age^2 in the prediction equation for bmi. If the relationship between bmi and age is indeed curvilinear, it would be unreasonable to assume that the conditional distribution of age given bmi is linear. One possibility is to determine what the relationship is between age and bmi given other predictors in the observed data (see, for example, [R] **mfp**) and include the appropriate functional terms of bmi in the prediction equation for age. Following White, Royston, and Wood (2011) to relax the linearity assumption, we use predictive mean matching instead of linear regression to impute age:

```
. mi impute chained
>   (pmm, knn(3) omit(hsgrad) incl((age^2)))   bmi
>   (pmm)                                       age
>   (logit, cond(if smokes==1) omit(i.smokes)) hightar
>   (logit)                                     smokes
>                                             = attack hsgrad female, replace

Conditional models:
          smokes: logit smokes bmi age attack hsgrad female
         hightar: logit hightar bmi age attack hsgrad female , cond(if smokes==1)
             bmi: pmm bmi i.smokes i.hightar age (age^2) attack female , knn(3)
             age: pmm age i.smokes i.hightar bmi attack hsgrad female

Performing chained iterations ...

Multivariate imputation                     Imputations =        5
Chained equations                                 added =        0
Imputed: m=1 through m=5                         updated =        5

Initialization: monotone                     Iterations =       50
                                                burn-in =       10

Conditional imputation:
  hightar: incomplete out-of-sample obs. replaced with value 0

             bmi: predictive mean matching
             age: predictive mean matching
         hightar: logistic regression
          smokes: logistic regression
```

	Observations per m			
Variable	Complete	Incomplete	Imputed	Total
bmi	130	24	24	154
age	124	30	30	154
hightar	135	19	19	154
smokes	140	14	14	154

```
(complete + incomplete = total; imputed is the minimum across m
of the number of filled-in observations.)
```

We included the expression term in parentheses in the include() option in the prediction equation for bmi.

◁

> ## Example 7: Imputing on subsamples

Suppose that in our primary logistic analysis of heart attacks, we are planning to investigate various interaction effects with respect to gender. The female variable is complete, so the best way to accommodate such interactions is to use the by() option to perform imputation separately for females and males.

We continue example 3. Before imputing missing values, let's review our conditional specifications for each group. We can use the `dryrun` option to see univariate conditional models that will be used during imputation without actually imputing data:

```
. use http://www.stata-press.com/data/r12/mheart8s0, clear
(Fictional heart attack data; bmi and age missing; arbitrary pattern)
. mi impute chained
>    (pmm, knn(3) omit(hsgrad)) bmi
>    (regress)                   age
>                                     = attack smokes hsgrad, by(female) dryrun
Performing setup for each by() group:
-> female = 0
Conditional models:
            age: regress age bmi attack smokes hsgrad
            bmi: pmm bmi age attack smokes , knn(3)
-> female = 1
Conditional models:
            age: regress age bmi attack smokes hsgrad
            bmi: pmm bmi age attack smokes , knn(3)
```

Conditional specifications are as we expected, so we can proceed to imputation.

```
. mi impute chained
>    (pmm, knn(3) omit(hsgrad)) bmi
>    (regress)                   age
>                                     = attack smokes hsgrad
>                                        , add(5) by(female, noreport) dots
-> female = 0
Performing chained iterations:
  imputing m=1 through m=5 ..... done
-> female = 1
Performing chained iterations:
  imputing m=1 through m=5 ..... done
Multivariate imputation                    Imputations =         5
Chained equations                               added =         5
Imputed: m=1 through m=5                       updated =         0

Initialization: monotone                     Iterations =        50
                                                burn-in =        10

              bmi: predictive mean matching
              age: linear regression
```

by() Variable	Observations per m			
	Complete	Incomplete	Imputed	Total
female = 0				
bmi	95	21	21	116
age	106	10	10	116
female = 1				
bmi	31	7	7	38
age	36	2	2	38
Overall				
bmi	126	28	28	154
age	142	12	12	154

(complete + incomplete = total; imputed is the minimum across m
 of the number of filled-in observations.)

To avoid longer output, we specified the `noreport` option within `by()` to suppress information about the setup and imputation steps that otherwise would have been reported for each group.

◁

Saved results

`mi impute chained` saves the following in `r()`:

Scalars
`r(M)`	total number of imputations
`r(M_add)`	number of added imputations
`r(M_update)`	number of updated imputations
`r(k_ivars)`	number of imputed variables
`r(burnin)`	number of burn-in iterations
`r(N_g)`	number of imputed groups (1 if `by()` is not specified)

Macros
`r(method)`	name of imputation method (`chained`)
`r(ivars)`	names of imputation variables
`r(uvmethods)`	names of univariate imputation methods
`r(init)`	type of initialization
`r(rseed)`	random-number seed
`r(by)`	names of variables specified within `by()`

Matrices
`r(N)`	number of observations in imputation sample in each group (per variable)
`r(N_complete)`	number of complete observations in imputation sample in each group (per variable)
`r(N_incomplete)`	number of incomplete observations in imputation sample in each group (per variable)
`r(N_imputed)`	number of imputed observations in imputation sample in each group (per variable)

Methods and formulas

Let $X_1, X_2, \ldots, X_p$ denote imputation variables ordered from the most observed to the least observed and let $\mathbf{Z}$ denote the set of complete independent variables. (If $X_1, X_2, \ldots, X_p$ are monotone-missing and neither `nomonotone` nor `nomonotonechk` is used, then `mi impute chained` uses monotone imputation; see *Methods and formulas* of [MI] **mi impute monotone** for details.)

With the default specification of prediction equations, the chained-equation algorithm proceeds as follows. First, at iteration $t = 0$, missing values are initialized using monotone imputation. That is, missing values of $X_i^{(0)}$, $i = 1, \ldots, p$, are simulated from conditional densities of the form

$$f_i(X_i | X_1^{(0)}, X_2^{(0)}, \ldots, X_{i-1}^{(0)}, \mathbf{Z}, \boldsymbol{\theta}_i) \qquad (2)$$

where the conditional density $f_i(\cdot)$ is determined according to the chosen univariate imputation method and $\boldsymbol{\theta}_i$ is its corresponding set of parameters with uniform prior; see *Methods and formulas* of chosen univariate imputation methods for details.

At iteration t, missing values of X_i for all $i = 1, \ldots, p$ are simulated from full conditionals, conditional densities of the form:

$$g_i(X_i | X_1^{(t)}, X_2^{(t)}, \ldots, X_{i-1}^{(t)}, X_{i+1}^{(t-1)}, \ldots, X_p^{(t-1)}, \mathbf{Z}, \phi_i) \qquad (3)$$

where again the conditional density $g_i(\cdot)$ is determined according to the chosen univariate imputation method and ϕ_i is its corresponding set of parameters with uniform prior.

The algorithm iterates for a prespecified number of iterations b, $t = 1, \ldots, b$, and a final set of imputed values is obtained from the last iteration. At each iteration, the imputation process consists of steps 1–3 described in *Methods and formulas* of each respective univariate imputation method's manual entry.

Each imputation is obtained independently by repeating (2) and (3).

Conditional specifications in (2) and (3) correspond to the default specification of prediction equations. With the custom specification, the sets of complete predictors $\mathbf{Z} = \mathbf{Z}_i$ and imputation variables may differ across univariate specifications, and prediction equations may additionally include functions of imputation variables.

In summary, `mi impute chained` follows the steps below to fill in missing values in $X_1, \ldots, X_p$:

1. `mi impute chained` first builds appropriate univariate imputation models using the supplied information about imputation methods, imputation variables $\mathbf{X}$, and complete predictors $\mathbf{Z}$. By default, fully conditional specification of prediction equations is used. The order in which imputation variables are listed is ignored unless the `orderasis` option is used. By default, `mi impute chained` imputes variables in order from the most observed to the least observed.

2. Initialize missing values at $t = 0$ using monotone imputation (2).

3. Perform the iterative procedure (3) for $t = 1, \ldots, b$, for the length of the burn-in period, to obtain imputed values. At each iteration t,

    3.1. Fit a univariate model for X_i to the observed data to obtain the estimates of ϕ_i. See step 1 in *Methods and formulas* of each respective univariate imputation method's manual entry for details.

    3.2. Fill in missing values of X_i according to the specified imputation model. See step 2 and step 3 in *Methods and formulas* of each respective univariate imputation method's manual entry for details.

    3.3. Repeat steps 3.1 and 3.2 for each imputation variable X_i, $i = 1, \ldots, p$.

4. Repeat steps 2 and 3 to obtain M multiple imputations.

The iterative procedure (3) may not always correspond to a genuine simulation of imputed values from their predictive distribution $f(\mathbf{X}_m | \mathbf{X}_o, \mathbf{Z})$ because the set of full conditionals $\{g_i : i = 1, 2, \ldots, p\}$ may not correspond to this distribution or, in fact, to any proper multivariate distribution. The extent to which this is a problem in practical applications is still an open research problem. Some limited simulation studies reported only minimal effect of such incompatibility on final MI estimates (for example, van Buuren et al. [2006]).

Acknowledgments

The `mi impute chained` command was inspired by the user-written command `ice` by Patrick Royston of the MRC Clinical Trials Unit and Ian White of the MRC Biostatistics Unit. We are indebted to them for their extensive work in the multiple-imputation area in Stata. We are also grateful to them for their comments and advice on `mi impute chained`.

References

Arnold, B. C., E. Castillo, and J. M. Sarabia. 1999. *Conditional Specification of Statistical Models*. New York: Springer.

——. 2001. Conditionally specified distributions: An introduction. *Statistical Science* 16: 249–274.

Gelfand, A. E., and A. M. F. Smith. 1990. Sampling-based approaches to calculating marginal densities. *Journal of the American Statistical Association* 85: 398–409.

Geman, S., and D. Geman. 1984. Stochastic relaxation, Gibbs distributions, and the Bayesian restoration of images. *IEEE Transactions on Pattern Analysis and Machine Intelligence* 6: 721–741.

Raghunathan, T. E., J. M. Lepkowski, J. Van Hoewyk, and P. Solenberger. 2001. A multivariate technique for multiply imputing missing values using a sequence of regression models. *Survey Methodology* 27: 85–95.

Royston, P. 2004. Multiple imputation of missing values. *Stata Journal* 4: 227–241.

——. 2005a. Multiple imputation of missing values: Update. *Stata Journal* 5: 188–201.

——. 2005b. Multiple imputation of missing values: Update of ice. *Stata Journal* 5: 527–536.

——. 2007. Multiple imputation of missing values: Further update of ice, with an emphasis on interval censoring. *Stata Journal* 7: 445–464.

——. 2009. Multiple imputation of missing values: Further update of ice, with an emphasis on categorical variables. *Stata Journal* 9: 466–477.

van Buuren, S. 2007. Multiple imputation of discrete and continuous data by fully conditional specification. *Statistical Methods in Medical Research* 16: 219–242.

van Buuren, S., H. C. Boshuizen, and D. L. Knook. 1999. Multiple imputation of missing blood pressure covariates in survival analysis. *Statistics in Medicine* 18: 681–694.

van Buuren, S., J. P. L. Brand, C. G. M. Groothuis-Oudshoorn, and D. B. Rubin. 2006. Fully conditional specification in multivariate imputation. *Journal of Statistical Computation and Simulation* 76: 1049–1064.

White, I. R., R. Daniel, and P. Royston. 2010. Avoiding bias due to perfect prediction in multiple imputation of incomplete categorical data. *Computational Statistics & Data Analysis* 54: 2267–2275.

White, I. R., P. Royston, and A. M. Wood. 2011. Multiple imputation using chained equations: Issues and guidance for practice. *Statistics in Medicine* 30: 377–399.

Also see

[MI] **mi impute** — Impute missing values

[MI] **mi impute monotone** — Impute missing values in monotone data

[MI] **mi impute mvn** — Impute using multivariate normal regression

[MI] **mi estimate** — Estimation using multiple imputations

[MI] **intro substantive** — Introduction to multiple-imputation analysis

[MI] **intro** — Introduction to mi

[MI] **Glossary**

Title

mi impute intreg — Impute using interval regression

Syntax

mi impute intreg *newivar* [*indepvars*] [*if*] [*weight*] [, *impute_options options*]

impute_options	Description
Main	
add(#)	specify number of imputations to add; required when no imputations exist
*replace	replace imputed values in existing imputations
rseed(#)	specify random-number seed
double	save imputed values in double precision; the default is to save them as float
by(*varlist*[, *byopts*])	impute separately on each group formed by *varlist*
Reporting	
dots	display dots as imputations are performed
noisily	display intermediate output
nolegend	suppress all table legends
Advanced	
force	proceed with imputation, even when missing imputed values are encountered
noupdate	do not perform mi update; see [MI] **noupdate option**

*add(#) is required when no imputations exist; add(#) or replace is required if imputations exist.
noupdate does not appear in the dialog box.

options	Description
Main	
noconstant	suppress constant term
*ll(*varname*)	lower limit for interval censoring
*ul(*varname*)	upper limit for interval censoring
offset(*varname_o*)	include *varname_o* in model with coefficient constrained to 1
conditional(*if*)	perform conditional imputation
bootstrap	estimate model parameters using sampling with replacement
Maximization	
maximize_options	control the maximization process; seldom used

You must mi set your data before using mi impute intreg; see [MI] **mi set**.
indepvars may contain factor variables; see [U] **11.4.3 Factor variables**.
aweights, fweights, iweights, and pweights are allowed; see [U] **11.1.6 weight**.

162

Menu

Statistics > Multiple imputation

Description

mi impute intreg fills in missing values of a continuous partially observed (censored) variable using an interval regression imputation method. You can perform separate imputations on different subsets of the data by using the by() option. You can also account for analytic, frequency, importance, and sampling weights.

Options

⌐ Main ⌐

noconstant; see [R] **estimation options**.

ll(*varname*) and ul(*varname*) specify variables containing the lower and upper limits for interval censoring. You must specify both. Nonmissing observations with equal values in ll() and ul() are fully observed observations with missing values in both ll() and ul() are unobserved (missing), and the remaining observations are partially observed (censored). Partially observed cases are left-censored when ll() contains missing, right-censored when ul() contains missing, and interval-censored when ll() < ul(). Fully observed cases are also known as point data; also see *Description* in [R] **intreg**. In addition to *newivar*, mi impute intreg fills in unobserved (missing) values of variables supplied in ll() and ul(); censored values remain unchanged.

add(), replace, rseed(), double, by(); see [MI] **mi impute**.

offset(*varname$_o$*); see [R] **estimation options**.

conditional(*if*) specifies that the imputation variable be imputed conditionally on observations satisfying *exp*; see [U] **11.1.3 if exp**. That is, missing values in a conditional sample, the sample identified by the *exp* expression, are imputed based only on data in that conditional sample. Missing values outside the conditional sample are replaced with a conditional constant, the value of the imputation variable in observations outside the conditional sample. As such, the imputation variable is required to be constant outside the conditional sample. Also, if any conditioning variables (variables involved in the conditional specification if *exp*) contain soft missing values (.), their missing values must be nested within missing values of the imputation variables. See *Conditional imputation* under *Remarks* in [MI] **mi impute**.

bootstrap specifies that posterior estimates of model parameters be obtained using sampling with replacement; that is, posterior estimates are estimated from a bootstrap sample. The default is to sample the estimates from the posterior distribution of model parameters or from the large-sample normal approximation of the posterior distribution. This option is useful when asymptotic normality of parameter estimates is suspect.

⌐ Reporting ⌐

dots, noisily, nolegend; see [MI] **mi impute**. noisily specifies that the output from the interval regression fit to the observed data be displayed. nolegend suppresses all legends that appear before the imputation table. Such legends include a legend about conditional imputation that appears when the conditional() option is specified and group legends that may appear when the by() option is specified.

⌐ Maximization ⌐_____

maximize_options; see [R] **intreg**. These options are seldom used.

⌐ Advanced ⌐_____

force; see [MI] **mi impute**.

The following option is available with mi impute but is not shown in the dialog box:

noupdate; see [MI] **noupdate option**.

Remarks

Remarks are presented under the following headings:

> *Univariate imputation using interval regression*
> *Using mi impute intreg*
> *Example*

See [MI] **mi impute** for a general description and details about options common to all imputation methods, *impute_options*. Also see [MI] **workflow** for general advice on working with mi.

Univariate imputation using interval regression

The interval regression imputation method can be used to fill in missing values of a continuous partially observed (censored) variable (Royston 2007). It is a parametric method that assumes an underlying normal model for the partially observed imputed variable (given other predictors). This method is based on the asymptotic approximation of the posterior predictive distribution of the missing data.

Partially observed data arise when instead of observing an actual value, we observe the range where that value can lie. Such data include interval-censored, left-censored, and right-censored data; see [R] **intreg** for a more detailed discussion of censored data.

Do not confuse censoring with truncation. Truncated data are observed and are known to be in a certain range. Censored data come from a mixture of a continuous distribution and point masses at censoring limits. Truncated data come from a continuous truncated distribution. See the technical note in *Remarks* of [R] **truncreg** for details. Use mi impute truncreg (see [MI] **mi impute truncreg**) to impute truncated data.

The imputation of censored data has certain unique characteristics. First, censored data are recorded in two variables containing the lower and the upper interval-censoring limits. So technically, there are two imputation variables. Second, in addition to complete observations (point data) and incomplete observations ("truly" missing data), there are partially complete (censored) observations for which only the lower and upper limits are known, not the values themselves. We can treat partially observed cases as "missing" and impute them along with other completely unobserved data, provided we respect their observed limits during imputation. As a result, we will end up with a single imputed variable where missing and partially observed cases are replaced with plausible values consistent with the observed censoring limits. See *Methods and formulas* for technical details.

In what follows, when referring to missing data (or missing observations) we will mean completely unobserved, truly missing data and when referring to incomplete data (or incomplete observations) we will mean both censored and truly missing data.

Using mi impute intreg

To accommodate the above characteristics, mi impute intreg requires modifications to the standard syntax of univariate imputation methods. First, mi impute intreg requires that variables containing interval-censoring limits be specified in the ll() and ul() options; see the description of ll() and ul() in *Options*. Second, mi impute intreg requires you to specify a new variable name *newivar* to store the resulting imputed values. mi impute intreg creates a new variable, *newivar*, and registers it as imputed.

The values of *newivar* are determined by ll() and ul(). Observations of *newivar* for which ll() and ul() are different or for which both contain soft missing are set to soft missing (.) and considered incomplete. Observations for which either ll() or ul() contains hard missing are set to the extended missing value .a and, as usual, are omitted from imputation. The remaining observations, corresponding to the observed point data, are complete.

After imputation, mi impute intreg stores imputed values in *newivar*. It also registers variables in ll() and ul() as passive (see mi register in [MI] **mi set**), if they are not already registered as passive, and replaces observations for which ll() and ul() both contain soft missing with the corresponding imputed values. That is, only missing data are replaced in these variables; censored data are not changed.

Later, you may decide to add more imputations or to revise your imputation model and replace existing imputations with new ones. In such cases, you do not need to provide a new variable name. You can reuse the name of the variable created previously by mi impute intreg. mi impute intreg will check that the variable is registered as imputed and that it is consistent in the observed data with the variables supplied in ll() and ul(). That is, the variable must have the same values as ll() and ul() in the observations where ll() and ul() are equal, and soft missing values in the remaining observations. If ll() or ul() contain hard missing values, the variable must contain hard missing values in the corresponding observations as well.

Example

We continue the example of imputing missing values of variable bmi from [MI] **mi impute pmm**. The primary analysis of interest is the logistic model investigating the effect of smoking adjusted for other predictors (including bmi) on heart attacks; see [MI] **intro substantive** for details.

The bmi variable is not censored in the original data. For the purpose of illustration, we use a version of the dataset in which the first three observations are censored:

```
. use http://www.stata-press.com/data/r12/mheartintreg
(Fictional heart attack data; BMI censored and missing)
. list lbmi ubmi in 1/10
```

	lbmi	ubmi
1.	.	22
2.	20	.
3.	30	31
4.	24.62917	24.62917
5.	22.52744	22.52744
6.	21.87975	21.87975
7.	17.77057	17.77057
8.	.	.
9.	23.47249	23.47249
10.	24.48916	24.48916

Rather than a single `bmi` variable, we have `lbmi` and `ubmi` variables containing lower and upper interval-censoring limits of BMI. The first observation is left-censored with an upper limit of 22, the second observation is right-censored with a lower limit of 20, and the third observation is interval-censored with the range [30, 31]. Observation 8, for which both `lbmi` and `ubmi` are missing, is missing.

Let's impute censored BMI values:

```
. mi set mlong

. mi impute intreg newbmi attack smokes age hsgrad female, add(20) ll(lbmi) ul(ubmi)

Univariate imputation                       Imputations =        20
Interval regression                               added =        20
Imputed: m=1 through m=20                       updated =         0

Limit: lower =         lbmi            Number missing =        22
       upper =         ubmi           Number censored =         3
                                             interval =         1
                                                 left =         1
                                                right =         1
```

	Observations per m			
Variable	Complete	Incomplete	Imputed	Total
newbmi	129	25	25	154

```
(complete + incomplete = total; imputed is the minimum across m
 of the number of filled-in observations.)
```

Following `mi impute intreg`, we provided a new variable name, `newbmi`, to contain imputed values. Because `newbmi` did not exist we did not need to register it before using `mi impute intreg`. We also specified the lower and upper interval-censoring limits in the `ll()` and `ul()` options. These options are required with `mi impute intreg`.

`mi impute intreg` reported that 25 incomplete BMI values were imputed. Among these incomplete observations, there are 22 missing observations and 3 censored observations (one interval-censored, one left-censored, and one right-censored).

Let's describe our `mi` data:

```
. mi describe, detail
   Style:  mlong
           last mi update 02apr2011 11:02:32, 0 seconds ago
   Obs.:   complete          129
           incomplete         25   (M = 20 imputations)
           ─────────────────────
           total             154
   Vars.:  imputed:  1; newbmi(25; 20*0)
           passive:  2; lbmi(23; 20*1) ubmi(23; 20*1)
           regular:  0
           system:   3; _mi_m _mi_id _mi_miss
           (there are 5 unregistered variables)
```

We used the `detail` option to also see missing-value counts in the imputed data.

According to `mi describe`, the new variable `newbmi` is registered as imputed and contains 25 incomplete observations in the original data. It does not contain incomplete values in any of the 20 imputations. `lbmi` and `ubmi` are registered as passive. Each of `lbmi` and `ubmi` contains 23 incomplete values in the original data and one incomplete value in each imputation. The 22 missing values for lbmi and ubmi are imputed. The incomplete value for each of these variables that is not imputed corresponds to a censored observation (left-censored observation 1 for lbmi and right-censored observation 2 for ubmi). mi impute intreg replaces only missing observations of lbmi and ubmi with imputed data and leaves censored observations unchanged.

As described in *Methods and formulas*, missing observations are simulated from an unrestricted normal distribution. So, the 22 imputed values may contain any value on the whole real line. This may not be desirable because the BMI measure is positive and, in fact, has a limited range.

To restrict imputed values to a certain range, we may replace lbmi and ubmi with lower and upper limits in observations for which these variables are missing. For example, let's restrict imputed values to be between 17 and 39, consistent with the observed range of BMI.

```
. use http://www.stata-press.com/data/r12/mheartintreg, clear
(Fictional heart attack data; BMI censored and missing)

. replace lbmi = 17 if lbmi==.
(23 real changes made)

. replace ubmi = 39 if ubmi==.
(23 real changes made)

. list lbmi ubmi in 1/10
```

	lbmi	ubmi
1.	17	22
2.	20	39
3.	30	31
4.	24.62917	24.62917
5.	22.52744	22.52744
6.	21.87975	21.87975
7.	17.77057	17.77057
8.	17	39
9.	23.47249	23.47249
10.	24.48916	24.48916

We replace missing lower limits with 17 and missing upper limits with 39 and proceed with imputation:

```
. mi set mlong

. mi impute intreg newbmi attack smokes age hsgrad female, add(20) ll(lbmi) ul(ubmi)
```

Univariate imputation		Imputations =	20
Interval regression		added =	20
Imputed: *m*=1 through *m*=20		updated =	0
Limit: lower =	lbmi	Number missing =	0
upper =	ubmi	Number censored =	25
		interval =	25
		left =	0
		right =	0

		Observations per *m*		
Variable	Complete	Incomplete	Imputed	Total
newbmi	129	25	25	154

(complete + incomplete = total; imputed is the minimum across *m*
 of the number of filled-in observations.)

All the incomplete observations are now interval-censored on [17, 39].

We can analyze these multiply imputed data using logistic regression with `mi estimate`:

```
. mi estimate: logit attack smokes age newbmi female hsgrad
  (output omitted )
```

In [MI] **mi impute truncreg**, we used `mi impute truncreg` to accommodate a restricted range of BMI during imputation. In the code above, we showed how to use `mi impute intreg` to ensure that imputed values are within a specified range. Which one should be used?

The answer to this question depends on our belief about the distribution of the imputation variable. If we believe that the underlying distribution of BMI is a normal distribution and we happened to only observe values within a certain range, then `mi impute intreg` should be used to impute BMI. We know, however, that BMI is positive and has an upper limit. As such, the assumption of a truncated distribution for BMI is more plausible, in which case `mi impute truncreg` should be used to impute its missing values.

Saved results

`mi impute intreg` saves the following in `r()`:

Scalars
`r(M)`	total number of imputations
`r(M_add)`	number of added imputations
`r(M_update)`	number of updated imputations
`r(N_miss)`	number of missing observations
`r(N_cens)`	number of censored observations
`r(N_lcens)`	number of left-censored observations
`r(N_rcens)`	number of right-censored observations
`r(N_intcens)`	number of interval-censored observations
`r(k_ivars)`	number of imputed variables (always 1)
`r(N_g)`	number of imputed groups (1 if by() is not specified)

Macros
`r(method)`	name of imputation method (intreg)
`r(ivars)`	names of imputation variables
`r(llname)`	name of variable containing lower interval-censoring limits
`r(ulname)`	name of variable containing upper interval-censoring limits
`r(rseed)`	random-number seed
`r(by)`	names of variables specified within by()

Matrices
`r(N)`	number of observations in imputation sample in each group
`r(N_complete)`	number of complete observations in imputation sample in each group
`r(N_incomplete)`	number of incomplete observations in imputation sample in each group
`r(N_imputed)`	number of imputed observations in imputation sample in each group

Methods and formulas

Consider a latent univariate variable $\mathbf{x}^u = (x_1^u, x_2^u, \ldots, x_n^u)'$ that follows a normal linear regression

$$x_i^u | \mathbf{z}_i \sim N(\mathbf{z}_i' \boldsymbol{\beta}, \sigma^2) \tag{1}$$

where $\mathbf{z}_i = (z_{i1}, z_{i2}, \ldots, z_{iq})'$ records values of predictors of $\mathbf{x}^u$ for observation i, $\boldsymbol{\beta}$ is the $q \times 1$ vector of unknown regression coefficients, and σ^2 is the unknown scalar variance. (When a constant is included in the model—the default—$z_{i1} = 1$, $i = 1, \ldots, n$.)

Instead of $\mathbf{x}^u$, we observe $(\mathbf{x}^{ll}, \mathbf{x}^{ul})$, where $x_j^{ll} = x_j^{ul} = x_j^u$ for point (observed) data $j \in \mathcal{C}$; $x_j^{ll} = -\infty$ and $x_j^{ul} < +\infty$ for left-censored data $j \in \mathcal{L}$; $x_j^{ll} > -\infty$ and $x_j^{ul} = +\infty$ for right-censored data $j \in \mathcal{R}$; $x_j^{ll} = -\infty$ and $x_j^{ul} = +\infty$ for missing data $j \in \mathcal{M}$. Observations from subset $\mathcal{C}$ are considered complete and the remaining observations are considered incomplete.

Let $\mathbf{x} = \mathbf{x}^u$ for observations in subset $\mathcal{C}$, and let $\mathbf{x}$ contain missing values in the remaining observations. We want to fill in missing values in $\mathbf{x}$. Consider the partition of $\mathbf{x} = (\mathbf{x}_o{'}, \mathbf{x}_m{'})$ into $n_0 \times 1$ and $n_1 \times 1$ vectors containing the complete and the incomplete observations. Consider a similar partition of $\mathbf{Z} = (\mathbf{Z}_o, \mathbf{Z}_m)$ into $n_0 \times q$ and $n_1 \times q$ submatrices.

`mi impute intreg` follows the steps below to fill in $\mathbf{x}_m$:

1. Fit an interval regression to the interval-censored data $(\mathbf{x}^{ll}, \mathbf{x}^{ul})$ to obtain the maximum likelihood estimates of parameters in (1), $\widehat{\boldsymbol{\theta}} = (\widehat{\boldsymbol{\beta}}', \ln\widehat{\sigma})'$, and their asymptotic sampling variance, $\widehat{\mathbf{U}}$. See [R] **intreg** for details.

2. Simulate new parameters, $\boldsymbol{\theta}_\star$, from the large-sample normal approximation, $N(\widehat{\boldsymbol{\theta}}, \widehat{\mathbf{U}})$, to its posterior distribution, assuming the noninformative prior $\Pr(\boldsymbol{\theta}) \propto \text{const}$.

3. Let $\mu_{\star i} = \mathbf{z}_i' \boldsymbol{\beta}_\star$. Obtain one set of imputed values, $\mathbf{x}_m^1$, by simulating from a truncated normal model with the density

$$f_{(x_i^{ll}, x_i^{ul})}(x|\mathbf{z}_i) = \frac{1}{\sigma_\star} \phi\left(\frac{x - \mu_{\star i}}{\sigma_\star}\right) \times \left\{ \Phi\left(\frac{x_i^{ul} - \mu_{\star i}}{\sigma_\star}\right) - \Phi\left(\frac{x_i^{ll} - \mu_{\star i}}{\sigma_\star}\right) \right\}^1 ,$$

$$x_i^{ll} < x < x_i^{ul}$$

for every incomplete observation $i \notin \mathcal{C}$. For missing observations $i \in \mathcal{M}$, when $x_i^{ll} = -\infty$ and $x_i^{ul} = +\infty$, the above density reduces to a normal density. Thus missing observations are simulated from the corresponding unrestricted normal distribution.

4. Repeat steps 2 and 3 to obtain M sets of imputed values, $\mathbf{x}_m^1, \mathbf{x}_m^2, \ldots, \mathbf{x}_m^M$.

Steps 2 and 3 above correspond to only approximate draws from the posterior predictive distribution of the missing data, $\Pr(\mathbf{x}_m|\mathbf{x}_o, \mathbf{Z}_o)$, because $\boldsymbol{\theta}_\star$ is drawn from the asymptotic approximation to its posterior distribution.

If weights are specified, a weighted regression model is fit to the observed data in step 1 (see [R] **intreg** for details). Also, in the case of `aweights`, $\sigma_\star$ is replaced with $\sigma_\star w_i^{-1/2}$ in step 3, where w_i is the analytic weight for observation i.

Reference

Royston, P. 2007. Multiple imputation of missing values: Further update of ice, with an emphasis on interval censoring. *Stata Journal* 7: 445–464.

Also see

[MI] **mi impute** — Impute missing values

[MI] **mi impute pmm** — Impute using predictive mean matching

[MI] **mi impute regress** — Impute using linear regression

[MI] **mi impute truncreg** — Impute using truncated regression

[MI] **mi estimate** — Estimation using multiple imputations

[MI] **intro** — Introduction to mi

[MI] **intro substantive** — Introduction to multiple-imputation analysis

Title

> **mi impute logit** — Impute using logistic regression

Syntax

mi impute logit *ivar* [*indepvars*] [*if*] [*weight*] [, *impute_options options*]

impute_options	Description
Main	
* add(*#*)	specify number of imputations to add; required when no imputations exist
* replace	replace imputed values in existing imputations
rseed(*#*)	specify random-number seed
double	save imputed values in double precision; the default is to save them as float
by(*varlist* [, *byopts*])	impute separately on each group formed by *varlist*
Reporting	
dots	display dots as imputations are performed
noisily	display intermediate output
nolegend	suppress all table legends
Advanced	
force	proceed with imputation, even when missing imputed values are encountered
noupdate	do not perform mi update; see [MI] **noupdate option**

* add(*#*) is required when no imputations exist; add(*#*) or replace is required if imputations exist.
noupdate does not appear in the dialog box.

options	Description
Main	
noconstant	suppress constant term
offset(*varname*)	include *varname* in model with coefficient constrained to 1
augment	perform augmented regression in the presence of perfect prediction
conditional(*if*)	perform conditional imputation
bootstrap	estimate model parameters using sampling with replacement
Maximization	
maximize_options	control the maximization process; seldom used

You must mi set your data before using mi impute logit; see [MI] **mi set**.

You must mi register *ivar* as imputed before using mi impute logit; see [MI] **mi set**.

indepvars may contain factor variables; see [U] **11.4.3 Factor variables**.

fweights, iweights, and pweights are allowed; see [U] **11.1.6 weight**.

171

Menu

Statistics > Multiple imputation

Description

mi impute logit fills in missing values of a binary variable by using a logistic regression imputation method. You can perform separate imputations on different subsets of the data by specifying the by() option. You can also account for frequency, importance, and sampling weights.

Options

─────┤ Main ├──

noconstant; see [R] **estimation options**.

add(), replace, rseed(), double, by(); see [MI] **mi impute**.

offset(*varname*); see [R] **estimation options**.

augment specifies that augmented regression be performed if perfect prediction is detected. By default, an error is issued when perfect prediction is detected. The idea behind the augmented-regression approach is to add a few observations with small weights to the data during estimation to avoid perfect prediction. See *The issue of perfect prediction during imputation of categorical data* under *Remarks* in [MI] **mi impute** for more information. augment is not allowed with importance weights.

conditional(*if*) specifies that the imputation variable be imputed conditionally on observations satisfying *exp*; see [U] **11.1.3 if exp**. That is, missing values in a conditional sample, the sample identified by the *exp* expression, are imputed based only on data in that conditional sample. Missing values outside the conditional sample are replaced with a conditional constant, the value of the imputation variable in observations outside the conditional sample. As such, the imputation variable is required to be constant outside the conditional sample. Also, if any conditioning variables (variables involved in the conditional specification if *exp*) contain soft missing values (.), their missing values must be nested within missing values of the imputation variables. See *Conditional imputation* under *Remarks* in [MI] **mi impute**.

bootstrap specifies that posterior estimates of model parameters be obtained using sampling with replacement; that is, posterior estimates are estimated from a bootstrap sample. The default is to sample the estimates from the posterior distribution of model parameters or from the large-sample normal approximation of the posterior distribution. This option is useful when asymptotic normality of parameter estimates is suspect.

─────┤ Reporting ├───

dots, noisily, nolegend; see [MI] **mi impute**. noisily specifies that the output from the logistic regression fit to the observed data be displayed. nolegend suppresses all legends that appear before the imputation table. Such legends include a legend about conditional imputation that appears when the conditional() option is specified and group legends that may appear when the by() option is specified.

─────┤ Maximization ├──

maximize_options; see [R] **logit**. These options are seldom used.

Advanced

`force`; see [MI] **mi impute**.

The following option is available with `mi impute` but is not shown in the dialog box:

`noupdate`; see [MI] **noupdate option**.

Remarks

Remarks are presented under the following headings:

> *Univariate imputation using logistic regression*
> *Using mi impute logit*

See [MI] **mi impute** for a general description and details about options common to all imputation methods, *impute_options*. Also see [MI] **workflow** for general advice on working with `mi`.

Univariate imputation using logistic regression

The logistic regression imputation method can be used to fill in missing values of a binary variable (for example, Rubin [1987]; Raghunathan et al. [2001]; and van Buuren [2007]). It is a parametric method that assumes an underlying logistic model for the imputed variable (given other predictors).

Unlike the linear regression method, the logistic imputation method is based on the asymptotic approximation of the posterior predictive distribution of the missing data. The actual posterior distribution of the logistic model parameters, β, does not have a simple form under the common noninformative prior distribution. Thus a large-sample normal approximation to the posterior distribution of β is used instead. Rubin (1987, 169) points out that although the actual posterior distribution may be far from normal (for example, when the number of observed cases is small or when the fraction of ones in the observed data is close to zero or one), the use of the normal approximation is common in practice.

Using mi impute logit

Continuing our heart attack example from [MI] **intro substantive** and [MI] **mi impute**, suppose that `hsgrad`, a binary variable recording whether subjects graduated from high school, contains missing values:

```
. use http://www.stata-press.com/data/r12/mheart2
(Fictional heart attack data; hsgrad missing)
. mi set mlong
. mi misstable summarize
```

				Obs<.		
Variable	Obs=.	Obs>.	Obs<.	Unique values	Min	Max
hsgrad	18		136	2	0	1

Thus we want to impute missing values of `hsgrad`, because `hsgrad` was one of the predictors in our logistic model (`logit attack smokes age bmi female hsgrad`). From our previous analysis of the heart attack data, we recall that `hsgrad` was not a significant predictor. So, we could have omitted `hsgrad` from the logistic model in the casewise-deletion analysis to avoid the reduction in sample size, and then imputing `hsgrad` would not have been needed. In general, the imputer rarely has such knowledge, and omitting `hsgrad` from the imputation model would prevent this predictor from being used in later analysis by the analyst (see, for example, *Imputation modeling* in [MI] **mi impute**). Thus we proceed with imputation.

We use `mi impute logit` to create 10 imputations of `hsgrad`:

```
. mi register imputed hsgrad
(18 m=0 obs. now marked as incomplete)

. mi impute logit hsgrad attack smokes age bmi female, add(10)

Univariate imputation                         Imputations =        10
Logistic regression                                 added =        10
Imputed: m=1 through m=10                          updated =         0
```

	Observations per *m*				
Variable	Complete	Incomplete	Imputed		Total
hsgrad	136	18	18		154

```
(complete + incomplete = total; imputed is the minimum across m
 of the number of filled-in observations.)
```

We can now use the imputed `hsgrad` in our analysis, for example,

```
. mi estimate: logit attack smokes age bmi female hsgrad
 (output omitted )
```

Saved results

`mi impute logit` saves the following in `r()`:

Scalars
`r(M)`	total number of imputations
`r(M_add)`	number of added imputations
`r(M_update)`	number of updated imputations
`r(k_ivars)`	number of imputed variables (always 1)
`r(pp)`	1 if perfect prediction detected, 0 otherwise
`r(N_g)`	number of imputed groups (1 if by() is not specified)

Macros
`r(method)`	name of imputation method (logit)
`r(ivars)`	names of imputation variables
`r(rseed)`	random-number seed
`r(by)`	names of variables specified within by()

Matrices
`r(N)`	number of observations in imputation sample in each group
`r(N_complete)`	number of complete observations in imputation sample in each group
`r(N_incomplete)`	number of incomplete observations in imputation sample in each group
`r(N_imputed)`	number of imputed observations in imputation sample in each group

Methods and formulas

Consider a univariate variable $\mathbf{x} = (x_1, x_2, \ldots, x_n)'$ that follows a logistic model

$$\Pr(x_i \neq 0 | \mathbf{z}_i) = \frac{\exp(\mathbf{z}_i'\boldsymbol{\beta})}{1 + \exp(\mathbf{z}_i'\boldsymbol{\beta})} \tag{1}$$

where $\mathbf{z}_i = (z_{i1}, z_{i2}, \ldots, z_{iq})'$ records values of predictors of $\mathbf{x}$ for observation i and $\boldsymbol{\beta}$ is the $q \times 1$ vector of unknown regression coefficients. (When a constant is included in the model—the default—$z_{i1} = 1$, $i = 1, \ldots, n$.)

$\mathbf{x}$ contains missing values that are to be filled in. Consider the partition of $\mathbf{x} = (\mathbf{x}_o', \mathbf{x}_m')$ into $n_0 \times 1$ and $n_1 \times 1$ vectors containing the complete and the incomplete observations. Consider a similar partition of $\mathbf{Z} = (\mathbf{Z}_o, \mathbf{Z}_m)$ into $n_0 \times q$ and $n_1 \times q$ submatrices.

mi impute logit follows the steps below to fill in $\mathbf{x}_m$:

1. Fit a logistic model (1) to the observed data $(\mathbf{x}_o, \mathbf{Z}_o)$ to obtain the maximum likelihood estimates, $\widehat{\boldsymbol{\beta}}$, and their asymptotic sampling variance, $\widehat{\mathbf{U}}$.

2. Simulate new parameters, $\boldsymbol{\beta}_\star$, from the large-sample normal approximation, $N(\widehat{\boldsymbol{\beta}}, \widehat{\mathbf{U}})$, to its posterior distribution assuming the noninformative prior $\Pr(\boldsymbol{\beta}) \propto \mathrm{const}$.

3. Obtain one set of imputed values, $\mathbf{x}_m^1$, by simulating from the logistic distribution:

$$\Pr(x_{i_m} = 1) = \exp(\mathbf{z}_{i_m}' \boldsymbol{\beta}_\star) / \left\{ 1 + \exp(\mathbf{z}_{i_m}' \boldsymbol{\beta}_\star) \right\}$$

for every missing observation i_m.

4. Repeat steps 2 and 3 to obtain M sets of imputed values, $\mathbf{x}_m^1, \mathbf{x}_m^2, \ldots, \mathbf{x}_m^M$.

Steps 2 and 3 above correspond to only approximate draws from the posterior predictive distribution of the missing data $\Pr(\mathbf{x}_m | \mathbf{x}_o, \mathbf{Z}_o)$ because $\boldsymbol{\beta}_\star$ is drawn from the asymptotic approximation to its posterior distribution.

If weights are specified, a weighted logistic regression model is fit to the observed data in step 1 (see [R] **logit** for details).

References

Raghunathan, T. E., J. M. Lepkowski, J. Van Hoewyk, and P. Solenberger. 2001. A multivariate technique for multiply imputing missing values using a sequence of regression models. *Survey Methodology* 27: 85–95.

Rubin, D. B. 1987. *Multiple Imputation for Nonresponse in Surveys.* New York: Wiley.

van Buuren, S. 2007. Multiple imputation of discrete and continuous data by fully conditional specification. *Statistical Methods in Medical Research* 16: 219–242.

Also see

Title

---
mi impute mlogit — Impute using multinomial logistic regression
---

Syntax

mi impute mlogit *ivar* [*indepvars*] [*if*] [*weight*] [, *impute_options options*]

impute_options	Description
Main	
* add(*#*)	specify number of imputations to add; required when no imputations exist
* replace	replace imputed values in existing imputations
rseed(*#*)	specify random-number seed
double	save imputed values in double precision; the default is to save them as float
by(*varlist* [, *byopts*])	impute separately on each group formed by *varlist*
Reporting	
dots	display dots as imputations are performed
noisily	display intermediate output
nolegend	suppress all table legends
Advanced	
force	proceed with imputation, even when missing imputed values are encountered
noupdate	do not perform mi update; see [MI] **noupdate option**

add(#*) is required when no imputations exist; add(*#*) or replace is required if imputations exist.
noupdate does not appear in the dialog box.

options	Description
Main	
noconstant	suppress constant term
baseoutcome(*#*)	specify value of *ivar* that will be the base outcome
augment	perform augmented regression in the presence of perfect prediction
conditional(*if*)	perform conditional imputation
bootstrap	estimate model parameters using sampling with replacement
Maximization	
maximize_options	control the maximization process; seldom used

You must mi set your data before using mi impute mlogit; see [MI] **mi set**.
You must mi register *ivar* as imputed before using mi impute mlogit; see [MI] **mi set**.
indepvars may contain factor variables; see [U] **11.4.3 Factor variables**.
fweights, iweights, and pweights are allowed; see [U] **11.1.6 weight**.

176

Menu

Statistics > Multiple imputation

Description

mi impute mlogit fills in missing values of a nominal variable by using the multinomial (polytomous) logistic regression imputation method. You can perform separate imputations on different subsets of the data by specifying the by() option. You can also account for frequency, importance, and sampling weights.

Options

<table><tr><td>Main</td></tr></table>

noconstant; see [R] **estimation options**.

add(), replace, rseed(), double, by(); see [MI] **mi impute**.

baseoutcome(#) specifies the value of *ivar* to be treated as the base outcome. The default is to choose the most frequent outcome.

augment specifies that augmented regression be performed if perfect prediction is detected. By default, an error is issued when perfect prediction is detected. The idea behind the augmented-regression approach is to add a few observations with small weights to the data during estimation to avoid perfect prediction. See *The issue of perfect prediction during imputation of categorical data* under *Remarks* in [MI] **mi impute** for more information. augment is not allowed with importance weights.

conditional(*if*) specifies that the imputation variable be imputed conditionally on observations satisfying *exp*; see [U] **11.1.3 if exp**. That is, missing values in a conditional sample, the sample identified by the *exp* expression, are imputed based only on data in that conditional sample. Missing values outside the conditional sample are replaced with a conditional constant, the value of the imputation variable in observations outside the conditional sample. As such, the imputation variable is required to be constant outside the conditional sample. Also, if any conditioning variables (variables involved in the conditional specification if *exp*) contain soft missing values (.), their missing values must be nested within missing values of the imputation variables. See *Conditional imputation* under *Remarks* in [MI] **mi impute**.

bootstrap specifies that posterior estimates of model parameters be obtained using sampling with replacement; that is, posterior estimates are estimated from a bootstrap sample. The default is to sample the estimates from the posterior distribution of model parameters or from the large-sample normal approximation of the posterior distribution. This option is useful when asymptotic normality of parameter estimates is suspect.

<table><tr><td>Reporting</td></tr></table>

dots, noisily, nolegend; see [MI] **mi impute**. noisily specifies that the output from the multinomial logistic regression fit to the observed data be displayed. nolegend suppresses all legends that appear before the imputation table. Such legends include a legend about conditional imputation that appears when the conditional() option is specified and group legends that may appear when the by() option is specified.

⌐ Maximization ⌐

maximize_options; see [R] **mlogit**. These options are seldom used.

⌐ Advanced ⌐

`force`; see [MI] **mi impute**.

The following option is available with `mi impute` but is not shown in the dialog box:

`noupdate`; see [MI] **noupdate option**.

Remarks

Remarks are presented under the following headings:

> *Univariate imputation using multinomial logistic regression*
> *Using mi impute mlogit*

See [MI] **mi impute** for a general description and details about options common to all imputation methods, *impute_options*. Also see [MI] **workflow** for general advice on working with `mi`.

Univariate imputation using multinomial logistic regression

The multinomial logistic regression imputation method can be used to fill in missing values of a nomial variable (for example, Raghunathan et al. [2001] and van Buuren [2007]). It is a parametric method that assumes an underlying multinomial logistic model for the imputed variable (given other predictors). Similarly to the logistic imputation method, this method is based on the asymptotic approximation of the posterior predictive distribution of the missing data.

Using mi impute mlogit

Consider the heart attack data introduced in [MI] **intro substantive** and discussed in [MI] **mi impute**. Suppose that we want our logistic model of interest to also include information about marital status (categorical variable `marstatus`)—`logit attack smokes age bmi female hsgrad i.marstatus`.

We first tabulate values of `marstatus`:

```
. use http://www.stata-press.com/data/r12/mheart3
(Fictional heart attack data; marstatus missing)

. tabulate marstatus, missing
```

Marital status: single, married, divorced	Freq.	Percent	Cum.
Single	53	34.42	34.42
Married	48	31.17	65.58
Divorced	46	29.87	95.45
.	7	4.55	100.00
Total	154	100.00	

From the output, the marstatus variable has three unique categories and seven missing observations. Because marstatus is a categorical variable, we use the multinomial logistic imputation method to fill in its missing values.

We mi set the data, register marstatus as an imputed variable, and then create 10 imputations by specifying the add(10) option with mi impute mlogit:

```
. mi set mlong

. mi register imputed marstatus
(7 m=0 obs. now marked as incomplete)

. mi impute mlogit marstatus attack smokes age bmi female hsgrad, add(10)

Univariate imputation                        Imputations =    10
Multinomial logistic regression                    added =    10
Imputed: m=1 through m=10                         updated =     0
```

	Observations per m			
Variable	Complete	Incomplete	Imputed	Total
marstatus	147	7	7	154

```
(complete + incomplete = total; imputed is the minimum across m
 of the number of filled-in observations.)
```

We can now analyze these multiply imputed data using logistic regression via mi estimate:

```
. mi estimate: logit attack smokes age bmi female hsgrad i.marstatus
  (output omitted )
```

Saved results

mi impute mlogit saves the following in r():

Scalars
r(M)	total number of imputations
r(M_add)	number of added imputations
r(M_update)	number of updated imputations
r(k_ivars)	number of imputed variables (always 1)
r(pp)	1 if perfect prediction detected, 0 otherwise
r(N_g)	number of imputed groups (1 if by() is not specified)

Macros
r(method)	name of imputation method (mlogit)
r(ivars)	names of imputation variables
r(rseed)	random-number seed
r(by)	names of variables specified within by()

Matrices
r(N)	number of observations in imputation sample in each group
r(N_complete)	number of complete observations in imputation sample in each group
r(N_incomplete)	number of incomplete observations in imputation sample in each group
r(N_imputed)	number of imputed observations in imputation sample in each group

Methods and formulas

Consider a univariate variable $\mathbf{x} = (x_1, x_2, \ldots, x_n)'$ that contains K categories (without loss of generality, let $k = 1$ be the base outcome) and follows a multinomial logistic model

$$
\Pr(x_i = k|\mathbf{z}_i) =
\begin{cases}
\dfrac{1}{1 + \sum_{l=2}^{K} \exp(\mathbf{z}_i'\boldsymbol{\beta}_l)}, & \text{if } k = 1 \\[3ex]
\dfrac{\exp(\mathbf{z}_i'\boldsymbol{\beta}_k)}{1 + \sum_{l=2}^{K} \exp(\mathbf{z}_i'\boldsymbol{\beta}_l)}, & \text{if } k > 1
\end{cases}
\tag{1}
$$

where $\mathbf{z}_i = (z_{i1}, z_{i2}, \ldots, z_{iq})'$ records values of predictors of $\mathbf{x}$ for observation i and $\boldsymbol{\beta}_l$ is the $q \times 1$ vector of unknown regression coefficients for outcome $l = 2, \ldots, K$. (When a constant is included in the model—the default—$z_{i1} = 1$, $i = 1, \ldots, n$.)

$\mathbf{x}$ contains missing values that are to be filled in. Consider the partition of $\mathbf{x} = (\mathbf{x}_o', \mathbf{x}_m')$ into $n_0 \times 1$ and $n_1 \times 1$ vectors containing the complete and the incomplete observations. Consider a similar partition of $\mathbf{Z} = (\mathbf{Z}_o, \mathbf{Z}_m)$ into $n_0 \times q$ and $n_1 \times q$ submatrices.

mi impute mlogit follows the steps below to fill in $\mathbf{x}_m$:

1. Fit a multinomial logistic model (1) to the observed data $(\mathbf{x}_o, \mathbf{Z}_o)$ to obtain the maximum likelihood estimates, $\widehat{\boldsymbol{\beta}} = (\widehat{\boldsymbol{\beta}}_2', \ldots, \widehat{\boldsymbol{\beta}}_K')'$, and their asymptotic sampling variance, $\widehat{\mathbf{U}}$.

2. Simulate new parameters, $\boldsymbol{\beta}_\star$, from the large-sample normal approximation, $N(\widehat{\boldsymbol{\beta}}, \widehat{\mathbf{U}})$, to its posterior distribution assuming the noninformative prior $\Pr(\boldsymbol{\beta}) \propto \text{const}$.

3. Obtain one set of imputed values, $\mathbf{x}_m^1$, by simulating from the multinomial logistic distribution: one of K categories is randomly assigned to a missing category, i_m, using the cumulative probabilities computed from (1) with $\boldsymbol{\beta}_l = \boldsymbol{\beta}_{\star l}$ and $\mathbf{z}_i = \mathbf{z}_{i_m}$.

4. Repeat steps 2 and 3 to obtain M sets of imputed values, $\mathbf{x}_m^1, \mathbf{x}_m^2, \ldots, \mathbf{x}_m^M$.

Steps 2 and 3 above correspond to only approximate draws from the posterior predictive distribution of the missing data $\Pr(\mathbf{x}_m|\mathbf{x}_o, \mathbf{Z}_o)$ because $\boldsymbol{\beta}_\star$ is drawn from the asymptotic approximation to its posterior distribution.

If weights are specified, a weighted multinomial logistic regression model is fit to the observed data in step 1 (see [R] **mlogit** for details).

References

Raghunathan, T. E., J. M. Lepkowski, J. Van Hoewyk, and P. Solenberger. 2001. A multivariate technique for multiply imputing missing values using a sequence of regression models. *Survey Methodology* 27: 85–95.

van Buuren, S. 2007. Multiple imputation of discrete and continuous data by fully conditional specification. *Statistical Methods in Medical Research* 16: 219–242.

Also see

Title

> **mi impute monotone** — Impute missing values in monotone data

Syntax

Default specification of prediction equations, basic syntax

    mi <u>imp</u>ute <u>mon</u>otone (*uvmethod*) *ivars* $\big[$ = *indepvars* $\big]$ $\big[$ *if* $\big]$ $\big[$ *weight* $\big]$ $\big[$, *impute_options options* $\big]$

Default specification of prediction equations, full syntax

    mi <u>imp</u>ute <u>mon</u>otone *lhs* $\big[$ = *indepvars* $\big]$ $\big[$ *if* $\big]$ $\big[$ *weight* $\big]$ $\big[$, *impute_options options* $\big]$

Custom specification of prediction equations

    mi <u>imp</u>ute <u>mon</u>otone *cmodels* $\big[$ *if* $\big]$ $\big[$ *weight* $\big]$, <u>cus</u>tom $\big[$ *impute_options options* $\big]$

where *lhs* is *lhs_spec* $\big[$ *lhs_spec* $\big[\dots \big]$ $\big]$ and *lhs_spec* is

    (*uvmethod* $\big[$ *if* $\big]$ $\big[$, *uvspec_options* $\big]$) *ivars*

cmodels is (*cond_spec*) $\big[$ (*cond_spec*) $\big[\dots \big]$ $\big]$ and a conditional specification, *cond_spec*, is

    uvmethod ivar $\big[$ *rhs_spec* $\big]$ $\big[$ *if* $\big]$ $\big[$, *uvspec_options* $\big]$

rhs_spec includes *varlist* and expressions of imputation variables bound in parentheses.

ivar(s) (or *newivar* if *uvmethod* is intreg) is the name(s) of the imputation variable(s).

uvspec_options are <u>asc</u>ontinuous, <u>noi</u>sily, and the method-specific *options* as described in the
    manual entry for each univariate imputation method.

uvmethod	Description
regress	linear regression for a continuous variable; [MI] **mi impute regress**
pmm	predictive mean matching for a continuous variable; [MI] **mi impute pmm**
truncreg	truncated regression for a continuous variable with a restricted range; [MI] **mi impute truncreg**
intreg	interval regression for a continuous censored variable; [MI] **mi impute intreg**
logit	logistic regression for a binary variable; [MI] **mi impute logit**
ologit	ordered logistic regression for an ordinal variable; [MI] **mi impute ologit**
mlogit	multinomial logistic regression for a nominal variable; [MI] **mi impute mlogit**
poisson	Poisson regression for a count variable; [MI] **mi impute poisson**
nbreg	negative binomial regression for an overdispersed count variable; [MI] **mi impute nbreg**

impute_options	Description
Main	
* add(#)	specify number of imputations to add; required when no imputations exist
* replace	replace imputed values in existing imputations
rseed(#)	specify random-number seed
double	save imputed values in double precision; the default is to save them as float
by(*varlist* [, *byopts*])	impute separately on each group formed by *varlist*
Reporting	
dots	display dots as imputations are performed
noisily	display intermediate output
nolegend	suppress all table legends
Advanced	
force	proceed with imputation, even when missing imputed values are encountered
noupdate	do not perform mi update; see [MI] **noupdate option**

* add(#) is required when no imputations exist; add(#) or replace is required if imputations exist.
noupdate does not appear in the dialog box.

options	Description
Main	
* custom	customize prediction equations of conditional specifications
augment	perform augmented regression in the presence of perfect prediction for all categorical imputation variables
bootstrap	estimate model parameters using sampling with replacement
Reporting	
dryrun	show conditional specifications without imputing data
verbose	show conditional specifications and impute data; implied when custom prediction equations are not specified
report	show report about each conditional specification
Advanced	
nomonotonechk	do not check whether variables follow a monotone-missing pattern

* custom is required when specifying customized prediction equations.

You must mi set your data before using mi impute monotone; see [MI] **mi set**.

You must mi register *ivars* as imputed before using mi impute monotone; see [MI] **mi set**.

indepvars and *rhs_spec* may contain factor variables; see [U] **11.4.3 Factor variables**.

fweights, aweights (regress, pmm, truncreg, and intreg only), iweights, and pweights are allowed; see [U] **11.1.6 weight**.

Menu

Statistics > Multiple imputation

Description

mi impute monotone fills in missing values in multiple variables by using a sequence of independent univariate conditional imputation methods. Variables to be imputed, *ivars*, must follow a monotone-missing pattern (see [MI] **intro substantive**). You can perform separate imputations on different subsets of the data by specifying the by() option. You can also account for frequency, analytic (with continuous variables only), importance, and sampling weights.

Options

⌐ Main ⌐

custom is required to build customized prediction equations within the univariate conditional specifications. Otherwise, the default specification of prediction equations is assumed.

add(), replace, rseed(), double, by(); see [MI] **mi impute**.

augment specifies that augmented regression be performed if perfect prediction is detected. By default, an error is issued when perfect prediction is detected. The idea behind the augmented-regression approach is to add a few observations with small weights to the data during estimation to avoid perfect prediction. See *The issue of perfect prediction during imputation of categorical data* under *Remarks* in [MI] **mi impute** for more information. augment is not allowed with importance weights. This option is equivalent to specifying augment within univariate specifications of all categorical imputation methods.

bootstrap specifies that posterior estimates of model parameters be obtained using sampling with replacement; that is, posterior estimates are estimated from a bootstrap sample. The default is to sample the estimates from the posterior distribution of model parameters or from the large-sample normal approximation of the posterior distribution. This option is useful when asymptotic normality of parameter estimates is suspect. This option is equivalent to specifying bootstrap within all univariate specifications.

The following options appear on a Specification dialog that appears when you click on the **Create ...** button on the **Main** tab.

uvspec_options are options specified within each univariate imputation method, *uvmethod*. *uvspec_options* include ascontinuous, noisily, and the method-specific *options* as described in the manual entry for each univariate imputation method.

ascontinuous specifies that categorical imputation variables corresponding to the current *uvmethod* be included as continuous in all prediction equations. This option is only allowed when *uvmethod* is logit, ologit, or mlogit.

noisily specifies that the output from the current univariate model fit to the observed data be displayed.

dots, noisily, nolegend; see [MI] **mi impute**. noisily specifies that the output from all univariate conditional models fit to the observed data be displayed. nolegend suppresses all imputation table legends which include a legend with the titles of the univariate imputation methods used, a legend about conditional imputation when conditional() is used within univariate specifications, and group legends when by() is specified.

dryrun specifies to show the conditional specifications that would be used to impute each variable without actually imputing data. This option is recommended for checking specifications of conditional models prior to imputation.

verbose specifies to show conditional specifications and impute data. verbose is implied when custom prediction equations are not specified.

report specifies to show a report about each univariate conditional specification. This option, in combination with dryrun, is recommended for checking specifications of conditional models prior to imputation.

force; see [MI] **mi impute**.

nomonotonechk specifies not to check that imputation variables follow a monotone-missing pattern. This option may be used to avoid potentially time-consuming checks. The monotonicity check may be time consuming when a large number of variables is being imputed. If you use nomonotonechk with a custom specification, make sure that you list the univariate conditional specifications in the order of monotonicity or you might obtain incorrect results.

The following option is available with mi impute but is not shown in the dialog box:

noupdate; see [MI] **noupdate option**.

Remarks

Remarks are presented under the following headings:

> *Multivariate imputation when a missing-data pattern is monotone*
> *First use*
> *Using mi impute monotone*
> *Default syntax of mi impute monotone*
> *The alternative syntax of mi impute monotone—custom prediction equations*
> *Examples of using default prediction equations*
> *Examples of using custom prediction equations*

See [MI] **mi impute** for a general description and details about options common to all imputation methods, *impute_options*. Also see [MI] **workflow** for general advice on working with mi.

Multivariate imputation when a missing-data pattern is monotone

When a pattern of missingness in multiple variables is monotone (or, more rigorously, when the missingness-modeling structure is monotone distinct), a multivariate imputation can be replaced with a set of conditional univariate imputations (Rubin 1987, 170–178). Let $X_1, X_2, \ldots, X_p$ be ordered such that if X_{1j} is missing, then X_{2j} is also missing, although X_2 may also be missing in other observations; if X_{2j} is missing, then X_{3j} is missing, although X_3 may also be missing in other observations; and so on. Then a simultaneous imputation of variables $X_1, X_2, \ldots, X_p$ according to a model, $f_{\mathbf{X}}(\cdot)$, and complete predictors (independent variables), $\mathbf{Z}$, is equivalent to the sequential conditional imputation

$$X_1^\star \sim f_1(X_1|\mathbf{Z})$$
$$X_2^\star \sim f_2(X_2|X_1^\star, \mathbf{Z})$$
$$\ldots$$
$$X_p^\star \sim f_p(X_p|X_1^\star, X_2^\star, \ldots, X_{p-1}^\star, \mathbf{Z})$$

(1)

where for brevity we omit conditioning on the model parameters. The univariate conditional imputation models $f_j(\cdot)$ can each be of a different type (normal, logistic, etc.), as is appropriate for imputing X_j.

The specification of a conditional imputation model $f_j(\cdot)$ includes an imputation method and a prediction equation relating an imputation variable to other explanatory variables. In what follows, we distinguish between the default specification in which the identities of the complete explanatory variables are the same for all imputed variables, and the custom specification in which the identities are allowed to differ.

Under the default specification, prediction equations of each imputation variable include all complete independent variables and all preceding imputation variables that have already been imputed. Under the custom specification, each prediction equation may include a subset of the predictors that would be used under the default specification. The custom specification implies nothing more than the assumption of conditional independence between certain imputation variables and certain sets of predictors.

Model (1) corresponds to the default specification. For example, consider imputation variables X_1, X_2, and X_3, ordered from the most observed to the least observed, and complete predictors Z_1 and Z_2. Under the default specification, the individual prediction equations are determined as follows. The most observed variable, X_1, is predicted from Z_1 and Z_2. The next most observed variable, X_2, is predicted from Z_1, Z_2, and previously imputed X_1. The least observed variable, X_3, is predicted from Z_1, Z_2, and previously imputed X_1 and X_2. (A constant is included in all prediction equations, by default.) We use the following notation to refer to the above sequence of prediction equations (imputation sequence): $X_1|Z_1, Z_2 \rightarrow X_2|X_1, Z_1, Z_2 \rightarrow X_3|X_1, X_2, Z_1, Z_2$.

A sequence such as $X_1|Z_1 \rightarrow X_2|X_1, Z_1, Z_2 \rightarrow X_3|X_1, Z_2$ would correspond to a custom specification. Here X_1 is assumed to be independent of Z_2 given Z_1, and X_3 is assumed to be independent of Z_1 and X_2 given X_1 and Z_2.

The monotone-distinct structure offers much flexibility in building a multivariate imputation model. It simplifies the often intractable multivariate imputation task to a set of simpler univariate imputation tasks. In addition, it accommodates imputation of a mixture of types of variables. So, what's the catch? The catch is that the pattern of missingness is rarely monotone in practice. There are types of data for which a monotone-missing data pattern can occur naturally (for example, follow-up measurements). Usually, however, this happens only by chance.

There are several ways to proceed if your data are not monotone missing. You can discard the observations that violate the monotone-missing pattern, especially if there are very few such observations. You can assume independence among the sets of variables to create independent monotone patterns. For example, the missingness pattern for X_1, X_2, X_3, X_4, X_5 may not be monotone, but it may be for X_1, X_3 and for X_2, X_4, X_5. If it is reasonable to assume independence between these two sets of variables, you can then impute each set separately by using monotone imputation. Other alternatives are to use certain techniques to complete the missing-data pattern to monotone (see, for example, Schafer 1997), to use an iterative sequential (fully conditional) imputation (see [MI] **mi impute chained**; Royston 2005, 2007, 2009; van Buuren, Boshuizen, and Knook 1999; Raghunathan et al. 2001), or to assume an explicit multivariate parametric model for the imputation variables (see [MI] **mi impute mvn**; Schafer 1997). Also see *Multivariate imputation* of [MI] **mi impute** for a general discussion of multivariate imputation.

Throughout this entry, we will assume that the considered imputation variables are monotone missing.

First use

Before we describe various uses of `mi impute monotone`, let's look at an example.

Consider the heart attack data examining the relationship between heart attack and smoking. The `age` and `bmi` variables contain missing values and follow a monotone-missing pattern. Recall multivariate imputation of `bmi` and `age` using `mi impute monotone` described in *Multivariate imputation* of [MI] **mi impute**:

```
. use http://www.stata-press.com/data/r12/mheart5s0
(Fictional heart attack data; bmi and age missing)

. mi impute monotone (regress) bmi age = attack smokes hsgrad female, add(10)

Conditional models:
                age: regress age attack smokes hsgrad female
                bmi: regress bmi age attack smokes hsgrad female

Multivariate imputation                    Imputations =        10
Monotone method                                  added =        10
Imputed: m=1 through m=10                       updated =         0

                bmi: linear regression
                age: linear regression
```

		Observations per m		
Variable	Complete	Incomplete	Imputed	Total
bmi	126	28	28	154
age	142	12	12	154

(complete + incomplete = total; imputed is the minimum across m
 of the number of filled-in observations.)

The `age` and `bmi` variables have monotone missingness, and so `mi impute monotone` is used to fill in missing values. Ten imputations are created (`add(10)` option). The linear regression imputation method (`regress`) is used to impute both continuous variables. The `attack`, `smokes`, `hsgrad`, and `female` variables are used as complete predictors (independent variables).

The conditional models legend shows that `age` (having the least number of missing values) is imputed first using the `regress` method, even though we specified `bmi` before `age` on the `mi impute` command. After that, `bmi` is imputed using the `regress` method and the previously imputed variable `age` and the other predictors.

The header and table output were described in detail in [MI] **mi impute**. The additional information above the imputation table is the legend describing what univariate imputation method was used to impute each variable. (If desired, this legend may be suppressed by specifying the `nolegend` option.)

Using mi impute monotone

Below we summarize general capabilities of `mi impute monotone`.

1. `mi impute monotone` requires that the specified imputation variables follow a monotone-missing pattern. If they do not, it will stop with an error:

   ```
   . mi impute monotone x1 x2 ...
   x1 x2: not monotone;
       imputation variables must have a monotone-missing structure;
       see mi misstable nested
   r(459);
   ```

   As indicated by the error message, we can use `mi misstable nested` to verify for ourselves that the imputation variables are not monotone missing. We could also use other features of `mi misstable` to investigate the pattern.

2. `mi impute monotone` offers two main syntaxes—one using the default prediction equations,

   ```
   . mi impute monotone ...
   ```

   and the other allowing customization of prediction equations,

   ```
   . mi impute monotone ..., custom ...
   ```

   We will refer to the two syntaxes as default and custom, respectively.

3. `mi impute monotone` allows specification of a global (outer) `if` condition,

   ```
   . mi impute monotone ... if exp ...
   ```

   and equation-specific (inner) `if` conditions,

   ```
   . mi impute monotone ... (... if exp ...) ...
   ```

   A global `if` is applied to all equations (conditional specifications). You may combine global and equation-specific `if` conditions:

   ```
   . mi impute monotone ... (... if exp ...) ... if exp ...
   ```

4. `mi impute monotone` allows specification of global weights, which are applied to all equations,

   ```
   . mi impute monotone ... [weight] ...
   ```

Use a combination of options `dryrun` and `report` to check the specification of each univariate imputation model prior to imputing data.

In the next two sections, we describe the use of `mi impute monotone` first using hypothetical situations and then using real examples.

Default syntax of mi impute monotone

We showed in *First use* an example of `mi impute monotone` with default prediction equations using the heart attack data. Here we provide more details about this default specification.

By default, `mi impute monotone` imputes missing values by using the full specification of prediction equations. It builds the corresponding univariate conditional imputation models based on the supplied information: *uvmethod*, the imputation method; *ivars*, the imputation variables; and *indepvars*, the complete predictors or independent variables.

Suppose that continuous variables x1, x2, and x3 contain missing values with a monotone-missing pattern. We want to impute these variables, and we decide to use the same univariate imputation method, say, linear regression, for all. We can do this by typing

```
. mi impute monotone (regress) x1 x2 x3 ...
```

The above corresponds to the first syntax diagram of mi impute monotone: *uvmethod* is regress and *ivars* is x1 x2 x3. Relating the above to the model notation used in (1), f_1, f_2, and f_3 represent linear regression imputation models and the prediction sequence is $X_1 \rightarrow X_2|X_1 \rightarrow X_3|X_2, X_1$.

If we have additional covariates containing no missing values (say, z1 and z2) that we want to include in the imputation model, we can do it by typing

```
. mi impute monotone (regress) x1 x2 x3 = z1 z2 ...
```

Now *indepvars* is z1 z2 and the prediction sequence is $X_1|Z_1, Z_2 \rightarrow X_2|X_1, Z_1, Z_2 \rightarrow X_3|X_2, X_1, Z_1, Z_2$. Independent variables are included in the prediction equations of all conditional models.

Suppose that we want to use a different imputation method for one of the variables—we want to impute x3 using predictive mean matching. We can do this by typing

```
. mi impute monotone (regress) x1 x2 (pmm) x3 = z1 z2 ...
```

The above corresponds to the second syntax diagram of mi impute monotone, a generalization of the first that accommodates differing imputation methods. The right-hand side of the equation is unchanged. z1 and z2 are included in all three prediction equations. The left-hand side now has two specifications: (regress) x1 x2 and (pmm) x3. In previous examples, we had only one left-hand-side specification, *lhs_spec*—(regress) x1 x2 x3. (Note that the number of left-hand-side specifications does not necessarily correspond to the number of conditional models; the latter is determined by the number of imputation variables.) In this example, x1 and x2 are imputed using linear regression, and x3 is imputed using predictive mean matching.

Now, instead of using the default one nearest neighbor with pmm, say that we want to use three, which requires pmm's knn(3) option. All method-specific options must be specified within the parentheses surrounding the method:

```
. mi impute monotone (regress) x1 x2 (pmm, knn(3)) x3 = z1 z2 ...
```

Under the default specification, you can list imputation variables in any order and mi impute monotone will determine the correct ordering that follows the monotone-missing pattern.

Suppose now we want to restrict the imputation sample for x2 to observations where z1 is one; also see *Imputing on subsamples* of [MI] **mi impute**. (We also omit pmm's knn() option here.) The corresponding syntax is

```
. mi impute monotone (regress) x1 (regress if z1==1) x2 (pmm) x3 = z1 z2 ...
```

If, in addition to the above, we want to impute all variables using an overall subsample where z3 is one, we can specify the global if z3==1 condition:

```
. mi impute monotone (regress) x1 (regress if z1==1) x2 (pmm) x3 = z1 z2
> if z3==1 ...
```

When any imputation variable is imputed using a categorical method, mi impute monotone automatically includes it as a factor variable in the prediction equations of other imputation variables. Suppose that x1 is a categorical variable and is imputed using the multinomial logistic method:

```
. mi impute monotone (mlogit) x1 (regress) x2 x3 ...
```

The above will result in the prediction sequence $X_1 \to X_2|\mathtt{i}.X_1 \to X_3|X_2,\mathtt{i}.X_1$ where $\mathtt{i}.X_1$ denotes the factors of X_1.

If you wish to include factor variables as continuous in prediction equations, you can use the `ascontinuous` option within a specification of the univariate imputation method for that variable:

```
. mi impute monotone (mlogit, ascontinuous) x1 (regress) x2 x3 ...
```

As we discussed in *The issue of perfect prediction during imputation of categorical data* of [MI] **mi impute**, perfect prediction often occurs during imputation of categorical variables. One way of dealing with it is to use the augmented-regression approach (White, Daniel, and Royston 2010), available through the `augment` option. For example, if perfect prediction occurs during imputation of `x1` in the above, you can specify `augment` within the method specification of `x1` to perform augmented regression:

```
. mi impute monotone (mlogit, augment) x1 (regress) x2 x3 ...
```

Alternatively, you can use the `augment` option with `mi impute monotone` to perform augmented regression for all categorical variables for which the issue of perfect prediction arises:

```
. mi impute monotone (mlogit) x1 (logit) x2 (regress) x3 ..., augment ...
```

The above command is equivalent to specifying `augment` within each specification of a univariate categorical imputation method:

```
. mi impute monotone (mlogit, augment) x1 (logit, augment) x2 (regress) x3 ...
```

Also see *Default prediction equations* in [MI] **mi impute chained** for other uses of the default syntax.

The alternative syntax of mi impute monotone—custom prediction equations

Consider the prediction sequence $X_1 \to X_2|X_1 \to X_3|X_2,X_1$. Suppose that we want to predict X_3 from X_1 rather than from X_1 and X_2. This could be achieved by simply imputing X_1 and X_2 and then X_3 given X_1 separately because of the implied assumption that X_3 and X_2 are independent given X_1. However, with a larger number of variables and more complicated prediction rules, separate imputations may not be appealing. So customization of the prediction equations is a good alternative.

You customize prediction equations using the custom syntax (the third syntax) of `mi impute monotone`. You must specify the `custom` option to notify `mi impute monotone` that you are specifying custom prediction equations.

Under the custom syntax, you specify a separate conditional imputation model for each imputation variable. The specification of a conditional model is the same as that for the chosen univariate imputation method, but the entire model must be bound in parentheses, for example,

```
. mi impute monotone (regress x1)
                      (regress x2 x1)
                      (regress x3 x1)
                              , custom ...
```

Here we have three conditional specifications: `(regress x1)`, `(regress x2 x1)`, and `(regress x3 x1)`. The corresponding prediction sequence is $X_1 \to X_2|X_1 \to X_3|X_1$. Prediction equations have the syntax *ivar* [*rhs_spec*].

When specifying custom prediction equations, you are required to list the conditional models in the correct order of missing monotonicity. `mi impute monotone` will issue an error if you are wrong:

```
mi impute monotone: incorrect equation order
    equations must be listed in the monotone-missing order of the imputation
    variables (from most observed to least observed);  x2(2) -> x1(5) -> x3(10)
r(198);
```

If we have additional covariates z1 and z2 containing no missing values, we can include them in the imputation model:

```
. mi impute monotone (regress x1 z1 z2)
                     (regress x2 x1 z1 z2)
                     (regress x3 x1 z1 z2), custom ...
```

To use the predictive mean matching method for x3, we simply change the method from regress to pmm in the last conditional specification:

```
. mi impute monotone (regress x1 z1 z2)
                     (regress x2 x1 z1 z2)
                     (pmm x3 x1 z1 z2), custom ...
```

To include more nearest neighbors in pmm, we specify the knn(3) option within the last conditional specification:

```
. mi impute monotone (regress x1 z1 z2)
                     (regress x2 x1 z1 z2)
                     (pmm x3 x1 z1 z2, knn(3)), custom ...
```

Under the custom syntax, you can also include expressions of previously imputed variables in prediction equations. For example, if you want to model x3 using main and squared effects of x1 (ignoring predictors z1 and z2), you can type

```
. mi impute monotone (regress x1)
                     (regress x2 x1)
                     (pmm x3 x1 (x1^2)), custom ...
```

Note that we bound the expression x1^2 in parentheses. Any expression may appear inside the parentheses.

Similar to the default specification, we can include equation-specific ifs,

```
. mi impute monotone (regress x1)
                     (regress x2 x1 if z1==1)
                     (pmm x3 x1), custom ...
```

and we can specify a global if,

```
. mi impute monotone (regress x1 z1 z2)
                     (regress x2 x1 z2 if z1==1)
                     (pmm x3 x1 z1 z2)
                                    if z3==1, custom ...
```

Suppose that one of the imputed variables is categorical. We can use the multinomial logistic method to impute its values:

```
. mi impute monotone (mlogit x1)
                     (regress x2 i.x1)
                     (regress x3 i.x1)
                            , custom ...
```

Also see *Link between mi impute chained and mi impute monotone* in [MI] **mi impute chained** for a discussion of custom syntaxes.

Examples of using default prediction equations

> Example 1: Different imputation methods

Recall the heart attack example from *First use*. If we wanted to impute age using predictive mean matching instead of linear regression, we could type

```
. use http://www.stata-press.com/data/r12/mheart5s0, clear
(Fictional heart attack data; bmi and age missing)
. mi impute monotone (regress) bmi (pmm) age = attack smokes hsgrad female,
> add(10)
Conditional models:
            age: pmm age attack smokes hsgrad female
            bmi: regress bmi age attack smokes hsgrad female

Multivariate imputation                        Imputations =       10
Monotone method                                      added =       10
Imputed: m=1 through m=10                           updated =        0
            bmi: linear regression
            age: predictive mean matching
```

	Observations per m			
Variable	Complete	Incomplete	Imputed	Total
bmi	126	28	28	154
age	142	12	12	154

```
(complete + incomplete = total; imputed is the minimum across m
 of the number of filled-in observations.)
```

As previously, we listed age and bmi in the reverse order here, and mi impute monotone determined the correct order of missing monotonicity.

◁

> Example 2: Imputing a variable on a subsample

Consider an mi set version of the heart attack data containing the indicator for smoking high-tar cigarettes (variable hightar):

```
. use http://www.stata-press.com/data/r12/mheart6s0
(Fictional heart attack data; bmi, age, and hightar missing)
. mi describe
  Style:  mlong
          last mi update 30mar2011 12:46:48, 1 day ago
  Obs.:   complete         124
          incomplete        30   (M = 0 imputations)
          ─────────────────────
          total            154
  Vars.:  imputed:  3; bmi(24) age(30) hightar(8)
          passive:  0
          regular:  4; attack smokes female hsgrad
          system:   3; _mi_m _mi_id _mi_miss
          (there are no unregistered variables)
```

mi describe reports that there are no imputations, three registered imputed variables (hightar is one of them), and four registered regular variables.

Next we use mi misstable nested to examine missing-data patterns in the data.

```
. mi misstable nested
    1.  hightar(8) -> bmi(24) -> age(30)
```

There is one monotone-missing pattern in the data. According to the output, missing values of hightar are nested within bmi, whose missing values are nested within age. So hightar, bmi, and age follow a monotone-missing pattern.

As before, to impute missing values of age and bmi, we use the regression method. The hightar variable is a binary variable, so we choose the logistic method to fill in its values (see [MI] **mi impute logit**). Because hightar records whether a subject smokes high-tar cigarettes, we use only those who smoke to impute its missing values. (If there were any missing values of hightar for the subjects who do not smoke, we would have replaced them with zeros.)

```
. mi impute monotone (reg) age bmi (logit if smokes) hightar
> = attack smokes hsgrad female, add(10)

Conditional models:
       hightar: logit hightar attack smokes hsgrad female if smokes
           bmi: regress bmi i.hightar attack smokes hsgrad female
           age: regress age bmi i.hightar attack smokes hsgrad female

note: smokes omitted because of collinearity

Multivariate imputation                 Imputations =       10
Monotone method                               added =       10
Imputed: m=1 through m=10                   updated =        0
            age: linear regression
            bmi: linear regression
        hightar: logistic regression
```

	Observations per m			
Variable	Complete	Incomplete	Imputed	Total
age	124	30	30	154
bmi	130	24	24	154
hightar	56	8	8	64

```
(complete + incomplete = total; imputed is the minimum across m
 of the number of filled-in observations.)
```

mi impute monotone reports which univariate conditional model was used to impute each variable. Because hightar has the least number of missing observations, it is imputed first using the specified complete predictors and using only observations for smokers. From the output, all incomplete values of each of the variables are imputed in all 10 imputations. Notice that because we restricted the imputation sample of hightar to smokers, the total number of observations reported for hightar is 64 and not 154.

It is safe to use the if restriction in the above because smokes does not contain any missing values and hightar does not contain any missing values in observations with smokes==0. Otherwise, the conditional() option should be used instead; see *Conditional imputation* of [MI] **mi impute** for details.

◁

Examples of using custom prediction equations

▷ Example 3: Using different sets of predictors within individual conditional models

Let's take a closer look at the conditional model for `hightar` used in the above example:

hightar: logit hightar attack smokes hsgrad female if (smokes)

Notice that predictor `smokes` is redundant in this model because it is collinear with the constant (included in the model by default) on the restricted sample of smokers. In fact, if we specify the `noisily` option (`noi` for short) within the `logit` specification to see the estimation results, we will notice that, as expected, `smokes` was omitted from the estimation model for `hightar`; that is, its coefficient is zero.

```
. mi impute monotone (reg) age bmi (logit if smokes, noi) hightar
> = attack smokes hsgrad female, replace
Conditional models:
         hightar: logit hightar attack smokes hsgrad female if smokes, noisily
            bmi: regress bmi i.hightar attack smokes hsgrad female
            age: regress age bmi i.hightar attack smokes hsgrad female

Running logit on observed data:

note: smokes omitted because of collinearity
Iteration 0:   log likelihood = -38.673263
Iteration 1:   log likelihood = -38.455029
Iteration 2:   log likelihood = -38.454991
Iteration 3:   log likelihood = -38.454991
```

```
Logistic regression                          Number of obs    =          56
                                             LR chi2(3)       =        0.44
                                             Prob > chi2      =      0.9326
Log likelihood = -38.454991                  Pseudo R2        =      0.0056
```

hightar	Coef.	Std. Err.	z	P>|z|	[95% Conf. Interval]	
attack	.0773715	.5630513	0.14	0.891	-1.026189	1.180932
smokes	0	(omitted)				
hsgrad	-.1663937	.5977995	-0.28	0.781	-1.338059	1.005272
female	-.3331926	.617736	-0.54	0.590	-1.543933	.8775477
_cons	.0138334	.6263152	0.02	0.982	-1.213722	1.241389

```
Multivariate imputation              Imputations =         10
Monotone method                            added =          0
Imputed: m=1 through m=10                updated =         10
            age: linear regression
            bmi: linear regression
        hightar: logistic regression
```

	Observations per m			
Variable	Complete	Incomplete	Imputed	Total
age	124	30	30	154
bmi	130	24	24	154
hightar	56	8	8	64

(complete + incomplete = total; imputed is the minimum across m
 of the number of filled-in observations.)

Although `mi impute` handles collinearity problems for us automatically, we can eliminate redundancy manually by removing `smokes` from the prediction equation for `hightar`. To do that, we need to specify custom prediction equations.

As discussed in *Using mi impute monotone*, custom prediction equations are available with `mi impute monotone` when the `custom` option is used. We also know that within this custom specification, we must fully specify prediction equations within each conditional model and must specify the conditional models in the monotone-missing order of the imputation variables.

Building such conditional models from scratch could be a tedious task except that we can use `mi impute monotone, dryrun` to display the conditional models with default prediction equations without performing the corresponding imputation:

```
. mi impute monotone (reg) age bmi (logit if smokes) hightar
> = attack smokes hsgrad female, dryrun

Conditional models:
        hightar: logit hightar attack smokes hsgrad female if smokes
            bmi: regress bmi i.hightar attack smokes hsgrad female
            age: regress age bmi i.hightar attack smokes hsgrad female
```

We can use these default conditional specifications as the basis for writing our own customized specifications. We will remove `smokes` from the predictor list for `hightar`:

```
. mi impute monotone (logit hightar attack hsgrad female if smokes)
>                    (regress bmi hightar attack smokes hsgrad female)
>                    (regress age bmi hightar attack smokes hsgrad female)
>                                                  , custom replace

Multivariate imputation                     Imputations =      10
Monotone method                                   added =       0
Imputed: m=1 through m=10                        updated =      10

        hightar: logistic regression
            bmi: linear regression
            age: linear regression
```

	Observations per *m*			
Variable	Complete	Incomplete	Imputed	Total
hightar	56	8	8	64
bmi	130	24	24	154
age	124	30	30	154

```
(complete + incomplete = total; imputed is the minimum across m
of the number of filled-in observations.)
```

◁

▷ Example 4: Including expressions of imputation variables in prediction equations

The distribution of `bmi` is slightly skewed. To take this into account, we can either use predictive mean matching to impute `bmi` or impute `bmi` on a logarithmic scale. We choose to impute the log of `bmi` here.

Following the steps described in *Imputing transformations of incomplete variables* of [MI] **mi impute**, we create a new variable, `lnbmi`, containing the log of `bmi` and register it as imputed. Here we also reset the number of imputations to zero.

```
. mi set M = 0
(10 imputations dropped; M = 0)
```

```
. mi unregister bmi

. generate lnbmi = ln(bmi)
(24 missing values generated)

. mi register imputed lnbmi
```

We are now ready to impute lnbmi. However, although we are imputing the log of bmi, we want to use bmi in the original scale when imputing age. To do that, we include exp(lnbmi) in the prediction equation for age. When including expressions in a custom specification, the expressions must appear in parentheses:

```
. mi impute monotone (logit hightar attack hsgrad female if smokes)
>                         (regress lnbmi hightar attack smokes hsgrad female)
>                         (regress age (exp(lnbmi)) hightar attack smokes hsgrad female)
>                                                        , custom add(10)

Multivariate imputation                  Imputations =       10
Monotone method                               added =       10
Imputed: m=1 through m=10                   updated =        0

           hightar: logistic regression
            lnbmi: linear regression
              age: linear regression
```

		Observations per *m*		
Variable	Complete	Incomplete	Imputed	Total
hightar	56	8	8	64
lnbmi	130	24	24	154
age	124	30	30	154

```
(complete + incomplete = total; imputed is the minimum across m
of the number of filled-in observations.)
```

If we also wanted to include a squared term for bmi in the conditional imputation model for age, we would type

```
. mi impute monotone
> (logit hightar attack hsgrad female if smokes)
> (regress lnbmi hightar attack smokes hsgrad female)
> (regress age (exp(lnbmi)) (exp(lnbmi)^2) hightar attack smokes hsgrad female)
>                                                        , custom replace
   (output omitted )
```

◁

Saved results

mi impute monotone saves the following in r():

Scalars
r(M)	total number of imputations
r(M_add)	number of added imputations
r(M_update)	number of updated imputations
r(k_ivars)	number of imputed variables
r(N_g)	number of imputed groups (1 if by() is not specified)

Macros
r(method)	name of imputation method (monotone)
r(ivars)	names of imputation variables
r(rseed)	random-number seed
r(uvmethods)	names of univariate conditional imputation methods
r(by)	names of variables specified within by()

Matrices
r(N)	number of observations in imputation sample in each group (per variable)
r(N_complete)	number of complete observations in imputation sample in each group (per variable)
r(N_incomplete)	number of incomplete observations in imputation sample in each group (per variable)
r(N_imputed)	number of imputed observations in imputation sample in each group (per variable)

Methods and formulas

Let $\mathbf{x}_{(i)} = (x_{i1}, x_{i2}, \ldots, x_{ip})$ be the ith observation containing values of the imputation variables ordered from the most observed to the least observed to form a monotone-missing data pattern. Let $\mathbf{z}_{(i)} = (z_{i1}, z_{i2}, \ldots, z_{iq})$ be the corresponding set of predictors of $\mathbf{x}_{(i)}$. Then, if the missingness-modeling structure is monotone distinct (imputation variables have monotone missingness and parameters of the conditional models are distinct as defined in Rubin [1987, 174]), the following decomposition holds:

$$f_{\mathbf{X}}(\mathbf{x}_{(i)}|\mathbf{z}_{(i)}, \boldsymbol{\theta}) = f_1(x_{i1}|\mathbf{z}_{(i)}, \boldsymbol{\theta}_1)f_2(x_{i2}|\mathbf{z}_{(i)}, x_{i1}, \boldsymbol{\theta}_2) \cdots f_p(x_{ip}|\mathbf{z}_{(i)}, x_{i1}, x_{i2}, \ldots, x_{i,p-1}, \boldsymbol{\theta}_p)$$

where the unknown parameters $\boldsymbol{\theta}_1, \ldots, \boldsymbol{\theta}_p$ are distinct, that is, $\Pr(\boldsymbol{\theta}) = \prod_{j=1}^{p} \Pr(\boldsymbol{\theta}_j)$. The monotone-distinct structure ensures that the univariate conditional models f_j do not depend on any unobserved values of variable $\mathbf{x}_j$ and the posterior distributions of $\boldsymbol{\theta}_j$ do not involve the imputed values of the previously filled-in variables $\mathbf{x}_1, \ldots, \mathbf{x}_{j-1}$. See Rubin (1987, 174–178) for a rigorous justification of the above decomposition.

The above allows substituting the imputation of $\mathbf{X}$ using the probability model $f_{\mathbf{X}}(\cdot)$ with a sequence of univariate conditional imputations of $\mathbf{x}_j$ using the probability models $f_j(\cdot)$. Note that f_j can be any proper imputation model (for example, linear regression or logistic regression).

mi impute monotone follows the steps below to fill in missing values in $\mathbf{x}_1, \ldots, \mathbf{x}_p$:

1. If the custom option is not used, mi impute monotone first builds univariate conditional models containing the default prediction equations using the supplied information about imputation methods, imputation variables $\mathbf{X}$, and complete predictors $\mathbf{Z}$. The order in which imputation variables are listed is irrelevant. The prediction equations are constructed as follows. Complete predictors *indepvars* are included first. The imputation variables are included next with each previously imputed variable added to the beginning of the prediction equation previously used.

   If the custom option is used, mi impute monotone uses the specified conditional models in the order supplied. The conditional models must be listed in the monotone-missing order of the corresponding imputation variables.

2. Fit univariate conditional models for each $\mathbf{x}_j$ to the observed data to obtain the estimates of $\boldsymbol{\theta}_j$, $j = 1, \ldots, p$. See step 1 in *Methods and formulas* of each respective univariate imputation method's manual entry for details.

3. Sequentially fill in missing values of $\mathbf{x}_1, \mathbf{x}_2, \ldots, \mathbf{x}_p$ according to the specified imputation model. See step 2 and step 3 in *Methods and formulas* of each respective univariate imputation method's manual entry for details.

4. Repeat step 3 to obtain M multiple imputations.

References

Raghunathan, T. E., J. M. Lepkowski, J. Van Hoewyk, and P. Solenberger. 2001. A multivariate technique for multiply imputing missing values using a sequence of regression models. *Survey Methodology* 27: 85–95.

Royston, P. 2005. Multiple imputation of missing values: Update. *Stata Journal* 5: 188–201.

——. 2007. Multiple imputation of missing values: Further update of ice, with an emphasis on interval censoring. *Stata Journal* 7: 445–464.

——. 2009. Multiple imputation of missing values: Further update of ice, with an emphasis on categorical variables. *Stata Journal* 9: 466–477.

Rubin, D. B. 1987. *Multiple Imputation for Nonresponse in Surveys*. New York: Wiley.

Schafer, J. L. 1997. *Analysis of Incomplete Multivariate Data*. Boca Raton, FL: Chapman & Hall/CRC.

van Buuren, S., H. C. Boshuizen, and D. L. Knook. 1999. Multiple imputation of missing blood pressure covariates in survival analysis. *Statistics in Medicine* 18: 681–694.

White, I. R., R. Daniel, and P. Royston. 2010. Avoiding bias due to perfect prediction in multiple imputation of incomplete categorical data. *Computational Statistics & Data Analysis* 54: 2267–2275.

Also see

Title

mi impute mvn — Impute using multivariate normal regression

Syntax

mi impute mvn *ivars* $\left[\, = indepvars \,\right]$ $\left[\, if \,\right]$ $\left[\,, impute\_options\ options \,\right]$

impute_options	Description
Main	
* add(*#*)	specify number of imputations to add; required when no imputations exist
* replace	replace imputed values in existing imputations
rseed(*#*)	specify random-number seed
double	save imputed values in double precision; the default is to save them as float
by(*varlist*[, *byopts*])	impute separately on each group formed by *varlist*
Reporting	
dots	display dots as imputations are performed
noisily	display intermediate output
nolegend	suppress all table legends
Advanced	
force	proceed with imputation, even when missing imputed values are encountered
noupdate	do not perform mi update; see [MI] **noupdate option**

* add(*#*) is required when no imputations exist; add(*#*) or replace is required if imputations exist. noupdate does not appear in the dialog box.

options	Description
Main	
noconstant	suppress constant term
MCMC options	
burnin(*#*)	specify number of iterations for the burn-in period; default is burnin(100)
burnbetween(*#*)	specify number of iterations between imputations; default is burnbetween(100)
prior(*prior_spec*)	specify a prior distribution; default is prior(uniform)
mcmconly	perform MCMC for the length of the burn-in period without imputing missing values
initmcmc(*init_mcmc*)	specify initial values for the MCMC procedure; default is initmcmc(em) using the EM estimates for initial values
wlfwgt(*matname*)	specify weights for the worst linear function
savewlf(*filename*[, ...])	save the worst linear function from each iteration in *filename*.dta
saveptrace(*fname*[, ...])	save MCMC parameter estimates from each iteration in *fname*.stptrace; see [MI] **mi ptrace**

Reporting

emlog	display iteration log from EM
emoutput	display intermediate output from EM estimation
mcmcdots	display dots as MCMC iterations are performed
alldots	display dots as intermediate iterations are performed
nolog	do not display information about the EM or MCMC procedures

Advanced

emonly[(em_options)]	perform EM estimation only

You must mi set your data before using mi impute mvn; see [MI] **mi set**.

You must mi register *ivars* as imputed before using mi impute mvn; see [MI] **mi set**.

indepvars may contain factor variables; see [U] **11.4.3 Factor variables**.

prior_spec	Description
uniform	use the uniform prior distribution; the default
jeffreys	use the Jeffreys noninformative prior distribution
ridge, df(#)	use a ridge prior distribution with degrees of freedom #

init_mcmc	Description
em[, *em_options*]	use EM to obtain starting values for MCMC; the default
initmatlist	supply matrices containing initial values for MCMC

em_options	Description
iterate(#)	specify the maximum number of iterations; default is iterate(100)
tolerance(#)	specify tolerance for the changes in parameter estimates; default is tolerance(1e-5)
init(*init_em*)	specify initial values for the EM algorithm; default is init(ac)
nolog	do not show EM iteration log
saveptrace(*fname*[, ...])	save EM parameter estimates from each iteration in *fname*.stptrace; see [MI] **mi ptrace**

init_em	Description
ac	use all available cases to obtain initial values for EM; the default
cc	use only complete cases to obtain initial values for EM
initmatlist	supply matrices containing initial values for EM

initmatlist is of the form *initmat* [*initmat* [...]]

initmat	Description
betas(# | *matname*)	specify coefficient vector; default is betas(0)
sds(# | *matname*)	specify standard deviation vector; default is sds(1)
vars(# | *matname*)	specify variance vector; default is vars(1)
corr(# | *matname*)	specify correlation matrix; default is corr(0)
cov(*matname*)	specify covariance matrix

In the above, # is understood to mean a vector containing all elements equal to #.

Menu

Statistics > Multiple imputation

Description

mi impute mvn fills in missing values of one or more continuous variables using multivariate normal regression. It accommodates arbitrary missing-value patterns. You can perform separate imputations on different subsets of the data by specifying the by() option. mi impute mvn uses an iterative Markov chain Monte Carlo (MCMC) method to impute missing values. See *Remarks* for details.

Options

⌐‾‾⌐ Main ⌐‾‾‾

noconstant; see [R] **estimation options**.

add(), replace, rseed(), double, by(); see [MI] **mi impute**.

⌐‾‾⌐ MCMC options ⌐‾‾

burnin(#) specifies the number of iterations for the initial burn-in period. The default is burnin(100). This option specifies the number of iterations necessary for the MCMC to reach approximate stationarity or, equivalently, to converge to a stationary distribution. The required length of the burn-in period will depend on the starting values used and the missing-data patterns observed in the data. It is important to examine the chain for convergence to determine an adequate length of the burn-in period prior to obtaining imputations; see *Convergence of the MCMC method* and examples 2 and 4. The provided default may be sufficient in many cases, but you are responsible for determining that sufficient iterations are performed.

burnbetween(#) specifies a number of iterations of the MCMC to perform between imputations, the purpose being to reduce correlation between sets of imputed values. The default is burnbetween(100). As with burnin(), you are responsible for determining that sufficient iterations are performed. See *Convergence of the MCMC method* and examples 2 and 4.

prior(*prior_spec*) specifies a prior distribution to be used by the MCMC procedure. The default is prior(uniform). The alternative prior distributions are useful when the default estimation of the parameters using maximum likelihood becomes unstable (for example, estimates on the boundary of the parameter space) and introducing some prior information about parameters stabilizes the estimation.

prior_spec is

　　　　u̲niform | j̲effreys | r̲idge, df(#)

　　uniform specifies the uniform (flat) prior distribution. Under this prior distribution, the posterior distribution is proportional to the likelihood function and thus the estimate of the posterior mode is the same as the maximum likelihood (ML) estimate.

　　jeffreys specifies the Jeffreys, noninformative prior distribution. This prior distribution can be used when there is no strong prior knowledge about the model parameters.

ridge, df(#) specifies a ridge, informative prior distribution with the degrees of freedom #. This prior introduces some information about the covariance matrix by smoothing the off-diagonal elements (correlations) toward zero. The degrees of freedom, df(), which may be noninteger, regulates the amount of smoothness—the larger this number, the closer the correlations are to zero. A ridge prior is useful to stabilize inferences about the mean parameters when the covariance matrix is poorly estimated, for example, when there are insufficient observations to estimate correlations between some variables reliably because of missing data, causing the estimated covariance matrix to become non–positive definite (see Schafer [1997, 155–157] for details).

mcmconly specifies that mi impute mvn run the MCMC for the length of the burn-in period and then stop. This option is useful in combination with savewlf() or saveptrace() to examine the convergence of the MCMC prior to imputation. No imputation is performed when mcmconly is specified, so add() or replace is not required with mi impute mvn, mcmconly, and they are ignored if specified. The mcmconly option is not allowed with emonly.

initmcmc() may be specified as initmcmc(em [, em_options]) or initmcmc(initmatlist).

initmcmc() specifies initial values for the regression coefficients and covariance matrix of the multivariate normal distribution to be used by the MCMC procedure. By default, initial values are obtained from the EM algorithm, initmcmc(em).

initmcmc(em[, em_options]) specifies that the initial values for the MCMC procedure be obtained from EM. You can control the EM estimation by specifying em_options. If the uniform prior is used, the initial estimates correspond to the ML estimates computed using EM. Otherwise, the initial values are the estimates of the posterior mode computed using EM.

em_options are

iterate(#) specifies the maximum number of EM iterations to perform. The default is iterate(100).

tolerance(#) specifies the convergence tolerance for the EM algorithm. The default is tolerance(1e-5). Convergence is declared once the maximum of the relative changes between two successive estimates of all model parameters is less than #.

init() may be specified as init(ac), init(cc), or init(matlist)

init() specifies initial values for the regression coefficients and covariance matrix of the multivariate normal distribution to be used by the EM algorithm. init(ac) is the default.

init(ac) specifies that initial estimates be obtained using all available cases. The initial values for regression coefficients are obtained from separate univariate regressions of each imputation variable on the independent variables. The corresponding estimates of the residual mean-squared error are used as the initial values for the diagonal entries of the covariance matrix (variances). The off-diagonal entries (correlations) are set to zero.

init(cc) specifies that initial estimates be obtained using only complete cases. The initial values for regression coefficients and the covariance matrix are obtained from a multivariate regression fit to the complete cases only.

init(initmatlist) specifies to use manually supplied initial values for the EM procedure and syntactically is identical to mcmcinit(initmatlist), described below, except that you specify init(initmatlist).

nolog suppresses the EM iteration log when emonly or emoutput is used.

saveptrace(fname[, replace]) specifies to save the parameter trace log from the EM algorithm to a file called fname.stptrace. If the file already exists, the replace suboption specifies to overwrite the existing file. See [MI] **mi ptrace** for details about the saved file and how to read it into Stata.

initmcmc(*initmatlist*), where *initmatlist* is

$$initmat \left[initmat \left[\ldots \right] \right]$$

specifies manually supplied initial values for the MCMC procedure.

initmat is

betas(*#* | *matname*) specifies initial values for the regression coefficients. The default is betas(0), implying a value of zero for all regression coefficients. If you specify betas(*#*), then *#* will be used as the initial value for all regression coefficients. Alternatively, you can specify the name of a Stata matrix, *matname*, containing values for each regression coefficient. *matname* must be conformable with the dimensionality of the specified model. That is, it can be one of the following dimensions: $p \times q$, $q \times p$, $1 \times pq$, or $pq \times 1$, where p is the number of imputation variables and q is the number of independent variables.

sds(*#* | *matname*) specifies initial values for the standard deviations (square roots of the diagonal elements of the covariance matrix). The default is sds(1), which sets all standard deviations and thus variances to one. If you specify sds(*#*), then the squared *#* will be used as the initial value for all variances. Alternatively, you can specify the name of a Stata matrix, *matname*, containing individual values. *matname* must be conformable with the dimensionality of the specified model. That is, it can be one of the following dimensions: $1 \times p$ or $p \times 1$, where p is the number of imputation variables. This option cannot be combined with cov() or vars(). The sds() option can be used in combination with corr() to provide initial values for the covariance matrix.

vars(*#* | *matname*) specifies initial values for variances (diagonal elements of the covariance matrix). The default is vars(1), which sets all variances to one. If you specify vars(*#*), then *#* will be used as the initial value for all variances. Alternatively, you can specify the name of a Stata matrix, *matname*, containing individual values. *matname* must be conformable with the dimensionality of the specified model. That is, it can be one of the following dimensions: $1 \times p$ or $p \times 1$, where p is the number of imputation variables. This option cannot be combined with cov() or sds(). The vars() option can be used in combination with corr() to provide initial values for the covariance matrix.

corr(*#* | *matname*) specifies initial values for the correlations (off-diagonal elements of the correlation matrix). The default is corr(0), which sets all correlations and, thus, covariances to zero. If you specify corr(*#*), then all correlation coefficients will be set to *#*. Alternatively, you can specify the name of a Stata matrix, *matname*, containing individual values. *matname* can be a square $p \times p$ matrix with diagonal elements equal to one or it can contain the corresponding lower (upper) triangular matrix in a vector of dimension $p(p+1)/2$, where p is the number of imputation variables. This option cannot be combined with cov(). The corr() option can be used in combination with sds() or vars() to provide initial values for the covariance matrix.

cov(*matname*) specifies initial values for the covariance matrix. *matname* must contain the name of a Stata matrix. *matname* can be a square $p \times p$ matrix or it can contain the corresponding lower (upper) triangular matrix in a vector of dimension $p(p+1)/2$, where p is the number of imputation variables. This option cannot be combined with corr(), sds(), or vars().

wlfwgt(*matname*) specifies the weights (coefficients) to use when computing the worst linear function (WLF). The coefficients must be saved in a Stata matrix, *matname*, of dimension $1 \times d$, where $d = pq + p(p+1)/2$, p is the number of imputation variables, and q is the number of predictors.

This option is useful when initial values from the EM estimation are supplied to data augmentation (DA) as matrices. This option can also be used to obtain the estimates of linear functions other than the default WLF.

savewlf(*filename*[, replace]) specifies to save the estimates of the WLF from each iteration of MCMC to a Stata dataset called *filename*.dta. If the file already exists, the replace suboption specifies to overwrite the existing file. This option is useful for monitoring convergence of the MCMC. savewlf() is allowed with initmcmc(em), when the initial values are obtained using the EM estimation, or with wlfwgt().

saveptrace(*fname*[, replace]) specifies to save the parameter trace log from the MCMC to a file called *fname*.stptrace. If the file already exists, the replace suboption specifies to overwrite the existing file. See [MI] **mi ptrace** for details about the saved file and how to read it into Stata. This option is useful for monitoring convergence of the MCMC.

⌐ Reporting ⌐_____

dots, noisily, nolegend; see [MI] **mi impute**. Also, noisily is a synonym for emoutput. nolegend suppresses group legends that may appear when the by() option is used. It is a synonym for by(, nolegend).

emlog specifies that the EM iteration log be shown. The EM iteration log is not displayed unless emonly or emoutput is specified.

emoutput specifies that the EM output be shown. This option is implied with emonly.

mcmcdots specifies to display all MCMC iterations as dots.

alldots specifies to display all intermediate iterations as dots in addition to the imputation dots. These iterations include the EM iterations and the MCMC burn-in iterations. This option implies mcmcdots.

nolog suppresses all output from EM or MCMC that is usually displayed by default.

⌐ Advanced ⌐_____

force; see [MI] **mi impute**.

emonly[(*em_options*)] specifies that mi impute mvn perform EM estimation and then stop. You can control the EM process by specifying *em_options*. This option is useful at the preliminary stage to obtain insight about the length of the burn-in period as well as to choose a prior specification. No imputation is performed, so add() or replace is not required with mi impute mvn, emonly, and they are ignored if specified. The emonly option is not allowed with mcmconly.

The following option is available with mi impute but is not shown in the dialog box:

noupdate; see [MI] **noupdate option**.

Remarks

Remarks are presented under the following headings:

> *Incomplete continuous data with arbitrary pattern of missing values*
> *Multivariate imputation using data augmentation*
> *Convergence of the MCMC method*
> *Using mi impute mvn*
> *Examples*

See [MI] **mi impute** for a general description and details about options common to all imputation methods, *impute_options*. Also see [MI] **workflow** for general advice on working with mi.

Incomplete continuous data with arbitrary pattern of missing values

As we described in detail in *Multivariate imputation* in [MI] **mi impute**, imputation of multiple variables with an arbitrary pattern of missing values is more challenging than when the missing-data pattern is monotone.

One approach for dealing with an arbitrary missing-value pattern is to assume an explicit tractable parametric model for the data and draw imputed values from the resulting distribution of the missing data given observed data. One of the more popular parametric models is the Gaussian normal model; see Rubin (1987) for other recommendations. Although a multivariate normal model is straightforward, difficulty arises in the simulation from the corresponding, more complicated, distribution of the missing data. One solution is to use one of the Bayesian iterative Markov chain Monte Carlo (MCMC) procedures to approximate the distribution of missing data.

Multivariate imputation using data augmentation

mi impute mvn uses data augmentation (DA) —an iterative MCMC procedure—to generate imputed values assuming an underlying multivariate normal model. For details about DA as a general MCMC procedure, see Gelman et al. (2004), Tanner and Wong (1987), and Li (1988), among others. For applications of DA to incomplete multivariate normal data, see, for example, Little and Rubin (2002) and Schafer (1997). Below we briefly describe the idea behind DA; see *Methods and formulas* for details.

Consider multivariate data $\mathbf{X} = (\mathbf{X}_o, \mathbf{X}_m)$, decomposed into the observed part $\mathbf{X}_o$ and the missing part $\mathbf{X}_m$, from a normal distribution $\Pr(\mathbf{X}|\theta) = N(\beta, \Sigma)$, where θ denotes the unknown model parameters (regression coefficients β and unique elements of the covariance matrix Σ). The goal is to replace missing values in $\mathbf{X}_m$ with draws from the distribution (or the predictive distribution in Bayesian terminology) of the missing data given observed data, $\Pr(\mathbf{X}_m|\mathbf{X}_o)$. The actual predictive distribution $\Pr(\mathbf{X}_m|\mathbf{X}_o)$ is difficult to draw from directly because of an underlying dependence on the posterior distribution of the unknown parameters θ, $\Pr(\theta|\mathbf{X}_o)$.

Originally, DA was used to approximate the posterior distribution of the model parameters, $\Pr(\theta|\mathbf{X}_o)$, in Bayesian applications with incomplete data. The idea of DA is to augment the observed data, $\mathbf{X}_o$, with the latent (unobserved) data, $\mathbf{X}_m$, such that the conditional posterior distribution $\Pr(\theta|\mathbf{X}_o, \mathbf{X}_m)$ becomes more tractable and easier to simulate from. Then the procedure becomes as follows. For a current $\theta^{(t)}$, draw $\mathbf{X}_m^{(t+1)}$ from its conditional predictive distribution given the observed data and θ, $\Pr(\mathbf{X}_m|\mathbf{X}_o, \theta^{(t)})$. Next draw $\theta^{(t+1)}$ from its conditional posterior distribution given the augmented data, $\Pr(\theta|\mathbf{X}_o, \mathbf{X}_m^{(t+1)})$. Continue to iterate until the sequence $\{(\mathbf{X}_m^{(t)}, \theta^{(t)}) : t = 1, 2, \ldots\}$, an MCMC sequence, converges to a stationary distribution $\Pr(\theta, \mathbf{X}_m|\mathbf{X}_o)$. This way a complicated task of simulating from $\Pr(\theta|\mathbf{X}_o)$ is replaced by a sequence of simpler simulation tasks of iteratively sampling from $\Pr(\theta|\mathbf{X}_o, \mathbf{X}_m)$ and $\Pr(\mathbf{X}_m|\mathbf{X}_o, \theta)$. How is this procedure related to imputation? The sequence $\{\mathbf{X}_m^{(t)} : t = 1, 2, \ldots\}$ contains draws from an approximate predictive distribution $\Pr(\mathbf{X}_m|\mathbf{X}_o)$, and thus $\mathbf{X}_m^{(t)}$'s are, in fact, imputations. The convergence of this procedure was studied by Li (1988).

The functional forms of the conditional distributions $\Pr(\theta|\mathbf{X}_o, \mathbf{X}_m)$ and $\Pr(\mathbf{X}_m|\mathbf{X}_o, \theta)$ are determined from the assumed distribution of the data, $\mathbf{X}$, and a prior distribution for the model parameters, θ, $\Pr(\theta)$. **mi impute mvn** assumes a normal distribution for the data and supports three prior distributions: uniform, Jeffreys, and ridge.

The prior distributions are categorized into noninformative (or also vague, diffuse, flat, reference) and informative prior distributions. The noninformative priors provide no extra information about model parameters beyond that already contained in the data. These priors are recommended when no strong prior knowledge is available about the parameters. Informative prior distributions are used when there is some a priori knowledge about the distribution of the parameters. For example, prior information about cancer mortality rates in a Poisson model can be assigned based on the available worldwide estimate. The uniform and Jeffreys priors are noninformative priors. The ridge prior is an informative prior.

The uniform prior assumes that all values of the parameters are equally probable. Under this prior specification, the posterior distribution of the parameters is equivalent to the likelihood function, and so the Bayesian and frequentist methods coincide. The Jeffreys prior is another widely used noninformative prior distribution, and with small samples, it may be preferable to the uniform prior. A ridge prior is often used when the estimated covariance matrix becomes singular (or nearly singular), as may occur with sparse missing data if there are not enough observations to estimate reliably all aspects of the covariance matrix. A ridge prior smooths the estimate of the covariance matrix toward a diagonal structure depending on the chosen degrees of freedom; the larger the degrees of freedom, the closer is the estimated covariance matrix to the diagonal matrix (see Schafer [1997, 155–157] for details).

Convergence of the MCMC method

For a brief overview of convergence of MCMC, see *Convergence of iterative methods* in [MI] **mi impute**.

The MCMC procedure DA is iterated until an MCMC sequence $\{(\mathbf{X}_m^{(t)}, \boldsymbol{\theta}^{(t)}) : t = 1, 2, \ldots\}$ converges to a stationary distribution. Unlike maximum likelihood, EM, or other optimization-based procedures, the DA procedure does not have a simple stopping rule that guarantees the convergence of the chain to a stationary distribution. Thus the question of how long to iterate to achieve convergence arises. In addition to determining convergence of MCMC, we must also investigate the serial dependence known to exist among the MCMC draws to obtain independent imputations.

Suppose that after an initial burn-in period, b, the sequence $\{(\mathbf{X}_m^{(b+t)}) : t = 1, 2, \ldots\}$ (imputations) can be regarded as an approximate sample from $\Pr(\mathbf{X}_m|\mathbf{X}_o)$. In general, this sample will not contain independent observations because the successive iterates of the MCMC tend to be correlated. To achieve independence among imputations, we can sample the chain. To do that, we need to determine the number of iterations, k, such that $\mathbf{X}_m^{(t)}$ and $\mathbf{X}_m^{(t+k)}$ are approximately independent. Then imputations can be obtained as the chain values of $\mathbf{X}_m$ from iterations $b, b + k, b + 2k, \ldots, b + mk$, where m is the required number of imputations. In our definition, b is the number of iterations necessary for the chain to achieve stationarity and k is the number of iterations between imputations necessary to achieve independent values of the chain.

Before we proceed, we notice that from the properties of MCMC, the convergence of the chain $\{(\mathbf{X}_m^{(t)}, \boldsymbol{\theta}^{(t)}) : t = 1, 2, \ldots\}$ to $\Pr(\boldsymbol{\theta}, \mathbf{X}_m|\mathbf{X}_o)$ is equivalent to the convergence of $\{(\boldsymbol{\theta}^{(t)}) : t = 1, 2, \ldots\}$ to $\Pr(\boldsymbol{\theta}|\mathbf{X}_o)$ or, alternatively, of $\{(\mathbf{X}_m^{(t)}) : t = 1, 2, \ldots\}$ to $\Pr(\mathbf{X}_m|\mathbf{X}_o)$. Because the parameter series are usually of lower dimension, we examine convergence using the series of parameter estimates rather than the series of imputations.

How to determine convergence and, in particular, to choose values for b and k, has received much attention in the MCMC literature. In practice, convergence is often examined visually from the trace and autocorrelation plots of the estimated parameters. Trace plots are plots of estimated parameters against iteration numbers. Long-term trends in trace plots and high serial dependence in autocorrelation plots are indicative of a slow convergence to stationarity. A value of b can be inferred from a trace plot as the earliest iteration after which the chain does not exhibit a visible trend and the parameter series stabilize, which is to say the fluctuations in values become more regular. A value of k can be chosen from autocorrelation plots as the lag k for which autocorrelations of all parameters decrease to zero. When the initial values are close to the posterior mode, the initial number of iterations, b, and number of iterations between imputations, k, will be similar. When the initial values are far off in the tails of the posterior distribution, the initial number of iterations will generally be larger.

In practice, when the number of parameters in the model is large, it may not be feasible to monitor the convergence of all the individual series. One solution is to find a function of the parameters that would be the slowest to converge to stationarity. The convergence of the series for this function will then be indicative of the convergence of other functions and, in particular, individual parameter series. Schafer (1997, 129–131) suggests the worst linear function (WLF), the function corresponding to the linear combination of the parameter estimates where the coefficients are chosen such that this function has the highest asymptotic rate of missing information; see *Method and formulas* for computational details. He found that when the observed-data posterior distribution is nearly normal, this function is among the slowest to approach stationarity. Thus we can determine b and k by monitoring the convergence of the WLF. When the observed-data posterior is not normal and some aspects of the model are poorly estimated, the WLF may not be the slowest to converge. In such cases, we recommend exploring convergence of other functions or of individual parameter series.

The number of iterations necessary for DA to converge depends on the rate of convergence of DA. The rate of convergence of DA mainly depends on the fractions of missing information and initial values. The higher the fractions of missing information and the farther the initial values are from the posterior mode, the slower the convergence, and thus the larger the number of iterations required. Initial values for the DA procedure can be obtained from the EM algorithm for incomplete data (for example, Dempster, Laird, and Rubin [1977]). In addition, the number of iterations necessary for the DA procedure to converge can be inferred based on the number of iterations that the EM algorithm took to converge (Schafer 1997).

The convergence of the chain and the required number of iterations can be also inferred by running multiple independent MCMC sequences using overdispersed initial values, that is, initial values from a distribution with greater variability than that of the posterior distribution (Gelman and Rubin 1992; Schafer 1997, 126–128). Then the number of iterations can be taken to be the largest iteration number for which the series in all the chains stabilize.

Although the graphical summaries described above are useful in checking convergence, they must be used with caution. They can be deceptive in cases when the observed-data posterior has an odd shape or has multiple modes, which may happen with small sample sizes or sparse missing data. Examination of the data and missing-data patterns, as well as the behavior of the EM algorithm, are highly recommended when investigating the MCMC convergence. How one checks for convergence will be shown in examples 2 and 4.

Using mi impute mvn

mi impute mvn imputes missing data using DA, an iterative MCMC method, assuming the multivariate normal distribution for the data. For the discussion of options, such as add() and replace, common to all imputation methods, see [MI] **mi impute**. Here we focus on the options and functionality specific to mi impute mvn.

The two main options are burnin() (which specifies the number of iterations necessary for the MCMC to converge, b) and burnbetween() (which specifies the number of iterations between imputations, k). We discussed how to choose these values in the previous section. By default, these values are arbitrarily set to be 100 each.

You can choose from the three prior specifications. You can use prior(uniform) (the default) to specify the uniform prior, prior(jeffreys) to specify the Jeffreys prior, or prior(ridge, df()) to specify a ridge prior. You must also choose the degrees of freedom with a ridge prior.

For initial values, mi impute mvn uses the estimates from the EM algorithm for incomplete data (initmcmc(em)). When the uniform prior distribution is used, the estimates obtained from EM are MLEs. Under other prior specifications, the estimates from EM correspond to the posterior mode of the respective posterior distribution of the model parameters. Using the estimates from EM as initial values in general accelerates the convergence of MCMC. To determine convergence, it may also be useful to try different sets of initial values. You can do this by creating Stata matrices containing the initial values and supplying them in the respective initmcmc() suboptions betas(), cov(), etc.

You can save the estimates of the WLF and parameter series from MCMC iterations by using the savewlf() and saveptrace() options. These options are useful when examining convergence of MCMC, as we will demonstrate in examples 2 and 4. You can use mi impute mvn to run the MCMC without imputing the data if you specify the mcmconly option. This option is useful in combination with savewlf() or saveptrace() when examining convergence of MCMC. When mcmconly is specified, the DA procedure is performed for the number of iterations as specified in burnin() and no imputations are performed.

You can also perform the EM estimation without MCMC iterations if you specify the emonly() option. This option is useful for detecting convergence problems prior to running MCMC. The number of iterations EM takes to converge can be used as an approximation for the burn-in period. Also, slow convergence of the EM algorithm can reveal problems with estimability of certain model parameters.

Examples

> ## Example 1: Monotone-missing data

Recall the heart attack example from *Multivariate imputation* in [MI] **mi impute**, where we used mi impute mvn to impute missing values for age and bmi that follow a monotone-missing pattern:

```
. use http://www.stata-press.com/data/r12/mheart5s0
(Fictional heart attack data; bmi and age missing)
. mi impute mvn age bmi = attack smokes hsgrad female, add(10)
Performing EM optimization:
note: 12 observations omitted from EM estimation because of all imputation
      variables missing
  observed log likelihood = -651.75868 at iteration 7
Performing MCMC data augmentation ...
```

Multivariate imputation			Imputations =	10
Multivariate normal regression			added =	10
Imputed: *m*=1 through *m*=10			updated =	0
Prior: uniform			Iterations =	1000
			burn-in =	100
			between =	100

		Observations per *m*		
Variable	Complete	Incomplete	Imputed	Total
age	142	12	12	154
bmi	126	28	28	154

```
(complete + incomplete = total; imputed is the minimum across m
 of the number of filled-in observations.)
```

In the above, we omitted the nolog option that was present in the example in [MI] **mi impute**.

In addition to the output reported by all imputation methods, `mi impute mvn` also provides some specific information.

As we previously explained, `mi impute mvn` uses an iterative MCMC technique to impute missing values. The two phases of `mi impute mvn` are 1) obtaining initial values (unless supplied directly) and 2) performing the MCMC procedure from which imputations are obtained. These two phases are noted in the output header.

In the above example, the initial values are obtained using the EM method (the default). We see from the output that EM converged in seven iterations. A note displayed thereafter reports that 12 observations contain missing values for both `bmi` and `age` and were omitted. The note is just explanatory and should not cause you concern. Those 12 observations would contribute nothing to the likelihood function even if they were included, although the algorithm would take longer to converge.

The estimates from EM are used as initial values for DA. The first part of the table header, containing the information about the method used and the number of imputations, was described in detail in [MI] **mi impute**. The second part of the table header is specific to `mi impute mvn`. From the output, a total of 1,000 iterations of MCMC are performed. The first 100 iterations (the default) are used for the burn-in period (`burn-in = 100`), the first imputation calculated from the last iteration; thereafter, each subsequent imputation is calculated after performing another 100 iterations. The default uniform prior is used for both the EM estimation and the MCMC procedure. Under this prior, the parameter estimates obtained are MLEs.

◁

▷ Example 2: Checking convergence of MCMC

In the above example, the monotone missingness of `age` and `bmi` as well as the quick convergence of EM suggest that the MCMC must converge rapidly. In fact, we know that under a monotone-missing pattern, no iterations are needed to obtain imputed values (see [MI] **mi impute monotone**). Let's examine the convergence of the MCMC procedure for the above heart attack data, the point being to see what quick convergence looks like.

As we discussed earlier, convergence is often assessed from the trace plots of the MCMC parameter estimates. Because of a possibly large number of estimated parameters, this approach may be tedious. Alternatively, we can plot the WLF for which the convergence is generally the slowest.

We use the `savewlf(wlf)` option to save estimates of the WLF to a Stata dataset called `wlf.dta`. To examine the convergence of MCMC, we do not need imputation, and so we use the `mcmconly` option to perform the MCMC procedure without subsequent imputation. We use a total of $1000 = 10 \times 100$ iterations (`burnin(1000)` option), corresponding to the length of the MCMC to obtain 10 imputations:

```
. mi impute mvn age bmi = attack smokes hsgrad female, mcmconly burnin(1000)
> rseed(2232) savewlf(wlf)

Performing EM optimization:
note: 12 observations omitted from EM estimation because of all imputation
      variables missing
   observed log likelihood = -651.75868 at iteration 7

Performing MCMC data augmentation ...

Note: no imputation performed.
```

We also specified the `rseed(2232)` option so that we can reproduce our results.

The created dataset contains three variables: `iter`, `m`, and `wlf`. The `iter` variable records iterations (the burn-in iterations are recorded as negative integers). The `m` variable records imputation numbers to which the iteration sequence corresponds (`m` contains 0 if `mcmconly` is used). The `wlf` variable records the WLF estimates.

```
. use wlf, clear

. describe

Contains data from wlf.dta
  obs:          1,000
  vars:             3                          2 Apr 2011 11:03
  size:        16,000
```

variable name	storage type	display format	value label	variable label
iter	long	%12.0g		
m	long	%12.0g		
wlf	double	%10.0g		

```
Sorted by:
```

We use the time-series commands `tsline` and `ac` (see [TS] **tsline** and [TS] **corrgram**) to plot the estimates and autocorrelations of `wlf` with respect to the iteration number. We first use `tsset` to set `iter` as the "time" variable and then use `tsline` to obtain a trace plot:

```
. tsset iter
        time variable:  iter, -999 to 0
               delta:  1 unit

. tsline wlf, ytitle(Worst linear function) xtitle(Burn-in period)
```

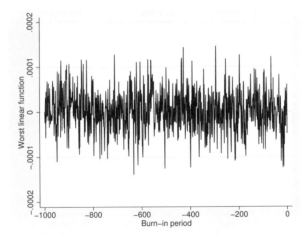

The graph shows no visible trend in the estimates of the WLF, just as we expected. Convergence of MCMC by the 100th iteration should be assured. In fact, taking into account the declared convergence of the EM algorithm in only seven iterations, we would be comfortable with using a much smaller burn-in period of, say, 10 iterations.

We next examine the autocorrelation in the WLF to obtain an idea of how many iterations to use between imputations to ensure their approximate independence:

```
. ac wlf, title(Worst linear function) ytitle(Autocorrelations)
> ciopts(astyle(none)) note("")
```

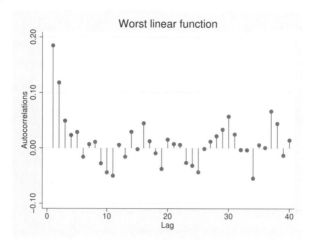

From the graphical output, the autocorrelations die off quickly. This suggests that we can use a smaller number, say, 10 or 20, rather than the default 100 iterations for the burn-between period.

◁

We considered an example with a monotone-missing pattern. mi impute mvn is designed to accommodate arbitrary missing-data patterns, so let's consider an example with them.

▷ Example 3: Arbitrary missing-data pattern

Consider data on house resale prices provided by the Albuquerque Board of Realtors and distributed by the Data and Story Library. You can find a detailed description of the data at http://lib.stat.cmu.edu/DASL/Stories/homeprice.html.

```
. use http://www.stata-press.com/data/r12/mhouses1993, clear
(Albuquerque Home Prices Feb15-Apr30, 1993)
. describe
Contains data from http://www.stata-press.com/data/r12/mhouses1993.dta
  obs:           117                    Albuquerque Home Prices
                                         Feb15-Apr30, 1993
  vars:            8                    19 Jun 2011 10:50
  size:        1,287                    (_dta has notes)
```

variable name	storage type	display format	value label	variable label
price	int	%8.0g		Sale price (hundreds)
sqft	int	%8.0g		Square footage of living space
age	byte	%10.0g		Home age (years)
nfeatures	byte	%8.0g		Number of certain features
ne	byte	%8.0g		Located in northeast (largest residential) sector of the city
custom	byte	%8.0g		Custom build
corner	byte	%8.0g		Corner location
tax	int	%10.0g		Tax amount (dollars)

```
Sorted by:
```

The dataset includes eight variables. The primary variable of interest is `price`, and other variables are used as its predictors.

We investigate the missing-data patterns of these data using `misstable`:

```
. misstable pattern

          Missing-value patterns
            (1 means complete)

                     |   Pattern
          Percent    |   1   2
        -------------+-----------
            56%      |   1   1
        -------------+-----------
             35      |   1   0
              7      |   0   0
              2      |   0   1
        -------------+-----------
            100%     |

Variables are   (1) tax   (2) age

. misstable nested
        1.  tax(10)
        2.  age(49)
```

We see from the output only 56% of observations are complete; the remaining 44% contain missing values of `age` or `tax`. The `tax` variable contains 10 missing values, and the `age` variable contains 49 missing values. `misstable nested` reports that missing values of `age` and `tax` are not nested because there are two statements describing the missing-value pattern; see [R] **misstable** for details.

Let's use `mi impute mvn` to impute missing values of `age` and `tax`. Before we do that, a quick examination of the data revealed that the distribution for `age` and `tax` are somewhat skewed. As such, we choose to impute the variables on a log-transformed scale.

Following the steps as described in *Imputing transformations of incomplete variables* of [MI] **mi impute**, we create new variables containing the log values,

```
. gen lnage = ln(age)
(49 missing values generated)
. gen lntax = ln(tax)
(10 missing values generated)
```

and register them as imputed variables,

```
. mi set mlong
. mi register imputed lnage lntax
(51 m=0 obs. now marked as incomplete)
. mi register regular price sqft nfeatures ne custom corner
```

We `mi set` our data as mlong and register the complete variables as regular. For the purpose of this analysis, we leave passive variables `age` and `tax` unregistered. (Note that all missing values of the created `lnage` and `lntax` variables are eligible for imputation; see [MI] **mi impute** for details.)

We now use `mi impute mvn` to impute values of `lnage` and `lntax`:

```
. mi impute mvn lnage lntax = price sqft nfeatures ne custom corner, add(20)
Performing EM optimization:
note: 8 observations omitted from EM estimation because of all imputation
      variables missing
   observed log likelihood =   112.1464 at iteration 48
Performing MCMC data augmentation ...
```

Multivariate imputation	Imputations =	20
Multivariate normal regression	added =	20
Imputed: m=1 through m=20	updated =	0
Prior: uniform	Iterations =	2000
	burn-in =	100
	between =	100

	Observations per m			
Variable	Complete	Incomplete	Imputed	Total
lnage	68	49	49	117
lntax	107	10	10	117

```
(complete + incomplete = total; imputed is the minimum across m
of the number of filled-in observations.)
```

◁

⊳ Example 4: Checking convergence of MCMC

In the above example, we arbitrarily created 20 imputations. The output is similar to that of the earlier example. Here the EM algorithm converges by the 48th iteration. This suggests that, again, the default 100 iterations for the burn-in period should be sufficient for the convergence of MCMC. Nevertheless, we choose to confirm this visually by repeating the steps from example 2.

We run the MCMC for a total of 2,000 iterations (as would be necessary to obtain 20 imputations) without imputing data and set the seed for reproducibility. We overwrite the existing `wlf.dta` file to contain the new estimates of the WLF by specifying `replace` within `savewlf()`:

```
. mi impute mvn lnage lntax = price sqft nfeatures ne custom corner,
> mcmconly burnin(2000) rseed(23) savewlf(wlf, replace)
Performing EM optimization:
note: 8 observations omitted from EM estimation because of all imputation
      variables missing
   observed log likelihood =   112.1464 at iteration 48
Performing MCMC data augmentation ...
Note: no imputation performed.
```

We generate the same graphs as in example 2, this time using the new estimates of the WLF:

```
. use wlf, clear
. tsset iter
        time variable:  iter, -1999 to 0
                delta:  1 unit
. tsline wlf, ytitle(Worst linear function) xtitle(Burn-in period)
```

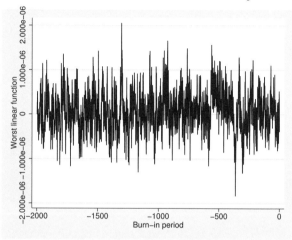

```
. ac wlf, title(Worst linear function) ytitle(Autocorrelations)
> ciopts(astyle(none)) note("")
```

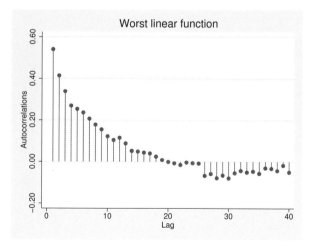

Compared with the earlier graphs, the time-series graphs do not reveal any apparent trend, but the autocorrelation dies out more slowly. The default values of 100 for the initial burn-in and between-imputation iterations should be sufficient.

◁

▷ Example 5: Alternative prior distribution

Consider some hypothetical data:

```
. use http://www.stata-press.com/data/r12/mvnexample0, clear
(Fictional data for -mi impute mvn-)
. mi describe
  Style:  mlong
          last mi update 30mar2011 12:46:49, 2 days ago
  Obs.:   complete            3
          incomplete         17  (M = 0 imputations)
          ─────────────────────
          total              20
  Vars.:  imputed:  3; x1(16) x2(5) x3(17)
          passive:  0
          regular:  0
          system:   3; _mi_m _mi_id _mi_miss
          (there are no unregistered variables)
```

Continuous normally distributed variables x1, x2, and x3 contain missing values. For illustration purposes, we consider an extreme case when some variables (x1 and x3 here) contain only a few complete observations.

We use mi impute mvn to impute missing values and create 30 imputations. Notice that in this example, we do not have complete predictors, and so the right-hand-side specification is empty:

```
. mi imp mvn x1-x3, add(30) rseed(332247)
Performing EM optimization:
note: 4 observations omitted from EM estimation because of all imputation
      variables missing
  observed log likelihood =  6.5368927 at iteration 100
  (EM did not converge)
Performing MCMC data augmentation ...
Iteration 92: variance-covariance matrix (Sigma) became not positive definite
posterior distribution is not proper
r(498);
```

mi impute mvn terminates with an error reporting that the estimated variance–covariance matrix became non–positive definite. mi impute mvn terminated because the posterior predictive distribution of missing data is not proper, but notice also that EM did not converge after the default 100 iterations.

There are two issues here. First, because EM did not converge after 100 iterations, we suspect that the default 100 iterations used for the burn-in period may not be large enough for MCMC to converge. Second, the observed missing-data pattern presents difficulties with estimating the covariance matrix reliably, which leads to a non–positive-definite estimate during the MCMC iteration.

The first issue may be resolved by increasing the maximum number of iterations for EM by using EM's iterate() suboption. Convergence of EM, however, does not guarantee convergence of the MCMC by the same number of iterations. For one, the convergence of EM is relative to the specified tolerance, and more stringent conditions may lead to a nonconvergent result. As such, we recommend that you always examine the obtained MCMC results.

The second issue is not surprising. Recall that x1 and x3 have very few complete observations. So the aspects of the covariance structure involving those variables (for example, the covariance between x1 and x2) are difficult to estimate reliably based on the information from the observed data only. The default uniform prior may not be viable here.

One solution is to introduce prior information to stabilize the estimation of the covariance matrix. We can do this by specifying a ridge prior using the prior() option. We introduce only a small amount of information by using a degrees of freedom value of 0.1:

```
. mi imp mvn x1-x3, add(30) prior(ridge, df(0.1)) rseed(332247)

Performing EM optimization:
note: 4 observations omitted from EM estimation because of all imputation
      variables missing
   observed log posterior =   -1.13422 at iteration 100
   (EM did not converge)

Performing MCMC data augmentation ...

Multivariate imputation               Imputations =        30
Multivariate normal regression              added =        30
Imputed: m=1 through m=30                  updated =         0

Prior: ridge, df=.1                       Iterations =      3000
                                           burn-in =       100
                                           between =       100
```

Variable	Observations per m			
	Complete	Incomplete	Imputed	Total
x1	4	16	16	20
x2	15	5	5	20
x3	3	17	17	20

```
(complete + incomplete = total; imputed is the minimum across m
 of the number of filled-in observations.)
```

This appears to be enough to alleviate the problem of a non–positive-definite estimate of the covariance matrix. Still, EM did not converge.

We will fix that and examine the resulting MCMC sequence. We will use the same random-number seed and this time save the WLF. Rather than imputing the data as before, we will simply run the MCMC for the same number of iterations it takes to obtain 30 imputations using the default settings, namely, $30 \times 100 = 3000$.

```
. mi imp mvn x1-x3, mcmconly prior(ridge, df(0.1)) initmcmc(em, iter(200) nolog)
> burnin(3000) savewlf(wlf, replace) rseed(332242)

Performing EM optimization:
note: 4 observations omitted from EM estimation because of all imputation variables
      missing
   observed log posterior = -1.1341806 at iteration 152

Performing MCMC data augmentation ...

Note: no imputation performed.
```

We increased the maximum number of iterations for the EM algorithm to 200; it converged in iteration 152.

We use the results from `wlf.dta` to obtain the trace and autocorrelation plots as we did in the earlier examples:

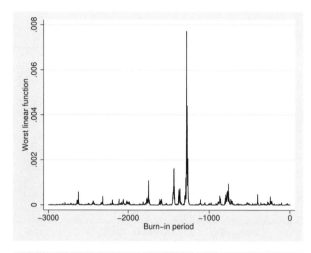

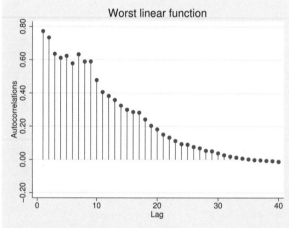

The serial correlation decreases slowly. There is no obvious trend in the WLF estimates, but we notice high variability and several spikes, some distinctive. The high variability and spikes are not surprising considering that certain model parameters could not be estimated reliably from the observed data and considering that we did not introduce enough prior information to obtain less variable estimates; we introduced only enough to achieve nonsingularity.

We could decrease the variability of the estimates by obtaining more data or introducing stronger prior information. For example, we could increase the number of degrees of freedom with a ridge prior to constrain the covariance matrix toward a diagonal structure:

```
. mi imp mvn x1-x3, replace prior(ridge, df(10)) burnin(300) rseed(332247)
  (output omitted)
```

If we create and examine the trace plots and autocorrelations of the WLF under the new prior specification, we find that variability of the estimates and serial dependence decrease greatly at a cost of bias if the prior assumptions are false.

◁

▷ Example 6: Saving all parameter series

The examples above used the WLF to monitor convergence of MCMC because in most applications it is sufficient. Although the WLF series often behave as the worst-case scenario, exceptions exist in practice. Sometimes, examining individual parameter series may be necessary.

We can save all parameter series from MCMC by using the `saveptrace()` option. These parameter series are saved in a parameter-trace file, a special file with extension `.stptrace`. Although the resulting file is not a Stata dataset, it can easily be loaded into Stata using `mi ptrace use`; see [MI] **mi ptrace** for details.

Let's look at several parameter series from the above example.

```
. use http://www.stata-press.com/data/r12/mvnexample0
. mi impute mvn x1-x3, mcmconly prior(ridge, df(0.1)) init(em, iter(200) nolog)
> burnin(3000) rseed(332247) saveptrace(parms)
  (output omitted)
```

We save all parameter series to a file called `parms` by using `stptrace(parms)`.

We first describe the contents of the `parms` file and then read it into Stata:

```
. mi ptrace describe parms
  file parms.stptrace created on 2 Apr 2011 11:07 contains 3,000 records
  (obs.) on
      m                           1 variable
      iter                        1 variable
      b[y, x]                     3 variables (3 x 1)
      v[y, y]                     6 variables (3 x 3, symmetric)
  where y and x are
      y:  (1) x1  (2) x2  (3) x3
      x:  (1) _cons
. mi ptrace use parms, clear
```

The output from `mi ptrace describe` reports that the file contains imputation numbers, iteration numbers, estimates of three regression coefficients (`b[x1,_cons]`, `b[x2,_cons]`, and `b[x3,_cons]`, which are effectively the means of x1, x2, and x3), and estimates of six covariances (`v[x1,x1]`, `v[x2,x1]`, `v[x2,x2]`, and so on).

Because x1 and x3 contain the least number of complete observations, we examine the series containing their variance and covariance estimates. We generate graphs separately for each series and then combine them in one graph by using `graph combine`; see [G-2] **graph combine**.

```
. tsset iter
        time variable:  iter, -2999 to 0
                delta:  1 unit
. tsline v_y1y1, name(gr1) nodraw ytitle(Var(x1)) xtitle("") ylabel(#4)
. tsline v_y3y1, name(gr2) nodraw ytitle(Cov(x3,x1)) xtitle("") ylabel(#4)
. tsline v_y3y3, name(gr3) nodraw ytitle(Var(x3)) xtitle("") ylabel(#4)
. graph combine gr1 gr2 gr3, xcommon cols(1) b1title(Iteration)
```

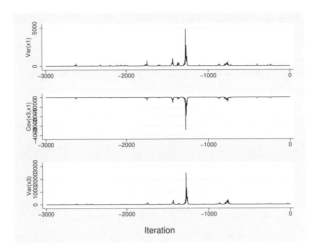

We repeat the same for the autocorrelation graphs:

```
. ac v_y1y1, ytitle(Var(x1)) xtitle("") ciopts(astyle(none)) note("")
> name(gr1, replace) nodraw ylabel(#4)
. ac v_y3y1, ytitle(Cov(x3,x1)) xtitle("") ciopts(astyle(none)) note("")
> name(gr2, replace) nodraw ylabel(#4)
. ac v_y3y3, ytitle(Var(x3)) xtitle("") ciopts(astyle(none)) note("")
> name(gr3, replace) nodraw ylabel(#4)
. graph combine gr1 gr2 gr3, xcommon cols(1) title(Autocorrelations) b1title(Lag)
```

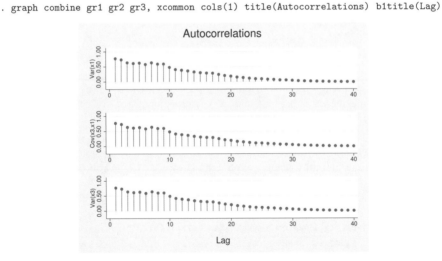

We can see that the trace plot and autocorrelations corresponding to the variance of x1 resemble the patterns of the earlier WLF estimates. We also notice that all series have high serial dependence within the first 20 iterations.

Again, if we switch to using a ridge prior with 10 degrees of freedom and repeat the steps above, the obtained trace plots will be more precise and more regular. The serial dependence in the series will be lower.

◁

Saved results

mi impute mvn saves the following in r():

Scalars
r(M)	total number of imputations
r(M_add)	number of added imputations
r(M_update)	number of updated imputations
r(k_ivars)	number of imputed variables
r(burnin)	number of burn-in iterations
r(burnbetween)	number of burn-between iterations
r(df_prior)	prior degrees of freedom (saved only with prior(ridge))
r(N_em)	number of observations used by EM (including omitted missing observations)
r(N_e_em)	number of observations used by EM in estimation (excluding omitted missing observations)
r(N_mis_em)	number of incomplete observations within the EM estimation sample
r(N_S_em)	number of unique missing-value patterns
r(niter_em)	number of iterations EM takes to converge
r(llobs_em)	observed log likelihood (saved with prior(uniform))
r(lpobs_em)	observed log posterior (saved with priors other than uniform)
r(converged_em)	convergence flag for EM
r(emonly)	1 if performed EM estimation only, 0 otherwise
r(mcmconly)	1 if performed MCMC only without imputing data, 0 otherwise
r(N_g)	number of imputed groups (1 if by() is not specified)

Macros
r(method)	name of imputation method (mvn)
r(ivars)	names of imputation variables
r(rseed)	random-number seed
r(prior)	prior distribution
r(init_mcmc)	type of initial values (em or user)
r(ivarsorder)	names of imputation variables in the order used in the computation
r(init_em)	type of initial values used by EM (ac, cc, or user)
r(by)	names of variables specified within by()

Matrices
r(N)	number of observations in imputation sample in each group (per variable)
r(N_complete)	number of complete observations in imputation sample in each group (per variable)
r(N_incomplete)	number of incomplete observations in imputation sample in each group (per variable)
r(N_imputed)	number of imputed observations in imputation sample in each group (per variable)
r(Beta0)	initial values for regression coefficients used by DA
r(Sigma0)	initial variance–covariance matrix used by DA
r(wlf_wgt)	coefficients for the WLF (saved with initmcmc(em) or if wlfwgt() is used)
r(Beta_em)	estimated regression coefficients from EM
r(Sigma_em)	estimated variance–covariance matrix from EM
r(Beta0_em)	initial values for regression coefficients used by EM
r(Sigma0_em)	initial variance–covariance matrix used by EM
r(N_pat)	minimum, average, and maximum numbers of observations per missing-value pattern

r(N_pat) and results with the _em suffix are saved only when the EM algorithm is used (with emonly or initmcmc(em)).

Methods and formulas

Let $\mathbf{x}_1, \mathbf{x}_2, \ldots, \mathbf{x}_N$ be a random sample from a p-variate normal distribution recording values of p imputation variables. Consider a multivariate normal regression

$$\mathbf{x}_i = \boldsymbol{\Theta}' \mathbf{z}_i + \boldsymbol{\epsilon}_i, \ i = 1, \ldots, N$$

where $\mathbf{z}_i$ is a $q \times 1$ vector of independent (complete) variables from observation i, $\boldsymbol{\Theta}$ is a $q \times p$ matrix of regression coefficients, and $\boldsymbol{\epsilon}_i$ is a $p \times 1$ vector of random errors from a p-variate normal distribution with a zero mean vector and a $p \times p$ positive-definite covariance matrix $\boldsymbol{\Sigma}$. We refer to $\boldsymbol{\Theta}$ and $\boldsymbol{\Sigma}$ as model parameters. Consider the partition $\mathbf{x}_i = (\mathbf{x}_{i(m)}, \mathbf{x}_{i(o)})$ corresponding to missing and observed values of imputation variables in observation i for $i = 1, \ldots, N$.

Methods and formulas are presented under the following headings:

Data augmentation
Prior distribution
Initial values: EM algorithm
Worst linear function

Data augmentation

mi impute mvn uses data augmentation (DA) to fill in missing values in $\mathbf{x}_i$ independently for each observation $i = 1, \ldots, N$. Data augmentation consists of two steps, an I step (imputation step) and a P step (posterior step), performed at each iteration $t = 0, 1, \ldots, T$. At iteration t of the I step, missing values in $\mathbf{x}_i$ are replaced with draws from the conditional posterior distribution of $\mathbf{x}_{i(m)}$ given observed data and current values of model parameters independently for each $i = 1, \ldots, N$. During the P step, new values of model parameters are drawn from their conditional posterior distribution given the observed data and the data imputed in the previous I step. Mathematically, this process can be described as follows:

I step:

$$
\mathbf{x}_{i(m)}^{(t+1)} \sim P\left(\mathbf{x}_{i(m)} | \mathbf{z}_i, \mathbf{x}_{i(o)}, \boldsymbol{\Theta}^{(t)}, \boldsymbol{\Sigma}^{(t)}\right), \; i = 1, \ldots, N \tag{1}
$$

P step:

$$
\begin{aligned}
\boldsymbol{\Sigma}^{(t+1)} &\sim P\left(\boldsymbol{\Sigma} | \mathbf{z}_i, \mathbf{x}_{i(o)}, \mathbf{x}_{i(m)}^{(t+1)}\right) \\
\boldsymbol{\Theta}^{(t+1)} &\sim P\left(\boldsymbol{\Theta} | \mathbf{z}_i, \mathbf{x}_{i(o)}, \mathbf{x}_{i(m)}^{(t+1)}, \boldsymbol{\Sigma}^{(t+1)}\right)
\end{aligned} \tag{2}
$$

The above two steps are repeated until the specified number of iterations, T, is reached. The total number of iterations, T, is determined by the length of the initial burn-in period, b, and the number of iterations between imputations, k. Specifically, $T = b + M_{new} \times k$, where M_{new} contains the number of added and updated imputations. mi impute mvn saves imputed values $\mathbf{x}_{i(m)}^{(t_1)}, \mathbf{x}_{i(m)}^{(t_2)}, \ldots, \mathbf{x}_{i(m)}^{(t_{M_{\text{new}}})}$ as final imputations, where iteration $t_i = b + (i-1)k$.

By default, mi impute mvn uses default values of 100 for b and k. These values may be adequate in some applications and may be too low in others. In general, b and k must be determined based on the properties of the observed Markov chain $\left(\mathbf{X}_m^{(1)}, \boldsymbol{\Theta}^{(1)}, \boldsymbol{\Sigma}^{(1)}\right), \left(\mathbf{X}_m^{(2)}, \boldsymbol{\Theta}^{(2)}, \boldsymbol{\Sigma}^{(2)}\right), \ldots$, where $\mathbf{X}_m^{(t)}$ denotes all values imputed at iteration t. b must be large enough so that the above chain converges to the stationary distribution $P(\mathbf{X}_m, \boldsymbol{\Theta}, \boldsymbol{\Sigma} | \mathbf{Z}, \mathbf{X}_o)$ by iteration $t = b$. k must be large enough so that random draws (imputations) $\mathbf{x}_{i(m)}^{(t_1)}, \mathbf{x}_{i(m)}^{(t_2)}, \ldots$ are approximately independent. See *Convergence of the MCMC method* for more details.

The functional form of the conditional posterior distributions (1) and (2) depends on the distribution of the data and a prior distribution of the model parameters. mi impute mvn assumes an improper uniform prior distribution for $\boldsymbol{\Theta}$ and an inverted Wishart distribution (Mardia, Kent, and Bibby 1979, 85) $W_p^{-1}(\Lambda, \lambda)$ for $\boldsymbol{\Sigma}$ under which the prior joint density function is

$$
f(\boldsymbol{\Theta}, \boldsymbol{\Sigma}) \propto |\boldsymbol{\Sigma}|^{-\left(\frac{\lambda+p+1}{2}\right)} \exp\left(-\frac{1}{2}\mathrm{tr}\Lambda^{-1}\boldsymbol{\Sigma}^{-1}\right)
$$

Under the multivariate normal model and the above prior distribution, the I and P steps become (Schafer 2008; Schafer 1997, 181–185) the following:

I step: $\mathbf{x}_{i(m)}^{(t+1)} \sim N_{p_i}\left(\boldsymbol{\mu}_{m\cdot o}^{(t)}, \boldsymbol{\Sigma}_{mm\cdot o}^{(t)}\right)$, $i = 1, \ldots, N$

P step: $\boldsymbol{\Sigma}^{(t+1)} \sim W^{-1}(\Lambda_{\star}^{(t+1)}, \lambda_{\star})$

$$\text{vec}\left(\boldsymbol{\Theta}^{(t+1)}\right) \sim N_{pq}\left\{\text{vec}\left(\widehat{\boldsymbol{\Theta}}^{(t+1)}\right), \boldsymbol{\Sigma}^{(t+1)} \otimes (\mathbf{Z}'\mathbf{Z})^{-1}\right\}$$

where p_i is the number of imputation variables containing missing values in observation i and $\otimes$ is the Kronecker product. Submatrices $\boldsymbol{\mu}_{m\cdot o}^{(t)}$ and $\boldsymbol{\Sigma}_{mm\cdot o}^{(t)}$ are the mean and variance of the conditional normal distribution of $\mathbf{x}_{i(m)}$ given $\mathbf{x}_{i(o)}$ based on $(\mathbf{x}_{i(m)}, \mathbf{x}_{i(o)}|\mathbf{z}_i) \sim N_p\left(\boldsymbol{\Theta}^{(t)'}\mathbf{z}_i, \boldsymbol{\Sigma}^{(t)}\right)$. See, for example, Mardia, Kent, and Bibby (1979, 63) for the corresponding formulas of the conditional mean and variance of the multivariate normal distribution. The matrix $\widehat{\boldsymbol{\Theta}}^{(t+1)} = (\mathbf{Z}'\mathbf{Z})^{-1}\mathbf{Z}'\mathbf{X}^{(t+1)}$ is the OLS estimate of the regression coefficients based on the augmented data $\mathbf{X}^{(t+1)} = (\mathbf{X}_o, \mathbf{X}_m^{(t+1)})$ from iteration t. The posterior cross-product matrix $\Lambda_{\star}^{(t+1)}$ and the posterior degrees of freedom $\lambda_{\star}$ are defined as follows:

$$\Lambda_{\star}^{(t+1)} = \left\{\Lambda^{-1} + (\mathbf{X}^{(t+1)} - \mathbf{Z}\widehat{\boldsymbol{\Theta}}^{(t+1)})'(\mathbf{X}^{(t+1)} - \mathbf{Z}\widehat{\boldsymbol{\Theta}}^{(t+1)})\right\}^{-1}$$

and

$$\lambda_{\star} = \lambda + N - q$$

Prior distribution

As we already mentioned, `mi impute mvn` assumes an improper uniform prior distribution for $\boldsymbol{\Theta}$ and an inverted Wishart distribution for $\boldsymbol{\Sigma}$ under which the prior joint density function is

$$f(\boldsymbol{\Theta}, \boldsymbol{\Sigma}) \propto |\boldsymbol{\Sigma}|^{-\left(\frac{\lambda + p + 1}{2}\right)} \exp\left(-\frac{1}{2}\text{tr}\Lambda^{-1}\boldsymbol{\Sigma}^{-1}\right)$$

Parameters of the inverted Wishart prior distribution, the prior cross-product matrix Λ, and the prior degrees of freedom λ are determined based on the requested prior distribution.

By default, `mi impute mvn` uses the uniform prior distribution under which $\lambda = -(p+1)$ and $\Lambda^{-1} = \mathbf{0}_{p \times p}$. Under the uniform prior, the log-likelihood and log-posterior functions are equivalent, and so the ML estimates of the parameters are equal to the posterior mode.

Under the noninformative Jeffreys prior distribution, $\lambda = 0$ and $\Lambda^{-1} = \mathbf{0}_{p \times p}$.

Under a ridge prior distribution, λ is equal to the user-specified value, and $\Lambda^{-1} = \lambda\boldsymbol{\Sigma}_{\star}$, where the diagonal matrix $\boldsymbol{\Sigma}_{\star}$ contains the diagonal elements of the estimate of the covariance matrix using all available cases. The variances (diagonal estimates) are the estimates of the mean squared error from regression of each imputation variable on the complete predictors. See Schafer (1997, 155–157) for details. With $\lambda = 0$, this prior specification reduces to the Jeffreys prior.

Initial values: EM algorithm

Initial values $\Theta^{(0)}$ and $\Sigma^{(0)}$ for DA are obtained from the EM algorithm for the incomplete multivariate normal data (for example, Dempster, Laird, and Rubin [1977], Little and Rubin [2002], Schafer [1997]). The EM algorithm iterates between the expectation step (E step) and the maximization step (M step) to maximize the log-likelihood (or log-posterior) function.

The observed-data log likelihood is

$$l_l(\Theta, \Sigma | \mathbf{X}_o) = \sum_{s=1}^{S} \sum_{i \in I(s)} \left\{ -0.5 \ln(|\Sigma_s|) - 0.5(\mathbf{x}_{i(o)} - \Theta_s' \mathbf{z}_i)' \Sigma_s^{-1} (\mathbf{x}_{i(o)} - \Theta_s' \mathbf{z}_i) \right\}$$

where S is the number of unique missing-value patterns, $I(s)$ is the set of observations from the same missing-value pattern s, and Θ_s and Σ_s are the submatrices of Θ and Σ that correspond to the imputation variables, which are observed in pattern s.

The observed-data log posterior is

$$l_p(\Theta, \Sigma | \mathbf{X}_o) = l_l(\Theta, \Sigma | \mathbf{X}_o) + \ln\{f(\Theta, \Sigma)\} = l_l(\Theta, \Sigma | \mathbf{X}_o) - \frac{\lambda + p + 1}{2} \ln(|\Sigma|) - \text{tr}(\Lambda^{-1} \Sigma^{-1})$$

The E step and M step of the EM algorithm are defined as follows (see Schafer [2008; 1997, 163–175] for details).

Let $T_1 = \sum_{i=1}^{N} \mathbf{z}_i \mathbf{x}_i'$ and $T_2 = \sum_{i=1}^{N} \mathbf{x}_i \mathbf{x}_i'$ denote the sufficient statistics for the multivariate normal model. Consider the submatrices $\Theta_{i(o)}$ and $\Theta_{i(m)}$ of Θ, and the submatrices $\Sigma_{i(mm)}$, $\Sigma_{i(mo)}$, and $\Sigma_{i(oo)}$ of Σ corresponding to the observed and missing columns of $\mathbf{x}_i$. Let $O(s)$ and $M(s)$ correspond to the column indexes of the observed and missing parts of $\mathbf{x}_i$ for each missing-values pattern s.

During the E step, the expectations $E(T_1)$ and $E(T_2)$ are computed with respect to the conditional distribution $\Pr(\mathbf{X}_m | \mathbf{X}_o, \Theta^{(t)}, \Sigma^{(t)})$ using the following relations:

$$E(x_{ij} | \mathbf{X}_o, \Theta^{(t)}, \Sigma^{(t)}) = \begin{cases} x_{ij}, & \text{for } j \in O(s) \\ x_{ij}^\star, & \text{for } j \in M(s) \end{cases}$$

and

$$E(x_{ij} x_{il} | \mathbf{X}_o, \Theta^{(t)}, \Sigma^{(t)}) = \begin{cases} x_{ij} x_{il}, & \text{for } j, l \in O(s) \\ x_{ij}^\star x_{il}, & \text{for } j \in M(s), l \in O(s) \\ c_{ij} + x_{ij}^\star x_{il}^\star, & \text{for } j, l \in M(s) \end{cases}$$

where $x_{ij}^\star$ is the jth element of the vector $\Theta_{i(m)}' \mathbf{z}_i + \Sigma_{i(mo)} \Sigma_{i(oo)}^{-1} \left(\mathbf{x}_{i(o)} - \Theta_{i(o)}' \mathbf{z}_i \right)$, and c_{ij} is the element of the matrix $\Sigma_{i(mm)} - \Sigma_{i(mo)} \Sigma_{i(oo)}^{-1} \Sigma_{i(mo)}'$.

During the M step, the model parameters are updated using the computed expectations of the sufficient statistics:

$$\Theta^{(t+1)} = (\mathbf{Z}'\mathbf{Z})^{-1} E(T_1)$$

$$\Sigma^{(t+1)} = \frac{1}{N + \lambda + p + 1} \left\{ E(T_2) - E(T_1)'(\mathbf{Z}'\mathbf{Z})^{-1} E(T_1) + \Lambda^{-1} \right\}$$

EM iterates between the E step and the M step until the maximum relative difference between the two successive values of all parameters is less than the default tolerance of 1e–5 (or the specified `tolerance()`).

Worst linear function

The worst linear function (WLF) is defined as follows (Schafer 1997, 129–131):

$$\xi(\theta) = \widehat{v}_1'(\theta - \widehat{\theta})$$

where θ and $\widehat{\theta}$ are column vectors of the unique model parameters and their respective EM estimates; $\widehat{v}_1 = \theta^{(t)} - \theta^{(t-1)}$, where $\theta^{(t)} = \widehat{\theta}$ and $\theta^{(t-1)}$ are the estimates from the last and one before the last iterations of the EM algorithm. This function is regarded to be the WLF because it has the highest asymptotic rate of missing information among all linear functions. This function is derived based on the convergence properties of the EM algorithm (see Schafer [1997, 55–59] for details).

References

Dempster, A. P., N. M. Laird, and D. B. Rubin. 1977. Maximum likelihood from incomplete data via the EM algorithm. *Journal of the Royal Statistical Society, Series B* 39: 1–38.

Gelman, A., J. B. Carlin, H. S. Stern, and D. B. Rubin. 2004. *Bayesian Data Analysis.* 2nd ed. London: Chapman & Hall/CRC.

Gelman, A., and D. B. Rubin. 1992. Inference from iterative simulation using multiple sequences. *Statistical Science* 7: 457–472.

Li, K.-H. 1988. Imputation using Markov chains. *Journal of Statistical Computation and Simulation* 30: 57–79.

Little, R. J. A., and D. B. Rubin. 2002. *Statistical Analysis with Missing Data.* 2nd ed. Hoboken, NJ: Wiley.

Mardia, K. V., J. T. Kent, and J. M. Bibby. 1979. *Multivariate Analysis.* London: Academic Press.

Rubin, D. B. 1987. *Multiple Imputation for Nonresponse in Surveys.* New York: Wiley.

Schafer, J. L. 1997. *Analysis of Incomplete Multivariate Data.* Boca Raton, FL: Chapman & Hall/CRC.

———. 2008. *NORM: Analysis of incomplete multivariate data under a normal model, Version 3.* Software package for R. University Park, PA: The Methodology Center, Pennsylvania State University.

Tanner, M. A., and W. H. Wong. 1987. The calculation of posterior distributions by data augmentation (with discussion). *Journal of the American Statistical Association* 82: 528–550.

Also see

[MI] **mi impute** — Impute missing values

[MI] **mi impute chained** — Impute missing values using chained equations

[MI] **mi impute monotone** — Impute missing values in monotone data

[MI] **mi estimate** — Estimation using multiple imputations

[MI] **intro substantive** — Introduction to multiple-imputation analysis

[MI] **intro** — Introduction to mi

[MI] **Glossary**

Stata Structural Equation Modeling Reference Manual

Title

> **mi impute nbreg** — Impute using negative binomial regression

Syntax

mi impute nbreg *ivar* [*indepvars*] [*if*] [*weight*] [, *impute_options options*]

impute_options	Description
Main	
* add(*#*)	specify number of imputations to add; required when no imputations exist
* replace	replace imputed values in existing imputations
rseed(*#*)	specify random-number seed
double	save imputed values in double precision; the default is to save them as float
by(*varlist*[, *byopts*])	impute separately on each group formed by *varlist*
Reporting	
dots	display dots as imputations are performed
noisily	display intermediate output
nolegend	suppress all table legends
Advanced	
force	proceed with imputation, even when missing imputed values are encountered
noupdate	do not perform mi update; see [MI] **noupdate option**

add(#*) is required when no imputations exist; add(*#*) or replace is required if imputations exist.
noupdate does not appear in the dialog box.

options	Description
Main	
noconstant	suppress constant term
dispersion(mean)	parameterization of dispersion; the default
dispersion(constant)	constant dispersion for all observations
exposure(*varname_e*)	include ln(*varname_e*) in model with coefficient constrained to 1
offset(*varname_o*)	include *varname_o* in model with coefficient constrained to 1
conditional(*if*)	perform conditional imputation
bootstrap	estimate model parameters using sampling with replacement
Maximization	
maximize_options	control the maximization process; seldom used

You must mi set your data before using mi impute nbreg; see [MI] **mi set**.

You must mi register *ivar* as imputed before using mi impute nbreg; see [MI] **mi set**.

indepvars may contain factor variables; see [U] **11.4.3 Factor variables**.

fweights, iweights, and pweights are allowed; see [U] **11.1.6 weight**.

Menu

Statistics > Multiple imputation

Description

mi impute nbreg fills in missing values of an overdispersed count variable using a negative binomial regression imputation method. You can perform separate imputations on different subsets of the data by specifying the by() option. You can also account for frequency, importance, and sampling weights.

Options

⌐ Main ¬

noconstant; see [R] **estimation options**.

add(), replace, rseed(), double, by(); see [MI] **mi impute**.

dispersion(mean | constant); see [R] **nbreg**.

exposure(*varname_e*), offset(*varname_o*); see [R] **estimation options**.

conditional(*if*) specifies that the imputation variable be imputed conditionally on observations satisfying *exp*; see [U] **11.1.3 if exp**. That is, missing values in a conditional sample, the sample identified by the *exp* expression, are imputed based only on data in that conditional sample. Missing values outside the conditional sample are replaced with a conditional constant, the value of the imputation variable in observations outside the conditional sample. As such, the imputation variable is required to be constant outside the conditional sample. Also, if any conditioning variables (variables involved in the conditional specification if *exp*) contain soft missing values (.), their missing values must be nested within missing values of the imputation variables. See *Conditional imputation* under *Remarks* in [MI] **mi impute**.

bootstrap specifies that posterior estimates of model parameters be obtained using sampling with replacement; that is, posterior estimates are estimated from a bootstrap sample. The default is to sample the estimates from the posterior distribution of model parameters or from the large-sample normal approximation of the posterior distribution. This option is useful when asymptotic normality of parameter estimates is suspect.

⌐ Reporting ¬

dots, noisily, nolegend; see [MI] **mi impute**. noisily specifies that the output from the negative binomial regression fit to the observed data be displayed. nolegend suppresses all legends that appear before the imputation table. Such legends include a legend about conditional imputation that appears when the conditional() option is specified and group legends that may appear when the by() option is specified.

⌐ Maximization ¬

maximize_options; see [R] **nbreg**. These options are seldom used.

⌐ Advanced ¬

force; see [MI] **mi impute**.

The following option is available with mi impute but is not shown in the dialog box:

noupdate; see [MI] **noupdate option**.

Remarks

Remarks are presented under the following headings:

Univariate imputation using negative binomial regression
Using mi impute nbreg

See [MI] **mi impute** for a general description and details about options common to all imputation methods, *impute_options*. Also see [MI] **workflow** for general advice on working with mi.

Univariate imputation using negative binomial regression

The negative binomial regression imputation method can be used to fill in missing values of an overdispersed count variable (Royston 2009). It is a parametric method that assumes an underlying negative binomial model (see [R] **nbreg**) for the imputed variable (given other predictors). This method is based on the asymptotic approximation of the posterior predictive distribution of the missing data.

Using mi impute nbreg

In [MI] **mi impute poisson**, we considered a version of the heart attack data containing a count variable, npreg, which records the number of pregnancies and is the only variable containing missing values. We imputed its missing values using mi impute poisson.

A Poisson model assumes that the mean and the variance are the same. In the presence of overdispersion, when the variance exceeds the mean, a negative binomial model is more appropriate. We can fit a negative binomial model for npreg to the observed data to see if there is any indication of overdispersion in the data.

```
. use http://www.stata-press.com/data/r12/mheartpois
(Fictional heart attack data; npreg missing)
. nbreg npreg attack smokes age bmi hsgrad if female==1, nolog
```

Negative binomial regression		Number of obs	=	35
		LR chi2(5)	=	1.69
Dispersion = mean		Prob > chi2	=	0.8903
Log likelihood = -54.638875		Pseudo R2	=	0.0152

npreg	Coef.	Std. Err.	z	P>|z|	[95% Conf. Interval]	
attack	.0551929	.4214484	0.13	0.896	-.7708309	.8812166
smokes	.0521987	.4182004	0.12	0.901	-.7674591	.8718565
age	-.0105877	.0174661	-0.61	0.544	-.0448206	.0236452
bmi	.0194787	.0489883	0.40	0.691	-.0765367	.115494
hsgrad	.5338139	.4972872	1.07	0.283	-.4408511	1.508479
_cons	-.0736959	1.551417	-0.05	0.962	-3.114417	2.967025
/lnalpha	-.7956602	.7987311			-2.361144	.769824
alpha	.4512832	.3604539			.0943122	2.159386

Likelihood-ratio test of alpha=0: chibar2(01) =   3.00 Prob>=chibar2 = 0.042

The estimate of the overdispersion parameter alpha is 0.45 with a 95% confidence interval of [0.094, 2.16]. The confidence interval does not include a value of 0 (no overdispersion), so there is slight overdispersion in the conditional distribution of nbreg in the observed data.

We now impute npreg using mi impute nbreg:

```
. mi set mlong
. mi register imputed npreg
(10 m=0 obs. now marked as incomplete)
. mi impute nbreg npreg attack smokes age bmi hsgrad, add(20) conditional(if female==1)
Univariate imputation                         Imputations =        20
Negative binomial regression                        added =        20
Imputed: m=1 through m=20                           updated =         0
Dispersion: mean
Conditional imputation:
  npreg: incomplete out-of-sample obs. replaced with value 0
```

	Observations per m			
Variable	Complete	Incomplete	Imputed	Total
npreg	144	10	10	154

```
(complete + incomplete = total; imputed is the minimum across m
of the number of filled-in observations.)
```

We specify the conditional() option to restrict imputation of npreg only to females; see *Conditional imputation* in [MI] **mi impute** for details.

We can analyze these multiply imputed data using logistic regression with mi estimate:

```
. mi estimate: logit attack smokes age bmi female hsgrad npreg
  (output omitted )
```

Saved results

mi impute nbreg saves the following in r():

Scalars
  r(M)   total number of imputations
  r(M_add)   number of added imputations
  r(M_update)   number of updated imputations
  r(k_ivars)   number of imputed variables (always 1)
  r(N_g)   number of imputed groups (1 if by() is not specified)

Macros
  r(method)   name of imputation method (nbreg)
  r(ivars)   names of imputation variables
  r(rseed)   random-number seed
  r(by)   names of variables specified within by()

Matrices
  r(N)   number of observations in imputation sample in each group
  r(N_complete)   number of complete observations in imputation sample in each group
  r(N_incomplete) number of incomplete observations in imputation sample in each group
  r(N_imputed)   number of imputed observations in imputation sample in each group

Methods and formulas

Consider a univariate variable $\mathbf{x} = (x_1, x_2, \ldots, x_n)'$ that follows a negative binomial model

$$\Pr(x_i = x | \mathbf{z}_i) = \frac{\Gamma(m_i + x)}{\Gamma(x+1)\Gamma(m_i)} \; p_i^{m_i}(1 - p_i)^x, \; x = 0, 1, 2, \ldots \tag{1}$$

where $m_i = m = 1/\alpha$, $p_i = 1/(1 + \alpha\mu_i)$ under mean-dispersion model and $m_i = \mu_i/\delta$, $p_i = p = 1/(1 + \delta)$ under constant-dispersion model, $\mu_i = \exp(\mathbf{z}_i'\boldsymbol{\beta} + \text{offset}_i)$, and $\alpha > 0$ and $\delta > 0$ are unknown dispersion parameters; see [R] **nbreg** for details. $\mathbf{z}_i = (z_{i1}, z_{i2}, \ldots, z_{iq})'$ records values of predictors of $\mathbf{x}$ for observation i and $\boldsymbol{\beta}$ is the $q \times 1$ vector of unknown regression coefficients. (When a constant is included in the model—the default—$z_{i1} = 1$, $i = 1, \ldots, n$.)

$\mathbf{x}$ contains missing values that are to be filled in. Consider the partition of $\mathbf{x} = (\mathbf{x}_o', \mathbf{x}_m')$ into $n_0 \times 1$ and $n_1 \times 1$ vectors containing the complete and the incomplete observations. Consider a similar partition of $\mathbf{Z} = (\mathbf{Z}_o, \mathbf{Z}_m)$ into $n_0 \times q$ and $n_1 \times q$ submatrices.

`mi impute nbreg` follows the steps below to fill in $\mathbf{x}_m$:

1. Fit a negative binomial regression model (1) to the observed data $(\mathbf{x}_o, \mathbf{Z}_o)$ to obtain the maximum likelihood estimates, $\widehat{\boldsymbol{\theta}} = (\widehat{\boldsymbol{\beta}}', \ln\widehat{\alpha})'$ under a mean-dispersion model or $\widehat{\boldsymbol{\theta}} = (\widehat{\boldsymbol{\beta}}', \ln\widehat{\delta})'$ under a constant-dispersion model, and their asymptotic sampling variance, $\widehat{\mathbf{U}}$.

2. Simulate new parameters, $\boldsymbol{\theta}_\star$, from the large-sample normal approximation, $N(\widehat{\boldsymbol{\theta}}, \widehat{\mathbf{U}})$, to its posterior distribution, assuming the noninformative prior $\Pr(\boldsymbol{\theta}) \propto \text{const}$.

3. Obtain one set of imputed values, $\mathbf{x}_m^1$, by simulating from a negative binomial distribution (1) with parameters set to their simulated values from step 2.

4. Repeat steps 2 and 3 to obtain M sets of imputed values, $\mathbf{x}_m^1, \mathbf{x}_m^2, \ldots, \mathbf{x}_m^M$.

Steps 2 and 3 above correspond to only approximate draws from the posterior predictive distribution of the missing data, $\Pr(\mathbf{x}_m|\mathbf{x}_o, \mathbf{Z}_o)$, because $\boldsymbol{\theta}_\star$ is drawn from the asymptotic approximation to its posterior distribution.

If weights are specified, a weighted negative binomial regression model is fit to the observed data in step 1 (see [R] **nbreg** for details).

Reference

Royston, P. 2009. Multiple imputation of missing values: Further update of ice, with an emphasis on categorical variables. *Stata Journal* 9: 466–477.

Also see

[MI] **mi impute** — Impute missing values

[MI] **mi impute poisson** — Impute using Poisson regression

[MI] **mi estimate** — Estimation using multiple imputations

[MI] **intro** — Introduction to mi

[MI] **intro substantive** — Introduction to multiple-imputation analysis

Title

> **mi impute ologit** — Impute using ordered logistic regression

Syntax

mi impute ologit *ivar* [*indepvars*] [*if*] [*weight*] [, *impute_options options*]

impute_options	Description
Main	
* add(*#*)	specify number of imputations to add; required when no imputations exist
* replace	replace imputed values in existing imputations
rseed(*#*)	specify random-number seed
double	save imputed values in double precision; the default is to save them as float
by(*varlist* [, *byopts*])	impute separately on each group formed by *varlist*
Reporting	
dots	display dots as imputations are performed
noisily	display intermediate output
nolegend	suppress all table legends
Advanced	
force	proceed with imputation, even when missing imputed values are encountered
noupdate	do not perform mi update; see [MI] **noupdate option**

* add(*#*) is required when no imputations exist; add(*#*) or replace is required if imputations exist.
noupdate does not appear in the dialog box.

options	Description
Main	
offset(*varname*)	include *varname* in model with coefficient constrained to 1
augment	perform augmented regression in the presence of perfect prediction
conditional(*if*)	perform conditional imputation
bootstrap	estimate model parameters using sampling with replacement
Maximization	
maximize_options	control the maximization process; seldom used

You must mi set your data before using mi impute ologit; see [MI] **mi set**.

You must mi register *ivar* as imputed before using mi impute ologit; see [MI] **mi set**.

indepvars may contain factor variables; see [U] **11.4.3 Factor variables**.

fweights, iweights, and pweights are allowed; see [U] **11.1.6 weight**.

Menu

Statistics > Multiple imputation

Description

mi impute ologit fills in missing values of an ordinal variable using an ordered logistic regression imputation method. You can perform separate imputations on different subsets of the data by specifying the by() option. You can also account for frequency, importance, and sampling weights.

Options

⌐──── Main ⌐──

add(), replace, rseed(), double, by(); see [MI] **mi impute**.

offset(*varname*); see [R] **estimation options**.

augment specifies that augmented regression be performed if perfect prediction is detected. By default, an error is issued when perfect prediction is detected. The idea behind the augmented-regression approach is to add a few observations with small weights to the data during estimation to avoid perfect prediction. See *The issue of perfect prediction during imputation of categorical data* under *Remarks* in [MI] **mi impute** for more information. augment is not allowed with importance weights.

conditional(*if*) specifies that the imputation variable be imputed conditionally on observations satisfying *exp*; see [U] **11.1.3 if exp**. That is, missing values in a conditional sample, the sample identified by the *exp* expression, are imputed based only on data in that conditional sample. Missing values outside the conditional sample are replaced with a conditional constant, the value of the imputation variable in observations outside the conditional sample. As such, the imputation variable is required to be constant outside the conditional sample. Also, if any conditioning variables (variables involved in the conditional specification if *exp*) contain soft missing values (.), their missing values must be nested within missing values of the imputation variables. See *Conditional imputation* under *Remarks* in [MI] **mi impute**.

bootstrap specifies that posterior estimates of model parameters be obtained using sampling with replacement; that is, posterior estimates are estimated from a bootstrap sample. The default is to sample the estimates from the posterior distribution of model parameters or from the large-sample normal approximation of the posterior distribution. This option is useful when asymptotic normality of parameter estimates is suspect.

⌐──── Reporting ⌐───

dots, noisily, nolegend; see [MI] **mi impute**. noisily specifies that the output from the ordered logistic regression fit to the observed data be displayed. nolegend suppresses all legends that appear before the imputation table. Such legends include a legend about conditional imputation that appears when the conditional() option is specified and group legends that may appear when the by() option is specified.

⌐──── Maximization ⌐──

maximize_options; see [R] **ologit**. These options are seldom used.

⌐ Advanced ⌐

`force`; see [MI] **mi impute**.

The following option is available with `mi impute` but is not shown in the dialog box:

`noupdate`; see [MI] **noupdate option**.

Remarks

Remarks are presented under the following headings:

Univariate imputation using ordered logistic regression
Using mi impute ologit

See [MI] **mi impute** for a general description and details about options common to all imputation methods, *impute_options*. Also see [MI] **workflow** for general advice on working with `mi`.

Univariate imputation using ordered logistic regression

The ordered logistic regression imputation method can be used to fill in missing values of an ordinal variable (for example, Raghunathan et al. [2001] and van Buuren [2007]). It is a parametric method that assumes an underlying logistic model for the imputed variable (given other predictors). Similarly to the logistic imputation method, this method is based on the asymptotic approximation of the posterior predictive distribution of the missing data.

Using mi impute ologit

Following the example from [MI] **mi impute mlogit**, we consider the heart attack data (for example, [MI] **intro substantive**, [MI] **mi impute**), where a logistic model of interest now includes information about alcohol consumption, variable `alcohol`—`logit attack smokes age bmi female hsgrad i.alcohol`.

```
. use http://www.stata-press.com/data/r12/mheart4
(Fictional heart attack data; alcohol missing)

. tabulate alcohol, missing
```

Alcohol consumption: none, <2 drinks/day, >=2 drinks/day	Freq.	Percent	Cum.
Do not drink	18	11.69	11.69
Less than 3 drinks/day	83	53.90	65.58
Three or more drinks/day	44	28.57	94.16
.	9	5.84	100.00
Total	154	100.00	

From the output, the `alcohol` variable has three unique ordered categories and nine missing observations. We use the ordered logistic imputation method to impute missing values of `alcohol`. We create 10 imputations by specifying the `add(10)` option:

```
. mi set mlong

. mi register imputed alcohol
(9 m=0 obs. now marked as incomplete)

. mi impute ologit alcohol attack smokes age bmi female hsgrad, add(10)
```

Univariate imputation	Imputations =	10
Ordered logistic regression	added =	10
Imputed: *m*=1 through *m*=10	updated =	0

		Observations per *m*		
Variable	Complete	Incomplete	Imputed	Total
alcohol	145	9	9	154

(complete + incomplete = total; imputed is the minimum across *m* of the number of filled-in observations.)

We can now analyze these multiply imputed data with logistic regression via `mi estimate`:

```
. mi estimate: logit attack smokes age bmi female hsgrad i.alcohol
  (output omitted )
```

Saved results

`mi impute ologit` saves the following in `r()`:

Scalars
`r(M)`	total number of imputations
`r(M_add)`	number of added imputations
`r(M_update)`	number of updated imputations
`r(k_ivars)`	number of imputed variables (always 1)
`r(pp)`	1 if perfect prediction detected, 0 otherwise
`r(N_g)`	number of imputed groups (1 if by() is not specified)

Macros
`r(method)`	name of imputation method (ologit)
`r(ivars)`	names of imputation variables
`r(rseed)`	random-number seed
`r(by)`	names of variables specified within by()

Matrices
`r(N)`	number of observations in imputation sample in each group
`r(N_complete)`	number of complete observations in imputation sample in each group
`r(N_incomplete)`	number of incomplete observations in imputation sample in each group
`r(N_imputed)`	number of imputed observations in imputation sample in each group

Methods and formulas

Consider a univariate variable $\mathbf{x} = (x_1, x_2, \ldots, x_n)'$ that contains K ordered categories and follows an ordered logistic model

$$\Pr(x_i = k | \mathbf{z}_i) = \Pr(\gamma_{k-1} < \mathbf{z}_i'\boldsymbol{\beta} + u \leq \gamma_k)$$
$$= \frac{1}{1 + \exp(-\gamma_k + \mathbf{z}_i'\boldsymbol{\beta})} - \frac{1}{1 + \exp(-\gamma_{k-1} + \mathbf{z}_i'\boldsymbol{\beta})} \qquad (1)$$

where $\mathbf{z}_i = (z_{i1}, z_{i2}, \ldots, z_{iq})'$ records values of predictors of $\mathbf{x}$ for observation i, $\boldsymbol{\beta}$ is the $q \times 1$ vector of unknown regression coefficients, and $\boldsymbol{\gamma} = (\gamma_1, \ldots, \gamma_{K-1})'$ are the unknown cutpoints with $\gamma_0 = -\infty$ and $\gamma_K = \infty$. (There is no constant in this model because its effect is absorbed into the cutpoints; see [R] **ologit** for details.)

$\mathbf{x}$ contains missing values that are to be filled in. Consider the partition of $\mathbf{x} = (\mathbf{x}'_o, \mathbf{x}'_m)$ into $n_0 \times 1$ and $n_1 \times 1$ vectors containing the complete and the incomplete observations. Consider a similar partition of $\mathbf{Z} = (\mathbf{Z}_o, \mathbf{Z}_m)$ into $n_0 \times q$ and $n_1 \times q$ submatrices.

`mi impute ologit` follows the steps below to fill in $\mathbf{x}_m$:

1. Fit an ordered logistic model (1) to the observed data $(\mathbf{x}_o, \mathbf{Z}_o)$ to obtain the maximum likelihood estimates, $\widehat{\boldsymbol{\theta}} = (\widehat{\boldsymbol{\beta}}', \widehat{\boldsymbol{\gamma}}')'$, and their asymptotic sampling variance, $\widehat{\mathbf{U}}$.

2. Simulate new parameters, $\boldsymbol{\theta}_\star$, from the large-sample normal approximation, $N(\widehat{\boldsymbol{\theta}}, \widehat{\mathbf{U}})$, to its posterior distribution assuming the noninformative prior $\Pr(\boldsymbol{\theta}) \propto \text{const}$.

3. Obtain one set of imputed values, $\mathbf{x}_m^1$, by simulating from an ordered logistic distribution as defined by (1): one of K categories is randomly assigned to a missing category, i_m, using the cumulative probabilities computed from (1) with $\boldsymbol{\beta} = \boldsymbol{\beta}_\star$, $\boldsymbol{\gamma} = \boldsymbol{\gamma}_\star$, and $\mathbf{z}_i = \mathbf{z}_{i_m}$.

4. Repeat steps 2 and 3 to obtain M sets of imputed values, $\mathbf{x}_m^1, \mathbf{x}_m^2, \ldots, \mathbf{x}_m^M$.

Steps 2 and 3 above correspond to only approximate draws from the posterior predictive distribution of the missing data, $\Pr(\mathbf{x}_m | \mathbf{x}_o, \mathbf{Z}_o)$, because $\boldsymbol{\theta}_\star$ is drawn from the asymptotic approximation to its posterior distribution.

If weights are specified, a weighted ordered logistic regression model is fit to the observed data in step 1 (see [R] **ologit** for details).

References

Raghunathan, T. E., J. M. Lepkowski, J. Van Hoewyk, and P. Solenberger. 2001. A multivariate technique for multiply imputing missing values using a sequence of regression models. *Survey Methodology* 27: 85–95.

van Buuren, S. 2007. Multiple imputation of discrete and continuous data by fully conditional specification. *Statistical Methods in Medical Research* 16: 219–242.

Also see

[MI] **mi impute** — Impute missing values

[MI] **mi impute mlogit** — Impute using multinomial logistic regression

[MI] **mi estimate** — Estimation using multiple imputations

[MI] **intro** — Introduction to mi

[MI] **intro substantive** — Introduction to multiple-imputation analysis

Title

---
mi impute pmm — Impute using predictive mean matching

Syntax

mi <u>imp</u>ute pmm *ivar* [*indepvars*] [*if*] [*weight*] [, *impute_options options*]

impute_options	Description
Main	
* add(*#*)	specify number of imputations to add; required when no imputations exist
* replace	replace imputed values in existing imputations
rseed(*#*)	specify random-number seed
double	save imputed values in double precision; the default is to save them as float
by(*varlist* [, *byopts*])	impute separately on each group formed by *varlist*
Reporting	
dots	display dots as imputations are performed
<u>noi</u>sily	display intermediate output
<u>noleg</u>end	suppress all table legends
Advanced	
force	proceed with imputation, even when missing imputed values are encountered
<u>noup</u>date	do not perform mi update; see [MI] **noupdate option**

add(#*) is required when no imputations exist; add(*#*) or replace is required if imputations exist.
noupdate does not appear in the dialog box.

options	Description
Main	
<u>nocons</u>tant	suppress constant term
knn(*#*)	specify # of closest observations (nearest neighbors) to draw from; default is knn(1)
<u>conditional</u>(*if*)	perform conditional imputation
<u>bootstrap</u>	estimate model parameters using sampling with replacement

You must mi set your data before using mi impute pmm; see [MI] **mi set**.
You must mi register *ivar* as imputed before using mi impute pmm; see [MI] **mi set**.
indepvars may contain factor variables; see [U] **11.4.3 Factor variables**.
aweights, fweights, iweights, and pweights are allowed; see [U] **11.1.6 weight**.

Menu

Statistics > Multiple imputation

Description

mi impute pmm fills in missing values of a continuous variable by using the predictive mean matching imputation method. You can perform separate imputations on different subsets of the data by specifying the by() option. You can also account for analytic, frequency, importance, and sampling weights.

Options

───────┐ Main └───

noconstant; see [R] **estimation options**.

add(), replace, rseed(), double, by(); see [MI] **mi impute**.

knn(#) specifies the number of closest observations (nearest neighbors) from which to draw imputed values. The default is to replace a missing value with the "closest" observation, knn(1). The closeness is determined based on the absolute difference between the linear prediction for the missing value and that for the complete values. The closest observation is the observation with the smallest difference. This option regulates the correlation among multiple imputations that affects the bias and the variability of the resulting multiple-imputation point estimates; see *Remarks* for details.

conditional(*if*) specifies that the imputation variable be imputed conditionally on observations satisfying *exp*; see [U] **11.1.3 if exp**. That is, missing values in a conditional sample, the sample identified by the *exp* expression, are imputed based only on data in that conditional sample. Missing values outside the conditional sample are replaced with a conditional constant, the value of the imputation variable in observations outside the conditional sample. As such, the imputation variable is required to be constant outside the conditional sample. Also, if any conditioning variables (variables involved in the conditional specification if *exp*) contain soft missing values (.), their missing values must be nested within missing values of the imputation variables. See *Conditional imputation* under *Remarks* in [MI] **mi impute**.

bootstrap specifies that posterior estimates of model parameters be obtained using sampling with replacement; that is, posterior estimates are estimated from a bootstrap sample. The default is to sample the estimates from the posterior distribution of model parameters or from the large-sample normal approximation of the posterior distribution. This option is useful when asymptotic normality of parameter estimates is suspect.

───────┐ Reporting └──

dots, noisily, nolegend; see [MI] **mi impute**. noisily specifies that the output from the linear regression fit to the observed data be displayed. nolegend suppresses all legends that appear before the imputation table. Such legends include a legend about conditional imputation that appears when the conditional() option is specified and group legends that may appear when the by() option is specified.

───────┐ Advanced └──

force; see [MI] **mi impute**.

The following option is available with `mi impute` but is not shown in the dialog box:

`noupdate`; see [MI] **noupdate option**.

Remarks

Remarks are presented under the following headings:

Univariate imputation using predictive mean matching
Using mi impute pmm

See [MI] **mi impute** for a general description and details about options common to all imputation methods, *impute_options*. Also see [MI] **workflow** for general advice on working with `mi`.

Univariate imputation using predictive mean matching

Either predictive mean matching (`pmm`) or normal linear regression (`regress`) imputation methods can be used to fill in missing values of a continuous variable (Rubin 1987; Schenker and Taylor 1996). Predictive mean matching may be preferable to linear regression when the normality of the underlying model is suspect.

Predictive mean matching (PMM) is a partially parametric method that matches the missing value to the observed value with the closest predicted mean (or linear prediction). It was introduced by Little (1988) based on Rubin's (1986) ideas applied to statistical file matching. PMM combines the standard linear regression and the nearest-neighbor imputation approaches. It uses the normal linear regression to obtain linear predictions. It then uses the linear prediction as a distance measure to form the set of nearest neighbors (possible donors) consisting of the complete values. Finally, it randomly draws an imputed value from this set. By drawing from the observed data, PMM preserves the distribution of the observed values in the missing part of the data, which makes it more robust than the fully parametric linear regression approach.

With PMM, you need to decide how many nearest neighbors to include in the set of possible donors. `mi impute pmm` defaults to one nearest neighbor, `knn(1)`. You may need to include more depending on your data. The number of nearest neighbors affects the correlation among imputations—the smaller the number, the higher the correlation. High correlation in turn increases the variability of the MI point estimates. Including too many possible donors may result in increased bias of the MI point estimates. Thus the number of nearest neighbors regulates the tradeoff between the bias and the variance of the point estimators in repeated sampling. The literature does not provide a definitive recommendation on how to choose this number in practice; see Schenker and Taylor (1996) for some insight into this issue.

Using mi impute pmm

Recall the heart attack data from *Univariate imputation* in [MI] **mi impute**. We wish to fit a logistic regression of `attack` on some predictors, one of which, `bmi`, has missing values. To avoid losing information contained in complete observations of the other predictors, we impute `bmi`.

We showed one way of imputing `bmi` in [MI] **mi impute regress**. Suppose, however, that we want to restrict the imputed values of `bmi` to be within the range observed for `bmi`. We can use the PMM imputation method to restrict the values. This method may also be preferable to the regression imputation of `bmi` because the distribution of `bmi` is slightly skewed.

```
. use http://www.stata-press.com/data/r12/mheart0
(Fictional heart attack data; bmi missing)
. mi set mlong
. mi register imputed bmi
(22 m=0 obs. now marked as incomplete)
. mi impute pmm bmi attack smokes age hsgrad female, add(20)
Univariate imputation                    Imputations =        20
Predictive mean matching                       added =        20
Imputed: m=1 through m=20                     updated =         0

                                    Nearest neighbors =         1
```

	Observations per _m_			
Variable	Complete	Incomplete	Imputed	Total
bmi	132	22	22	154

```
(complete + incomplete = total; imputed is the minimum across m
 of the number of filled-in observations.)
```

By default, mi impute pmm uses one nearest neighbor to draw from. That is, it replaces missing values with an observed value whose linear prediction is the closest to that of the missing value. Using only one nearest neighbor may result in high variability of the MI estimates. You can increase the number of nearest neighbors from which the imputed value is drawn by specifying the knn() option. For example, we use 5 below:

```
. mi impute pmm bmi attack smokes age hsgrad female, replace knn(5)
Univariate imputation                    Imputations =        20
Predictive mean matching                       added =         0
Imputed: m=1 through m=20                     updated =        20

                                    Nearest neighbors =         5
```

	Observations per _m_			
Variable	Complete	Incomplete	Imputed	Total
bmi	132	22	22	154

```
(complete + incomplete = total; imputed is the minimum across m
 of the number of filled-in observations.)
```

You can now refit the logistic model and examine the effect of using more neighbors:

```
. mi estimate: logit attack smokes age bmi hsgrad female
  (output omitted )
```

See [MI] **mi impute**, [MI] **mi impute regress**, and [MI] **mi estimate** for more details.

Saved results

mi impute pmm saves the following in r():

Scalars
r(M)	total number of imputations
r(M_add)	number of added imputations
r(M_update)	number of updated imputations
r(knn)	number of k nearest neighbors
r(k_ivars)	number of imputed variables (always 1)
r(N_g)	number of imputed groups (1 if by() is not specified)

Macros
r(method)	name of imputation method (pmm)
r(ivars)	names of imputation variables
r(rseed)	random-number seed
r(by)	names of variables specified within by()

Matrices
r(N)	number of observations in imputation sample in each group
r(N_complete)	number of complete observations in imputation sample in each group
r(N_incomplete)	number of incomplete observations in imputation sample in each group
r(N_imputed)	number of imputed observations in imputation sample in each group

Methods and formulas

mi impute pmm follows the steps as described in *Methods and formulas* of [MI] **mi impute regress** with the exception of step 3.

Consider a univariate variable $\mathbf{x} = (x_1, x_2, \ldots, x_n)'$ that follows a normal linear regression model

$$x_i | \mathbf{z}_i \sim N(\mathbf{z}_i'\boldsymbol{\beta}, \sigma^2) \tag{1}$$

where $\mathbf{z}_i = (z_{i1}, z_{i2}, \ldots, z_{iq})'$ records values of predictors of $\mathbf{x}$ for observation i, $\boldsymbol{\beta}$ is the $q \times 1$ vector of unknown regression coefficients, and σ^2 is the unknown scalar variance. (Note that when a constant is included in the model—the default—$z_{i1} = 1$, $i = 1, \ldots, n$.)

$\mathbf{x}$ contains missing values that are to be filled in. Consider the partition of $\mathbf{x} = (\mathbf{x}_o', \mathbf{x}_m')$ into $n_0 \times 1$ and $n_1 \times 1$ vectors containing the complete and the incomplete observations. Consider a similar partition of $\mathbf{Z} = (\mathbf{Z}_o, \mathbf{Z}_m)$ into $n_0 \times q$ and $n_1 \times q$ submatrices.

mi impute pmm follows the steps below to fill in $\mathbf{x}_m$ (for simplicity, we omit the conditioning on the observed data in what follows):

1. Fit a regression model (1) to the observed data $(\mathbf{x}_o, \mathbf{Z}_o)$ to obtain estimates $\widehat{\boldsymbol{\beta}}$ and $\widehat{\sigma}^2$ of the model parameters.

2. Simulate new parameters $\boldsymbol{\beta}_\star$ and $\sigma_\star^2$ from their joint posterior distribution under the conventional noninformative improper prior $\Pr(\boldsymbol{\beta}, \sigma^2) \propto 1/\sigma^2$. This is done in two steps:

$$\sigma_\star^2 \sim \widehat{\sigma}^2(n_0 - q)/\chi_{n_0-q}^2$$
$$\boldsymbol{\beta}_\star | \sigma_\star^2 \sim N\left\{\widehat{\boldsymbol{\beta}}, \sigma_\star^2(\mathbf{Z}_o'\mathbf{Z}_o)^{-1}\right\}$$

3. Generate the imputed values, $\mathbf{x}_m^1$, as follows. Let $\widehat{x}_i$ be the linear prediction of $\mathbf{x}$ based on predictors $\mathbf{Z}$ for observation i. Then for any missing observation i of $\mathbf{x}$, $x_i = x_{j_{\min}}$, where $j_{\min}$ is randomly drawn from the set of indices $\{i_1, i_2, \ldots, i_k\}$ corresponding to

the first k minimums determined based on the absolute differences between the linear prediction for incomplete observation i and linear predictions for all complete observations, $|\widehat{x}_i - \widehat{x}_j|$, $j \in obs$. For example, if $k = 1$ (the default), $j_{\min}$ is determined based on $|\widehat{x}_i - \widehat{x}_{j_{\min}}| = \min_{j \in \text{obs}} |\widehat{x}_i - \widehat{x}_j|$.

4. Repeat steps 2 and 3 to obtain M sets of imputed values, $\mathbf{x}_m^1, \mathbf{x}_m^2, \ldots, \mathbf{x}_m^M$.

If weights are specified, a weighted linear regression model is fit to the observed data in step 1 (see [R] **regress** for details).

References

Little, R. J. A. 1988. Missing-data adjustments in large surveys. *Journal of Business and Economic Statistics* 6: 287–296.

Rubin, D. B. 1986. Statistical matching using file concatenation with adjusted weights and multiple imputations. *Journal of Business and Economic Statistics* 4: 87–94.

——. 1987. *Multiple Imputation for Nonresponse in Surveys*. New York: Wiley.

Schenker, N., and J. M. G. Taylor. 1996. Partially parametric techniques for multiple imputation. *Computational Statistics & Data Analysis* 22: 425–446.

Also see

[MI] **mi impute** — Impute missing values

[MI] **mi impute regress** — Impute using linear regression

[MI] **mi impute truncreg** — Impute using truncated regression

[MI] **mi impute intreg** — Impute using interval regression

[MI] **mi estimate** — Estimation using multiple imputations

[MI] **intro** — Introduction to mi

[MI] **intro substantive** — Introduction to multiple-imputation analysis

Title

mi impute poisson — Impute using Poisson regression

Syntax

mi <u>imp</u>ute poisson *ivar* [*indepvars*] [*if*] [*weight*] [, *impute_options options*]

impute_options	Description
Main	
* add(#)	specify number of imputations to add; required when no imputations exist
* replace	replace imputed values in existing imputations
rseed(#)	specify random-number seed
double	save imputed values in double precision; the default is to save them as float
by(*varlist* [, *byopts*])	impute separately on each group formed by *varlist*
Reporting	
dots	display dots as imputations are performed
<u>noi</u>sily	display intermediate output
<u>nolegend</u>	suppress all table legends
Advanced	
force	proceed with imputation, even when missing imputed values are encountered
<u>noup</u>date	do not perform mi update; see [MI] **noupdate option**

*add(#) is required when no imputations exist; add(#) or replace is required if imputations exist.
noupdate does not appear in the dialog box.

options	Description
Main	
<u>nocons</u>tant	suppress constant term
<u>exp</u>osure(*varname_e*)	include ln(*varname_e*) in model with coefficient constrained to 1
<u>off</u>set(*varname_o*)	include *varname_o* in model with coefficient constrained to 1
<u>cond</u>itional(*if*)	perform conditional imputation
<u>boot</u>strap	estimate model parameters using sampling with replacement
Maximization	
maximize_options	control the maximization process; seldom used

You must mi set your data before using mi impute poisson; see [MI] **mi set**.

You must mi register *ivar* as imputed before using mi impute poisson; see [MI] **mi set**.

indepvars may contain factor variables; see [U] **11.4.3 Factor variables**.

fweights, iweights, and pweights are allowed; see [U] **11.1.6 weight**.

Menu

Statistics > Multiple imputation

Description

mi impute poisson fills in missing values of a count variable using a Poisson regression imputation method. You can perform separate imputations on different subsets of the data by specifying the by() option. You can also account for frequency, importance, and sampling weights.

Options

_____⌐ Main ⌐_____

noconstant; see [R] **estimation options**.

add(), replace, rseed(), double, by(); see [MI] **mi impute**.

exposure($varname_e$), offset($varname_o$); see [R] **estimation options**.

conditional(_if_) specifies that the imputation variable be imputed conditionally on observations satisfying _exp_; see [U] **11.1.3 if exp**. That is, missing values in a conditional sample, the sample identified by the _exp_ expression, are imputed based only on data in that conditional sample. Missing values outside the conditional sample are replaced with a conditional constant, the value of the imputation variable in observations outside the conditional sample. As such, the imputation variable is required to be constant outside the conditional sample. Also, if any conditioning variables (variables involved in the conditional specification if _exp_) contain soft missing values (.), their missing values must be nested within missing values of the imputation variables. See _Conditional imputation_ under _Remarks_ in [MI] **mi impute**.

bootstrap specifies that posterior estimates of model parameters be obtained using sampling with replacement; that is, posterior estimates are estimated from a bootstrap sample. The default is to sample the estimates from the posterior distribution of model parameters or from the large-sample normal approximation of the posterior distribution. This option is useful when asymptotic normality of parameter estimates is suspect.

_____⌐ Reporting ⌐_____

dots, noisily, nolegend; see [MI] **mi impute**. noisily specifies that the output from the Poisson regression fit to the observed data be displayed. nolegend suppresses all legends that appear before the imputation table. Such legends include a legend about conditional imputation that appears when the conditional() option is specified and group legends that may appear when the by() option is specified.

_____⌐ Maximization ⌐_____

_maximize_options_; see [R] **poisson**. These options are seldom used.

_____⌐ Advanced ⌐_____

force; see [MI] **mi impute**.

The following option is available with mi impute but is not shown in the dialog box:

noupdate; see [MI] **noupdate option**.

Remarks

Remarks are presented under the following headings:

Univariate imputation using Poisson regression
Using mi impute poisson

See [MI] **mi impute** for a general description and details about options common to all imputation methods, *impute_options*. Also see [MI] **workflow** for general advice on working with mi.

Univariate imputation using Poisson regression

The Poisson regression imputation method can be used to fill in missing values of a count variable (for example, Raghunathan et al. [2001] and van Buuren [2007]). It is a parametric method that assumes an underlying Poisson model for the imputed variable (given other predictors). For imputation of overdispersed count variables, see [MI] **mi impute nbreg**. The Poisson method is based on the asymptotic approximation of the posterior predictive distribution of the missing data.

Using mi impute poisson

To illustrate the use of mi impute poisson, we continue with our heart attack data analysis example in [MI] **intro substantive** and consider an additional predictor, npreg, which records the number of pregnancies:

```
. use http://www.stata-press.com/data/r12/mheartpois
(Fictional heart attack data; npreg missing)
. misstable summarize
```

					Obs<.	
Variable	Obs=.	Obs>.	Obs<.	Unique values	Min	Max
npreg	10		144	6	0	5

```
. tab female if npreg==.
```

Gender	Freq.	Percent	Cum.
Male	7	70.00	70.00
Female	3	30.00	100.00
Total	10	100.00	

According to misstable summarize, npreg is the only variable containing missing values, and it has 10 out of 154 observations missing. The tabulation of missing values of npreg by gender reveals that most missing values (7) correspond to males.

In this example, we could replace missing npreg for males with 0 and proceed with complete-data analysis, disregarding the remaining three missing observations. Instead, as an illustration, we use mi impute poisson to impute missing values of npreg. Our dataset is not declared yet, so we use mi set to declare it. We also use mi register to register npreg as the imputed variable before using mi impute poisson:

```
. mi set mlong

. mi register imputed npreg
(10 m=0 obs. now marked as incomplete)

. mi impute poisson npreg attack smokes age bmi hsgrad, add(20) conditional(if female==1)
Univariate imputation                    Imputations =        20
Poisson regression                             added =        20
Imputed: m=1 through m=20                     updated =         0

Conditional imputation:
  npreg: incomplete out-of-sample obs. replaced with value 0
```

		Observations per _m_		
Variable	Complete	Incomplete	Imputed	Total
npreg	144	10	10	154

```
(complete + incomplete = total; imputed is the minimum across m
of the number of filled-in observations.)
```

The npreg variable is relevant to females only, so we used the conditional() option to restrict imputation to observations with female==1; see *Conditional imputation* in [MI] **mi impute**.

We can analyze these multiply imputed data using logistic regression with mi estimate:

```
. mi estimate: logit attack smokes age bmi female hsgrad npreg
  (output omitted )
```

Saved results

mi impute poisson saves the following in r():

Scalars
r(M)	total number of imputations
r(M_add)	number of added imputations
r(M_update)	number of updated imputations
r(k_ivars)	number of imputed variables (always 1)
r(N_g)	number of imputed groups (1 if by() is not specified)

Macros
r(method)	name of imputation method (poisson)
r(ivars)	names of imputation variables
r(rseed)	random-number seed
r(by)	names of variables specified within by()

Matrices
r(N)	number of observations in imputation sample in each group
r(N_complete)	number of complete observations in imputation sample in each group
r(N_incomplete)	number of incomplete observations in imputation sample in each group
r(N_imputed)	number of imputed observations in imputation sample in each group

Methods and formulas

Consider a univariate variable $\mathbf{x} = (x_1, x_2, \ldots, x_n)'$ that follows a Poisson model

$$\Pr(x_i = x | \mathbf{z}_i) = \frac{e^{-\lambda_i} \lambda_i^x}{x!}, \ x = 0, 1, 2, \ldots \tag{1}$$

where $\lambda_i = \exp(\mathbf{z}_i'\boldsymbol{\beta}+\text{offset}_i)$ (see [R] **poisson**), $\mathbf{z}_i = (z_{i1}, z_{i2}, \ldots, z_{iq})'$ records values of predictors of $\mathbf{x}$ for observation i and $\boldsymbol{\beta}$ is the $q \times 1$ vector of unknown regression coefficients. (When a constant is included in the model—the default—$z_{i1} = 1$, $i = 1, \ldots, n$.)

$\mathbf{x}$ contains missing values that are to be filled in. Consider the partition of $\mathbf{x} = (\mathbf{x}_o', \mathbf{x}_m')$ into $n_0 \times 1$ and $n_1 \times 1$ vectors containing the complete and the incomplete observations. Consider a similar partition of $\mathbf{Z} = (\mathbf{Z}_o, \mathbf{Z}_m)$ into $n_0 \times q$ and $n_1 \times q$ submatrices.

`mi impute poisson` follows the steps below to fill in $\mathbf{x}_m$:

1. Fit a Poisson regression model (1) to the observed data $(\mathbf{x}_o, \mathbf{Z}_o)$ to obtain the maximum likelihood estimates, $\widehat{\boldsymbol{\beta}}$, and their asymptotic sampling variance, $\widehat{\mathbf{U}}$.

2. Simulate new parameters, $\boldsymbol{\beta}_\star$, from the large-sample normal approximation, $N(\widehat{\boldsymbol{\beta}}, \widehat{\mathbf{U}})$, to its posterior distribution assuming the noninformative prior $\Pr(\boldsymbol{\beta}) \propto \text{const}$.

3. Obtain one set of imputed values, $\mathbf{x}_m^1$, by simulating from a Poisson distribution (1) with $\lambda_i = \lambda_{i_m} = \exp(\mathbf{z}_{i_m}'\boldsymbol{\beta}_\star + \text{offset}_{i_m})$.

4. Repeat steps 2 and 3 to obtain M sets of imputed values $\mathbf{x}_m^1, \mathbf{x}_m^2, \ldots, \mathbf{x}_m^M$.

Steps 2 and 3 above correspond to only approximate draws from the posterior predictive distribution of the missing data, $\Pr(\mathbf{x}_m | \mathbf{x}_o, \mathbf{Z}_o)$, because $\boldsymbol{\beta}_\star$ is drawn from the asymptotic approximation to its posterior distribution.

If weights are specified, a weighted Poisson regression model is fit to the observed data in step 1 (see [R] **poisson** for details).

References

Raghunathan, T. E., J. M. Lepkowski, J. Van Hoewyk, and P. Solenberger. 2001. A multivariate technique for multiply imputing missing values using a sequence of regression models. *Survey Methodology* 27: 85–95.

van Buuren, S. 2007. Multiple imputation of discrete and continuous data by fully conditional specification. *Statistical Methods in Medical Research* 16: 219–242.

Also see

Title

mi impute regress — Impute using linear regression

Syntax

mi impute regress *ivar* [*indepvars*] [*if*] [*weight*] [, *impute_options options*]

impute_options	Description
Main	
* add(#)	specify number of imputations to add; required when no imputations exist
* replace	replace imputed values in existing imputations
rseed(#)	specify random-number seed
double	save imputed values in double precision; the default is to save them as float
by(varlist[, byopts])	impute separately on each group formed by varlist
Reporting	
dots	display dots as imputations are performed
noisily	display intermediate output
nolegend	suppress all table legends
Advanced	
force	proceed with imputation, even when missing imputed values are encountered
noupdate	do not perform mi update; see [MI] **noupdate option**

*add(#) is required when no imputations exist; add(#) or replace is required if imputations exist.
noupdate does not appear in the dialog box.

options	Description
Main	
noconstant	suppress constant term
conditional(if)	perform conditional imputation
bootstrap	estimate model parameters using sampling with replacement

You must mi set your data before using mi impute regress; see [MI] **mi set**.
You must mi register *ivar* as imputed before using mi impute regress; see [MI] **mi set**.
indepvars may contain factor variables; see [U] **11.4.3 Factor variables**.
aweights, fweights, iweights, and pweights are allowed; see [U] **11.1.6 weight**.

Menu

Statistics > Multiple imputation

Description

mi impute regress fills in missing values of a continuous variable using the Gaussian normal regression imputation method. You can perform separate imputations on different subsets of the data by specifying the by() option. You can also account for analytic, frequency, importance, and sampling weights.

Options

⌐ Main ⌐

noconstant; see [R] **estimation options**.

add(), replace, rseed(), double, by(); see [MI] **mi impute**.

conditional(*if*) specifies that the imputation variable be imputed conditionally on observations satisfying *exp*; see [U] **11.1.3 if exp**. That is, missing values in a conditional sample, the sample identified by the *exp* expression, are imputed based only on data in that conditional sample. Missing values outside the conditional sample are replaced with a conditional constant, the value of the imputation variable in observations outside the conditional sample. As such, the imputation variable is required to be constant outside the conditional sample. Also, if any conditioning variables (variables involved in the conditional specification if *exp*) contain soft missing values (.), their missing values must be nested within missing values of the imputation variables. See *Conditional imputation* under *Remarks* in [MI] **mi impute**.

bootstrap specifies that posterior estimates of model parameters be obtained using sampling with replacement; that is, posterior estimates are estimated from a bootstrap sample. The default is to sample the estimates from the posterior distribution of model parameters or from the large-sample normal approximation of the posterior distribution. This option is useful when asymptotic normality of parameter estimates is suspect.

⌐ Reporting ⌐

dots, noisily, nolegend; see [MI] **mi impute**. noisily specifies that the output from a linear regression fit to the observed data be displayed. nolegend suppresses all legends that appear before the imputation table. Such legends include a legend about conditional imputation that appears when the conditional() option is specified and group legends that may appear when the by() option is specified.

⌐ Advanced ⌐

force; see [MI] **mi impute**.

The following option is available with mi impute but is not shown in the dialog box:

noupdate; see [MI] **noupdate option**.

Remarks

Remarks are presented under the following headings:

> *Univariate imputation using linear regression*
> *Using mi impute regress*

See [MI] **mi impute** for a general description and details about options common to all imputation methods, *impute_options*. Also see [MI] **workflow** for general advice on working with mi.

Univariate imputation using linear regression

When a continuous variable contains missing values, a linear regression imputation method (or predictive mean matching; see [MI] **mi impute pmm**) can be used to fill in missing values (Rubin 1987; Schenker and Taylor 1996). The linear regression method is a fully parametric imputation method that relies on the normality of the model. Thus the imputation variable may need to be transformed from the original scale to meet the normality assumption prior to using mi impute regress.

The linear regression method is perhaps the most popular method for imputing quantitative variables. It is superior to other imputation methods when the underlying normal model holds. However, it can be more sensitive to violations of this assumption than other nonparametric and partially parametric imputation methods, such as predictive mean matching. For example, Schenker and Taylor (1996) studied the sensitivity of the regression method to the misspecification of the regression function and error distribution. They found that this method still performs well in the presence of heteroskedasticity and when the error distribution is heavier-tailed than the normal. However, it resulted in increased bias and variances under a misspecified regression function.

Using mi impute regress

Recall the heart attack data from *Univariate imputation* of [MI] **mi impute**. We wish to fit a logistic regression of attack on some predictors, one of which (bmi) has missing values. To avoid losing information contained in complete observations of the other predictors, we impute bmi.

The distribution of BMI is slightly skewed to the right, so we choose to fill in missing values of BMI on a log-transformed scale here. To do that, we need to create a new variable, lnbmi, containing the log of bmi and impute it:

```
. use http://www.stata-press.com/data/r12/mheart0
(Fictional heart attack data; bmi missing)
. generate lnbmi = ln(bmi)
(22 missing values generated)
. mi set mlong
. mi register imputed lnbmi
(22 m=0 obs. now marked as incomplete)
```

Following the steps in *Imputing transformations of incomplete variables* of [MI] **mi impute**, we create the imputed variable lnbmi containing the log of bmi and register it as imputed. We omitted the step of eliminating possible ineligible missing values in lnbmi because bmi ranges from 17 to 38 and we do not anticipate any extra (algebraic) missing from the operation ln(bmi).

We now use mi impute to impute missing values of lnbmi. We create 20 imputations and specify a random-number seed for reproducibility:

```
. mi impute regress lnbmi attack smokes age hsgrad female, add(20) rseed(2232)
Univariate imputation                    Imputations =        20
Linear regression                              added =        20
Imputed: m=1 through m=20                     updated =         0
```

	Observations per m			
Variable	Complete	Incomplete	Imputed	Total
lnbmi	132	22	22	154

```
(complete + incomplete = total; imputed is the minimum across m
 of the number of filled-in observations.)
```

From the output, all 22 incomplete values of lnbmi are imputed.

We want to use BMI in its original scale in the analysis. To do that, we need to replace bmi with exponentiated lnbmi. Because bmi now is a function of the imputed variable, it becomes a passive variable:

```
. mi register passive bmi
. quietly mi passive: replace bmi = exp(lnbmi)
```

Finally, we fit the logistic regression:

```
. mi estimate, dots: logit attack smokes age bmi hsgrad female
Imputations (20):
.........10.........20 done

Multiple-imputation estimates          Imputations       =          20
Logistic regression                    Number of obs     =         154
                                       Average RVI       =      0.0385
                                       Largest FMI       =      0.1610
DF adjustment:    Large sample         DF:       min     =      753.42
                                                 avg     =   144093.21
                                                 max     =   425800.51
Model F test:     Equal FMI            F(  5,47946.1)    =        3.75
Within VCE type:       OIM             Prob > F          =      0.0022
```

attack	Coef.	Std. Err.	t	P>|t|	[95% Conf. Interval]	
smokes	1.23917	.3629712	3.41	0.001	.52775	1.950589
age	.0355306	.0154994	2.29	0.022	.0051521	.0659091
bmi	.1192117	.0495795	2.40	0.016	.0218814	.216542
hsgrad	.1863652	.4075066	0.46	0.647	-.6123403	.9850706
female	-.0996965	.4189836	-0.24	0.812	-.9208917	.7214986
_cons	-5.868808	1.723036	-3.41	0.001	-9.248163	-2.489453

We obtain results comparable with those from [MI] **intro substantive**.

Saved results

mi impute regress saves the following in r():

Scalars
r(M)	total number of imputations
r(M_add)	number of added imputations
r(M_update)	number of updated imputations
r(k_ivars)	number of imputed variables (always 1)
r(N_g)	number of imputed groups (1 if by() is not specified)

Macros
r(method)	name of imputation method (regress)
r(ivars)	names of imputation variables
r(rseed)	random-number seed
r(by)	names of variables specified within by()

Matrices
r(N)	number of observations in imputation sample in each group
r(N_complete)	number of complete observations in imputation sample in each group
r(N_incomplete)	number of incomplete observations in imputation sample in each group
r(N_imputed)	number of imputed observations in imputation sample in each group

Methods and formulas

Consider a univariate variable $\mathbf{x} = (x_1, x_2, \ldots, x_n)'$ that follows a normal linear regression model

$$x_i | \mathbf{z}_i \sim N(\mathbf{z}_i' \boldsymbol{\beta}, \sigma^2) \tag{1}$$

where $\mathbf{z}_i = (z_{i1}, z_{i2}, \ldots, z_{iq})'$ records values of predictors of $\mathbf{x}$ for observation i, $\boldsymbol{\beta}$ is the $q \times 1$ vector of unknown regression coefficients, and σ^2 is the unknown scalar variance. (Note that when a constant is included in the model—the default—$z_{i1} = 1$, $i = 1, \ldots, n$.)

$\mathbf{x}$ contains missing values that are to be filled in. Consider the partition of $\mathbf{x} = (\mathbf{x}_o', \mathbf{x}_m')$ into $n_0 \times 1$ and $n_1 \times 1$ vectors containing the complete and the incomplete observations. Consider a similar partition of $\mathbf{Z} = (\mathbf{Z}_o, \mathbf{Z}_m)$ into $n_0 \times q$ and $n_1 \times q$ submatrices.

`mi impute regress` follows the steps below to fill in $\mathbf{x}_m$ (for simplicity, we omit the conditioning on the observed data in what follows):

1. Fit a regression model (1) to the observed data $(\mathbf{x}_o, \mathbf{Z}_o)$ to obtain estimates $\widehat{\boldsymbol{\beta}}$ and $\widehat{\sigma}^2$ of the model parameters.

2. Simulate new parameters $\boldsymbol{\beta}_\star$ and $\sigma_\star^2$ from their joint posterior distribution under the conventional noninformative improper prior $\Pr(\boldsymbol{\beta}, \sigma^2) \propto 1/\sigma^2$. This is done in two steps:

$$\sigma_\star^2 \sim \widehat{\sigma}^2 (n_0 - q) / \chi_{n_0 - q}^2$$
$$\boldsymbol{\beta}_\star | \sigma_\star^2 \sim N \left\{ \widehat{\boldsymbol{\beta}}, \sigma_\star^2 (\mathbf{Z}_o' \mathbf{Z}_o)^{-1} \right\}$$

3. Obtain one set of imputed values, $\mathbf{x}_m^1$, by simulating from $N(\mathbf{Z}_m \boldsymbol{\beta}_\star, \sigma_\star^2 I_{n_1 \times n_1})$.

4. Repeat steps 2 and 3 to obtain M sets of imputed values, $\mathbf{x}_m^1, \mathbf{x}_m^2, \ldots, \mathbf{x}_m^M$.

Steps 2 and 3 above correspond to simulating from the posterior predictive distribution of the missing data $\Pr(\mathbf{x}_m | \mathbf{x}_o, \mathbf{Z}_o)$ (for example, see Gelman et al. [2004, 355–358]).

If weights are specified, a weighted linear regression model is fit to the observed data in step 1 (see [R] **regress** for details). Also, in the case of `aweights`, $\sigma_\star^2 I_{n_1 \times n_1}$ is replaced with $\sigma_\star^2 \mathbf{W}_{n_1 \times n_1}^{-1}$ in step 3, where $\mathbf{W} = \text{diag}(w_i)$ and w_i is the analytic weight for observation i.

References

Gelman, A., J. B. Carlin, H. S. Stern, and D. B. Rubin. 2004. *Bayesian Data Analysis*. 2nd ed. London: Chapman & Hall/CRC.

Rubin, D. B. 1987. *Multiple Imputation for Nonresponse in Surveys*. New York: Wiley.

Schenker, N., and J. M. G. Taylor. 1996. Partially parametric techniques for multiple imputation. *Computational Statistics & Data Analysis* 22: 425–446.

Also see

[MI] **mi impute** — Impute missing values

[MI] **mi impute pmm** — Impute using predictive mean matching

[MI] **mi impute truncreg** — Impute using truncated regression

[MI] **mi impute intreg** — Impute using interval regression

[MI] **mi estimate** — Estimation using multiple imputations

[MI] **intro** — Introduction to mi

[MI] **intro substantive** — Introduction to multiple-imputation analysis

Title

mi impute truncreg — Impute using truncated regression

Syntax

mi impute truncreg *ivar* [*indepvars*] [*if*] [*weight*] [, *impute_options options*]

impute_options	Description
Main	
add(#)	specify number of imputations to add; required when no imputations exist
*replace	replace imputed values in existing imputations
rseed(#)	specify random-number seed
double	save imputed values in double precision; the default is to save them as float
by(*varlist*[, *byopts*])	impute separately on each group formed by *varlist*
Reporting	
dots	display dots as imputations are performed
noisily	display intermediate output
nolegend	suppress all table legends
Advanced	
force	proceed with imputation, even when missing imputed values are encountered
noupdate	do not perform mi update; see [MI] **noupdate option**

*add(#) is required when no imputations exist; add(#) or replace is required if imputations exist.
noupdate does not appear in the dialog box.

options	Description
Main	
noconstant	suppress constant term
ll(*varname* | #)	lower limit for left-truncation
ul(*varname* | #)	upper limit for right-truncation
offset(*varname_o*)	include *varname_o* in model with coefficient constrained to 1
conditional(*if*)	perform conditional imputation
bootstrap	estimate model parameters using sampling with replacement
Maximization	
maximize_options	control the maximization process; seldom used

You must mi set your data before using mi impute truncreg; see [MI] **mi set**.
You must mi register *ivar* as imputed before using mi impute truncreg; see [MI] **mi set**.
indepvars may contain factor variables; see [U] **11.4.3 Factor variables**.
aweights, fweights, iweights, and pweights are allowed; see [U] **11.1.6 weight**.

Menu

Statistics > Multiple imputation

Description

mi impute truncreg fills in missing values of a continuous variable with a restricted range using a truncated regression imputation method. You can perform separate imputations on different subsets of the data by specifying the by() option. You can also account for analytic, frequency, importance, and sampling weights.

Options

———⌐ Main ⌐——

noconstant; see [R] **estimation options**.

add(), replace, rseed(), double, by(); see [MI] **mi impute**.

ll(*varname* | #) and ul(*varname* | #) indicate the lower and upper limits for truncation, respectively. You may specify one or both. Observations with $ivar \leq$ ll() are left-truncated, observations with $ivar \geq$ ul() are right-truncated, and the remaining observations are not truncated.

offset(*varname$_o$*); see [R] **estimation options**.

conditional(*if*) specifies that the imputation variable be imputed conditionally on observations satisfying *exp*; see [U] **11.1.3 if exp**. That is, missing values in a conditional sample, the sample identified by the *exp* expression, are imputed based only on data in that conditional sample. Missing values outside the conditional sample are replaced with a conditional constant, the value of the imputation variable in observations outside the conditional sample. As such, the imputation variable is required to be constant outside the conditional sample. Also, if any conditioning variables (variables involved in the conditional specification if *exp*) contain soft missing values (.), their missing values must be nested within missing values of the imputation variables. See *Conditional imputation* under *Remarks* in [MI] **mi impute**.

bootstrap specifies that posterior estimates of model parameters be obtained using sampling with replacement; that is, posterior estimates are estimated from a bootstrap sample. The default is to sample the estimates from the posterior distribution of model parameters or from the large-sample normal approximation of the posterior distribution. This option is useful when asymptotic normality of parameter estimates is suspect.

———⌐ Reporting ⌐——

dots, noisily, nolegend; see [MI] **mi impute**. noisily specifies that the output from the truncated regression fit to the observed data be displayed. nolegend suppresses all legends that appear before the imputation table. Such legends include a legend about conditional imputation that appears when the conditional() option is specified and group legends that may appear when the by() option is specified.

———⌐ Maximization ⌐——

maximize_options; see [R] **truncreg**. These options are seldom used.

Advanced
force; see [MI] **mi impute**.

The following option is available with mi impute but is not shown in the dialog box:

noupdate; see [MI] **noupdate option**.

Remarks

Remarks are presented under the following headings:

Univariate imputation using truncated regression
Using mi impute truncreg

See [MI] **mi impute** for a general description and details about options common to all imputation methods, *impute_options*. Also see [MI] **workflow** for general advice on working with mi.

Univariate imputation using truncated regression

The truncated regression imputation method can be used to fill in missing values of a continuous variable with a restricted range (for example, Raghunathan et al. [2001] and Schafer [1997, 203]). It is a parametric method that assumes an underlying truncated normal model for the imputed variable (given other predictors). This method is based on the asymptotic approximation of the posterior predictive distribution of the missing data.

Similar to estimation, it is important to distinguish between truncation and censoring when imputing continuous variables with a limited range. Truncation arises when the distribution of a variable of interest is restricted to a certain range—a truncated distribution. The probability that the variable takes on values outside that range is zero. Truncated data may arise naturally (for example, SAT section scores may not exceed 800) or may be the result of a particular study design (for example, only subjects with income below a certain threshold are of interest in the study). See [R] **truncreg** for more details.

Use mi impute intreg (see [MI] **mi impute intreg**) to impute continuous partially observed (censored) variables.

Using mi impute truncreg

In [MI] **mi impute pmm**, we used predictive mean matching to impute missing values of bmi (used as a predictor in the logistic analysis of heart attacks as described in [MI] **intro substantive**), restricting imputed values to be within the observed range of bmi.

mi impute pmm imputes missing values of bmi, replacing them only with values already observed in the data. Suppose that, instead, we want to allow imputed bmi values to take on any value within a certain range. We can achieve this by using mi impute truncreg.

```
. use http://www.stata-press.com/data/r12/mheart0
(Fictional heart attack data; bmi missing)

. summarize bmi
```

Variable	Obs	Mean	Std. Dev.	Min	Max
bmi	132	25.24136	4.027137	17.22643	38.24214

The observed range of bmi in our data is between roughly 17 and 39.

We impute `bmi` from a normal distribution truncated at (17, 39):

```
. mi set mlong
. mi register imputed bmi
(22 m=0 obs. now marked as incomplete)
. mi impute truncreg bmi attack smokes age hsgrad female, add(20) ll(17) ul(39)
Univariate imputation                          Imputations =       20
Truncated regression                                 added =       20
Imputed: m=1 through m=20                           updated =        0

Limit: lower =             17        Number truncated =        0
       upper =             39                     left =        0
                                                 right =        0
```

	Observations per *m*			
Variable	Complete	Incomplete	Imputed	Total
bmi	132	22	22	154

```
(complete + incomplete = total; imputed is the minimum across m
 of the number of filled-in observations.)
```

`mi impute truncreg` reports in the output header the truncation limits used (17 and 39 in our example). If the `ll()` and `ul()` options are not specified, the truncation limits are displayed as `-inf` and `+inf`, respectively, and the imputation model becomes equivalent to that using (unrestricted) normal linear regression.

`mi impute truncreg` also reports numbers of truncated observations. In our example, all values of `bmi` lie between 17 and 39, so there are no truncated observations. Truncated observations are not used during estimation; see [R] **truncreg**.

Rather than restricting `bmi` to the observed range during imputation, it may be reasonable to assume a wider range that is still consistent with the observed dataset. It may also be reasonable to use different ranges for males and females. For example, considering the observed ages, suppose that we assume a normal distribution for `bmi` truncated at (14, 55) for females and at (17, 50) for males.

To accommodate varying ranges, we first create variables containing gender-specific truncation limits:

```
. qui mi xeq: gen lbmi = cond(female==1, 14, 17)
. qui mi xeq: gen ubmi = cond(female==1, 55, 50)
```

The declared style of our `mi` data is `mlong`, so it is not necessary to use the `mi xeq` prefix for generating new variables. It is good practice, however, to use `mi`-specific commands so that your data manipulation is appropriate no matter what the `mi` style is; see [MI] **mi xeq** and [MI] **styles** for details.

We now replace the existing imputations with new ones, which account for varying ranges of bmi among males and females:

```
. mi impute truncreg bmi attack smokes age hsgrad female, replace ll(lbmi) ul(ubmi)
Univariate imputation              Imputations =    20
Truncated regression                     added =     0
Imputed: m=1 through m=20              updated =    20

Limit: lower =       lbmi     Number truncated =     0
       upper =       ubmi                 left =     0
                                         right =     0
```

		Observations per m			
Variable	Complete	Incomplete	Imputed		Total
bmi	132	22	22		154

(complete + incomplete = total; imputed is the minimum across m of the number of filled-in observations.)

We can analyze these multiply imputed data using logistic regression with mi estimate:

```
. mi estimate: logit attack smokes age bmi female hsgrad
  (output omitted )
```

Saved results

mi impute truncreg saves the following in r():

Scalars
r(M)	total number of imputations
r(M_add)	number of added imputations
r(M_update)	number of updated imputations
r(k_ivars)	number of imputed variables (always 1)
r(N_trunc)	number of truncated observations
r(N_ltrunc)	number of left-truncated observations
r(N_rtrunc)	number of right-truncated observations
r(ll)	lower truncation limit (if ll(#) is specified)
r(ul)	upper truncation limit (if ul(#) is specified)
r(N_g)	number of imputed groups (1 if by() is not specified)

Macros
r(method)	name of imputation method (truncreg)
r(ivars)	names of imputation variables
r(llopt)	contents of ll(), if specified
r(ulopt)	contents of ul(), if specified
r(rseed)	random-number seed
r(by)	names of variables specified within by()

Matrices
r(N)	number of observations in imputation sample in each group
r(N_complete)	number of complete observations in imputation sample in each group
r(N_incomplete)	number of incomplete observations in imputation sample in each group
r(N_imputed)	number of imputed observations in imputation sample in each group

Methods and formulas

Consider a univariate variable $\mathbf{x} = (x_1, x_2, \ldots, x_n)'$ that follows a truncated normal model with the density

$$f_{(a,b)}(x|\mathbf{z}_i) = \frac{\frac{1}{\sigma}\phi\left(\frac{x-\mu_i}{\sigma}\right)}{\Phi\left(\frac{b-\mu_i}{\sigma}\right) - \Phi\left(\frac{a-\mu_i}{\sigma}\right)}, \; a < x < b \qquad (1)$$

where $\phi(\cdot)$ and $\Phi(\cdot)$ are the standard normal density and cumulative distribution functions, respectively, $\mu_i = \mathbf{z}_i'\boldsymbol{\beta}$, $\mathbf{z}_i = (z_{i1}, z_{i2}, \ldots, z_{iq})'$ records values of predictors of $\mathbf{x}$ for observation i, $\boldsymbol{\beta}$ is the $q \times 1$ vector of unknown regression coefficients, σ^2 is the unknown scalar variance, and a and b are the respective known lower and upper truncation limits; also see [R] **truncreg**. (When a constant is included in the model—the default—$z_{i1} = 1$, $i = 1, \ldots, n$.)

$\mathbf{x}$ contains missing values that are to be filled in. Consider the partition of $\mathbf{x} = (\mathbf{x}_o', \mathbf{x}_m')$ into $n_0 \times 1$ and $n_1 \times 1$ vectors containing the complete and the incomplete observations. Consider a similar partition of $\mathbf{Z} = (\mathbf{Z}_o, \mathbf{Z}_m)$ into $n_0 \times q$ and $n_1 \times q$ submatrices.

`mi impute truncreg` follows the steps below to fill in $\mathbf{x}_m$:

1. Fit a truncated regression (1) to the observed data $(\mathbf{x}_o, \mathbf{Z}_o)$ to obtain the maximum likelihood estimates, $\widehat{\boldsymbol{\theta}} = (\widehat{\boldsymbol{\beta}}', \ln\widehat{\sigma})'$, and their asymptotic sampling variance, $\widehat{\mathbf{U}}$.

2. Simulate new parameters, $\boldsymbol{\theta}_\star$, from the large-sample normal approximation, $N(\widehat{\boldsymbol{\theta}}, \widehat{\mathbf{U}})$, to its posterior distribution assuming the noninformative prior $\Pr(\boldsymbol{\theta}) \propto \text{const}$.

3. Obtain one set of imputed values, $\mathbf{x}_m^1$, by simulating from a truncated normal model (1) with parameters set to their simulated values from step 2: $\boldsymbol{\beta} = \boldsymbol{\beta}_\star$ and $\sigma = \sigma_\star$.

4. Repeat steps 2 and 3 to obtain M sets of imputed values, $\mathbf{x}_m^1, \mathbf{x}_m^2, \ldots, \mathbf{x}_m^M$.

Steps 2 and 3 above correspond to only approximate draws from the posterior predictive distribution of the missing data, $\Pr(\mathbf{x}_m|\mathbf{x}_o, \mathbf{Z}_o)$, because $\boldsymbol{\theta}_\star$ is drawn from the asymptotic approximation to its posterior distribution.

If weights are specified, a weighted regression model is fit to the observed data in step 1 (see [R] **truncreg** for details). Also, in the case of `aweights`, $\sigma_\star$ is replaced with $\sigma_\star w_i^{-1/2}$ in step 3, where w_i is the analytic weight for observation i.

References

Raghunathan, T. E., J. M. Lepkowski, J. Van Hoewyk, and P. Solenberger. 2001. A multivariate technique for multiply imputing missing values using a sequence of regression models. *Survey Methodology* 27: 85–95.

Schafer, J. L. 1997. *Analysis of Incomplete Multivariate Data*. Boca Raton, FL: Chapman & Hall/CRC.

Also see

[MI] **mi impute** — Impute missing values

[MI] **mi impute pmm** — Impute using predictive mean matching

[MI] **mi impute regress** — Impute using linear regression

[MI] **mi impute intreg** — Impute using interval regression

[MI] **mi estimate** — Estimation using multiple imputations

[MI] **intro** — Introduction to mi

[MI] **intro substantive** — Introduction to multiple-imputation analysis

Title

mi merge — Merge mi data

Syntax

mi merge 1:1 *varlist* using *filename* [, *options*]

mi merge m:1 *varlist* using *filename* [, *options*]

mi merge 1:m *varlist* using *filename* [, *options*]

mi merge m:m *varlist* using *filename* [, *options*]

options	Description
Main	
generate(*newvar*)	create *newvar* recording how observations matched
nolabel	do not copy value-label definitions from using
nonotes	do not copy notes from using
force	allow string/numeric variable type mismatch without error
Results	
assert(*results*)	require observations to match as specified
keep(*results*)	results to keep
noreport	do not display result summary table
noupdate	see [MI] **noupdate option**

Notes:

1. Jargon:
    match variables = *varlist*, variables on which match performed
    master = data in memory
    using = data on disk (*filename*)

2. Master must be mi set; using may be mi set.

3. mi merge is syntactically and logically equivalent to merge (see [D] **merge**).

4. mi merge syntactically differs from merge in that the nogenerate, sorted, keepusing(), update, and replace options are not allowed. Also, no _merge variable is created unless the generate() option is specified.

5. *filename* must be enclosed in double quotes if *filename* contains blanks or other special characters.

Menu

Statistics > Multiple imputation

257

Description

mi merge is merge for mi data; see [D] **merge** for a description of merging datasets.

It is recommended that the match variables (*varlist* in the syntax diagram) not include imputed or passive variables, or any varying or super-varying variables. If they do, the values of the match variables in $m = 0$ will be used to control the merge even in $m = 1$, $m = 2$, ..., $m = M$. Thus $m = 0$, $m = 1$, ..., $m = M$ will all be merged identically, and there will continue to be a one-to-one correspondence between the observations in $m = 0$ with the observations in each of $m > 0$.

Options

    ⌐ Main ⌐

generate(*newvar*) creates new variable *newvar* containing the match status of each observation in the resulting data. The codes are 1, 2, and 3 from the table below.

nolabel prevents copying the value-label definitions from the using data to the master. Even if you do not specify this option, label definitions from the using never replace those of the master.

nonotes prevents any notes in the using from being incorporated into the master; see [D] **notes**.

force allows string/numeric variable type mismatches, resulting in missing values from the using dataset. If omitted, mi merge issues an error message; if specified, mi merge issues a warning message.

    ⌐ Results ⌐

assert(*results*) specifies how observations should match. If results are not as you expect, an error message will be issued and the master data left unchanged.

Code	Word	Description
1	master	observation appeared in master only
2	using	observation appeared in using only
3	match	observation appeared in both

(Numeric codes and words are equivalent; you may use either.)

assert(match) specifies that all observations in both the master and the using are expected to match, and if that is not so, an error message is to be issued. assert(match master) means that all observations match or originally appeared only in the master. See [D] **merge** for more information.

keep(*results*) specifies which observations are to be kept from the merged dataset. keep(match) would specify that only matches are to be kept.

noreport suppresses the report that mi merge ordinarily presents.

noupdate in some cases suppresses the automatic mi update this command might perform; see [MI] **noupdate option**.

Remarks

Use mi merge when you would use merge if the data were not mi.

Remarks are presented under the following headings:

    Merging with non-mi data
    Merging with mi data
    Merging with mi data containing overlapping variables

Merging with non-mi data

Assume that file `ipats.dta` contains data on the patients in the ICU of a local hospital. The data are `mi set`, $M = 5$, and missing values have been imputed. File `nurses.dta` contains information on nurses and is not `mi` data. You wish to add the relevant nurse information to each patient. Type

```
. use ipats, clear
. mi merge m:1 nurseid using nurses, keep(master)
```

The resulting data are still `mi set` with $M = 5$. The new variables are unregistered.

Merging with mi data

Now assume the same situation as above except this time `nurses.dta` is `mi` data. Some of the nurse variables have missing values, and those values have been imputed. M is 6. To combine the datasets, you type the same as you would have typed before:

```
. use ipats, clear
. mi merge m:1 nurseid using nurses, keep(master)
```

Remember, $M = 5$ in `ipats.dta` and $M = 6$ in `nurses.dta`. The resulting data have $M = 6$, the larger value. There are missing values in the patient variables in $m = 6$, so we need to either impute them or drop the extra imputation by typing `mi set M = 5`.

Merging with mi data containing overlapping variables

Now assume the situation as directly above but this time `nurses.dta` contains variables other than `nurseid` that also appear in `ipats.dta`. Such variables—variables in common that are not used as matching variables—are called overlapping variables. Assume `seniornurse` is such a variable. Let's imagine that `seniornurse` has no missing values and is unregistered in `ipats.dta`, but does have missing values and is registered as imputed in `nurses.dta`.

You will want `seniornurse` registered as imputed if merging `nurses.dta` adds new observations that have `seniornurse` equal to missing. On the other hand, if none of the added observations has `seniornurse` equal to missing, then you will want the variable left unregistered. And that is exactly what `mi merge` does. That is,

- Variables unique to the master will be registered according to how they were registered in the master.

- Variables unique to the using will be registered according to how they were registered in the using.

- Variables that overlap will be registered according to how they were in the master if there are no unmatched using observations in the final result.

- If there are such unmatched using observations in the final result, then the unique variables that do not contain missing in the unmatched-and-kept observations will be registered according to how they were registered in the master. So will all variables registered as imputed in the master.

- Variables that do contain missing in the unmatched-and-kept observations will be registered as imputed if they were registered as imputed in the using data or as passive if they were registered as passive in the using data.

Thus variables might be registered differently if we typed

```
. mi merge m:1 nurseid using nurses, keep(master)
```

rather than

```
. mi merge m:1 nurseid using nurses, gen(howmatch)
. keep if howmatch==3
```

If you want to keep the matched observations, it is better to specify merge's keep() option.

Saved results

mi merge saves the following in r():

Scalars
    r(N_master) number of observations in $m=0$ in master
    r(N_using)   number of observations in $m=0$ in using
    r(N_result) number of observations in $m=0$ in result

    r(M_master) number of imputations (M) in master
    r(M_using)   number of imputations (M) in using
    r(M_result) number of imputations (M) in result

Macros
    r(newvars)   new variables added

Thus values in the resulting data are

$$N = \text{\# of observations in } m = 0$$
$$= \texttt{r(N\_result)}$$

$$k = \text{\# of variables}$$
$$= k\_master + \text{`:word count `r(newvars)''}$$

$$M = \text{\# of imputations}$$
$$= \texttt{max(r(M\_master), r(M\_using))}$$
$$= \texttt{r(M\_result)}$$

Also see

[MI] **intro** — Introduction to mi

[D] **merge** — Merge datasets

[MI] **mi append** — Append mi data

Title

mi misstable — Tabulate pattern of missing values

Syntax

mi <u>misstable</u> <u>summ</u>arize [*varlist*] [*if*] [, *options*]

mi <u>misstable</u> <u>pat</u>terns [*varlist*] [*if*] [, *options*]

mi <u>misstable</u> tree [*varlist*] [*if*] [, *options*]

mi <u>misstable</u> <u>nest</u>ed [*varlist*] [*if*] [, *options*]

options	Description
Main	
exmiss	treat .a, .b, ..., .z as missing
m(#)	run misstable on $m = \#$; default is $m = 0$
other_options	see [R] **misstable** (generate() is not allowed; exok is assumed)
nopreserve	programmer's option; see [P] **nopreserve option**

Menu

Statistics > Multiple imputation

Description

mi misstable runs misstable on $m = 0$ or on $m = \#$ if the m(#) option is specified. misstable makes tables to help in understanding the pattern of missing values in your data; see [R] **misstable**.

Options

⌐ Main ⌐

exmiss specifies that the extended missing values, .a, .b, ..., .z, are to be treated as missing. misstable treats them as missing by default and has the exok option to treat them as nonmissing. mi misstable turns that around and has the exmiss option.

In the mi system, extended missing values that are recorded in imputed variables indicate values not to be imputed and thus are, in a sense, not missing, or more accurately, missing for a good and valid reason.

The exmiss option is intended for use with the patterns, tree, and nested subcommands. You may specify exmiss with the summarize subcommand, but the option is ignored because summarize reports both extended and system missing in separate columns.

m(#) specifies the imputation dataset on which misstable is to be run. The default is $m = 0$, the original data.

other_options are allowed; see [R] **misstable**.

Remarks

See [R] **misstable**.

Saved results

See [R] **misstable**.

Also see

[MI] **intro** — Introduction to mi

[R] **misstable** — Tabulate missing values

[MI] **mi varying** — Identify variables that vary across imputations

Title

mi passive — Generate/replace and register passive variables

Syntax

mi <u>pass</u>ive: { <u>g</u>enerate | egen | replace } ...

mi <u>pass</u>ive: by *varlist*: { <u>g</u>enerate | egen | replace } ...

The full syntax is

mi <u>pass</u>ive[, *options*]: [by *varlist* [(*varlist*)] :] { <u>g</u>enerate | egen | replace } ...

options	Description
<u>noup</u>date	see [MI] **noupdate option**
nopreserve	do not first preserve

Also see [D] **generate** and [D] **egen**.

Menu

Statistics > Multiple imputation

Description

mi passive creates and registers passive variables or replaces the contents of existing passive variables.

More precisely, mi passive executes the specified generate, egen, or replace command on each of $m = 0$, $m = 1$, ..., $m = M$; see [D] **generate** and [D] **egen**. If the command is generate or egen, then mi passive registers the new variable as passive. If the command is replace, then mi passive verifies that the variable is already registered as passive.

Options

<u>noup</u>date in some cases suppresses the automatic mi update this command might perform; see [MI] **noupdate option**.

nopreserve is a programmer's option. It specifies that mi passive is not to preserve the data if it ordinarily would. This is used by programmers who have already preserved the data before calling mi passive.

Remarks

Remarks are presented under the following headings:

> *mi passive basics*
> *mi passive works with the by prefix*
> *mi passive works fastest with the wide style*
> *mi passive and super-varying variables*
> *Renaming passive variables*
> *Dropping passive variables*
> *Update passive variables when imputed values change*
> *Alternatives to mi passive*

mi passive basics

A passive variable is a variable that is a function of imputed variables or of other passive variables. For instance, if variable age were imputed and you created lnage from it, the lnage variable would be passive. The right way to create lnage is to type

. mi passive: generate lnage = ln(age)

Simply typing

. generate lnage = ln(age)

is not sufficient because that would create lnage in the $m = 0$ data, and age, being imputed, varies across m. There are situations where omitting the mi passive prefix would be almost sufficient, namely, when the data are mlong or flong style, but even then you would need to follow up by typing mi register passive lnage.

To create passive variables or to change the values of existing passive variables, use mi passive. Passive variables cannot be super-varying; see *mi passive and super-varying variables.*

mi passive works with the by prefix

You can use mi passive with the by prefix. For instance, you can type

. mi passive: by person: generate totaltodate = sum(amount)

You do not need to sort the data before issuing either of these commands, nor are you required to specify by's sort option. mi passive handles sorting issues for you.

Use by's parenthetical syntax to specify the order within by, if that is necessary. For instance,

. mi passive: by person (time): generate lastamount = amount[_n-1]

Do not omit the parenthetical time and instead attempt to sort the data yourself:

. sort person time
. mi passive: by person: generate lastamount = amount[_n-1]

Sorting the data yourself will work if your data happen to be wide style; it will not work in general.

mi passive works fastest with the wide style

mi passive works with any style, but it works fastest when the data are wide style. If you are going to issue multiple mi passive commands, you can usually speed execution by first converting your data to the wide style; see [MI] **mi convert**.

mi passive and super-varying variables

You should be careful not to mistakenly use mi passive to create super-varying variables. Super-varying variables cannot be passive variables because the values of a super-varying variable differ not only in the incomplete observations but also in the complete observations across imputations.

As noted in [MI] **mi set**, super-varying variables should never be registered. If a super-varying variable is registered as passive, it will be converted to a varying variable. All complete observations of the super-varying variable in each imputation will be replaced with their values from $m = 0$.

mi passive registers the created variable as passive. Even if the command you use with mi passive creates a super-varying variable, mi passive will convert it to varying, as described above.

You can use mi passive with any function that produces values that solely depend on values within the observation. In general, you cannot use mi passive with functions that produce values that depend on groups of observations.

For example, most egen functions result in super-varying variables. In such cases, you should use mi xeq: egen to create them and leave them unregistered; see [MI] **mi xeq**. You might thus conclude that you should never use mi passive with egen. That is not true, but it is nearly true. You may use mi passive with egen's rowmean() function, for instance, because it produces values that depend only on one observation at a time.

Renaming passive variables

Use mi rename (see [MI] **mi rename**) to rename all variables, not just passive variables:

. mi rename *oldname newname*

rename (see [D] **rename**) is insufficient for renaming passive variables regardless of the style of your data.

Dropping passive variables

Use drop (see [D] **drop**) to drop variables (or observations), but run mi update (see [MI] **mi update**) afterward.

. drop *var_or_vars*

. mi update

This advice applies for all variables, not just passive ones.

Update passive variables when imputed values change

Passive variables are not automatically updated when the values of the underlying imputed variables change.

If imputed values change or if you add more imputations, you must update or re-create the passive variables. If you have several passive variables, we suggest you make a do-file to create them. You can run the do-file again whenever necessary. A do-file to create lnage and totaltodate might read

```
                                                          begin cr_passive.do
  use mydata, clear

  capture drop lnage
  capture drop totaltodate
  mi update

  mi passive: generate lnage = ln(age)
  mi passive: by person (time): generate totaltodate = sum(amount)
                                                          end cr_passive.do
```

Alternatives to mi passive

mi passive can run any generate, replace, or egen command. If that is not sufficient to create the variable you need, you will have to create the variable for yourself. Here is how you do that:

1. If your data are wide or mlong, use mi convert (see [MI] **mi convert**) to convert them to one of the fully long styles, flong or flongsep, and then continue with the appropriate step below.

2. If your data are flong, mi system variable _mi_m records m. Create your new variable by using standard Stata commands, but do that by _mi_m. After creating the variable, mi register it as passive; see [MI] **mi set**.

3. If your data are flongsep, create the new variable in each of the $m = 0$, $m = 1$, ..., $m = M$ datasets, and then register the result. Start by working with a copy of your data:

   . mi copy *newname*

   The data in memory at this point correspond to $m = 0$. Create the new variable and then save the data:

   . *(create new_variable)*

   . save *newname*, replace

   Now use the $m = 1$ data and repeat the process:

   . use _1_*newname*

   . *(create new_variable)*

   . save _1_*newname*, replace

   Repeat for $m = 2$, $m = 3$, ..., $m = M$.

   At this point, the new variable is created but not yet registered. Reload the original $m = 0$ data, register the new variable as passive, and run mi update (see [MI] **mi update**):

   . use *newname*

   . register passive *new_variable*

   . mi update

Finally, copy the result back to your original flongsep data,

```
. mi copy name, replace
```

or if you started with mlong, flong, or wide data, then convert the data back to your preferred style:

```
. mi convert original_style
```

Either way, erase the *newname* flongsep dataset collection:

```
. mi erase newname
```

The third procedure can be tedious and error-prone if M is large. We suggest that you make a do-file to create the variable and then run it on each of the $m = 0$, $m = 1$, ..., $m = M$ datasets:

```
. mi copy newname

. do mydofile

. save newname, replace

. forvalues m=1(1)20 {            // we assume M=20
>      use _'m'_newname
>      do mydofile
>      save _'m'_newname, replace
> }

. use newname

. register passive new_variable

. mi update
```

Also see

[MI] **intro** — Introduction to mi

[MI] **mi reset** — Reset imputed or passive variables

[MI] **mi xeq** — Execute command(s) on individual imputations

Title

> **mi predict** — Obtain multiple-imputation predictions

Syntax

Obtain multiple-imputation linear predictions

> mi predict [*type*] *newvar* [*if*] using *miestfile* [, *predict_options options*]

Obtain multiple-imputation nonlinear predictions

> mi predictnl [*type*] *newvar* = *pnl_exp* [*if*] using *miestfile* [, *pnl_options options*]

miestfile.ster contains estimation results previously saved by mi estimate, saving(*miestfile*); see [MI] **mi estimate**.

pnl_exp is any valid Stata expression and may also contain calls to two special functions unique to predictnl: predict() and xb(); see [R] **predictnl** for details.

predict_options	Description
Predict options	
xb	calculate linear prediction; the default
stdp	calculate standard error of the prediction
nooffset	ignore any offset() or exposure() variable
equation(*eqno*)	specify equations after multiple-equation commands

pnl_options	Description
Predict options	
se(*newvar*)	create *newvar* containing standard errors
variance(*newvar*)	create *newvar* containing variances
wald(*newvar*)	create *newvar* containing the Wald test statistic
p(*newvar*)	create *newvar* containing the significance level (*p*-value) of the Wald test
ci(*newvars*)	create *newvars* containing lower and upper confidence intervals
level(#)	set confidence level; default is level(95)
bvariance(*newvar*)	create *newvar* containing between-imputation variances
wvariance(*newvar*)	create *newvar* containing within-imputation variances
df(*newvar*)	create *newvar* containing MI degrees of freedom
nosmall	do not apply small-sample correction to degrees of freedom
rvi(*newvar*)	create *newvar* containing relative variance increases
fmi(*newvar*)	create *newvar* containing fractions of missing information
re(*newvar*)	create *newvar* containing relative efficiencies

Advanced

<u>iter</u>ate(#)	maximum iterations for finding optimal step size to compute completed-data numerical derivatives of *pnl_exp*; default is 100
force	calculate completed-data standard errors, etc., even when possibly inappropriate

options	Description
MI options	
<u>ni</u>mputations(#)	specify number of imputations to use in computation; default is to use all saved imputations
<u>imp</u>utations(*numlist*)	specify which imputations to use in computation
<u>est</u>imations(*numlist*)	specify which estimation results to use in computation
esample(*varname*)	restrict the prediction to the estimation subsample identified by a binary variable *varname*
storecompleted	store completed-data predictions in the imputed data; available only in the flong and flongsep styles
Reporting	
replay	replay command-specific results from each individual estimation in *miestfile*.ster
cmdlegend	display the command legend
noupdate	do not perform mi update; see [MI] **noupdate option**
<u>noerr</u>notes	suppress error notes associated with failed estimation results in *miestfile*.ster
showimputations	show imputations saved in *miestfile*.ster

noupdate, noerrnotes, and showimputations do not appear in the dialog box.

Menu

Statistics > Multiple imputation

Description

mi predict using *miestfile* is for use after mi estimate, saving(*miestfile*): ... to obtain multiple-imputation (MI) linear predictions or their standard errors.

mi predictnl using *miestfile* is for use after mi estimate, saving(*miestfile*): ... to obtain MI (possibly) nonlinear predictions, their standard errors, and other statistics, including statistics specific to MI.

MI predictions, their standard errors, and other statistics are obtained by applying Rubin's combination rules observationwise to the completed-data predictions, predictions computed for each imputation (White, Royston, and Wood 2011). The results are stored in the original data ($m = 0$). See [R] **predict** and [R] **predictnl** for details about the computation of the completed-data predictions.

mi predict and mi predictnl may change the sort order of the data.

Options

Predict options

xb, stdp, nooffset, equation(*eqno*); see [R] **predict**.

se(*newvar*), variance(*newvar*), wald(*newvar*), p(*newvar*), ci(*newvars*), level(*#*); see [R] **predictnl**. These options store the specified MI statistics in variable *newvar* in the original data ($m = 0$). level() is relevant in combination with ci() only. If storecompleted is specified, then *newvar* contains the respective completed-data estimates in the imputed data ($m > 0$). Otherwise, *newvar* is missing in the imputed data.

bvariance(*newvar*) adds *newvar* of storage type *type*, where for each i in the prediction sample, *newvar*[i] contains the estimated between-imputation variance of *pnl_exp*[i]. storecompleted has no effect on bvariance().

wvariance(*newvar*) adds *newvar* of storage type *type*, where for each i in the prediction sample, *newvar*[i] contains the estimated within-imputation variance of *pnl_exp*[i]. storecompleted has no effect on wvariance().

df(*newvar*) adds *newvar* of storage type *type*, where for each i in the prediction sample, *newvar*[i] contains the estimated MI degrees of freedom of *pnl_exp*[i]. If storecompleted is specified, then *newvar* in the imputed data will contain the complete-data degrees of freedom as saved by mi estimate. In the absence of the complete-data degrees of freedom or if nosmall is used, then *newvar* is missing in the imputed data, even if storecompleted is specified.

nosmall specifies that no small-sample correction be made to the degrees of freedom. By default, the small-sample correction of Barnard and Rubin (1999) is used. This option has an effect on the results stored by p(), ci(), df(), fmi(), and re().

rvi(*newvar*) adds *newvar* of storage type *type*, where for each i in the prediction sample, *newvar*[i] contains the estimated relative variance increase of *pnl_exp*[i]. storecompleted has no effect on rvi().

fmi(*newvar*) adds *newvar* of storage type *type*, where for each i in the prediction sample, *newvar*[i] contains the estimated fraction of missing information of *pnl_exp*[i]. storecompleted has no effect on fmi().

re(*newvar*) adds *newvar* of storage type *type*, where for each i in the prediction sample, *newvar*[i] contains the estimated relative efficiency of *pnl_exp*[i]. storecompleted has no effect on re().

MI options

nimputations(*#*) specifies that the first # imputations be used; # must be $2 \leq \# \leq M$. The default is to use all imputations, M. Only one of nimputations(), imputations(), or estimations() may be specified.

imputations(*numlist*) specifies which imputations to use. The default is to use all of them. *numlist* must contain at least two numbers corresponding to the imputations saved in *miestfile*.ster. You can use the showimputations option to display imputations currently saved in *miestfile*.ster. Only one of nimputations(), imputations(), or estimations() may be specified.

estimations(*numlist*) does the same thing as imputations(*numlist*), but this time the imputations are numbered differently. Say that *miestfile*.ster was created by mi estimate and mi estimate was told to limit itself to imputations 1, 3, 5, and 9. With imputations(), the imputations are still numbered 1, 3, 5, and 9. With estimations(), they are numbered 1, 2, 3, and 4. Usually, one does not specify a subset of imputations when using mi estimate, and so usually, the imputations() and estimations() options are identical. The specified *numlist* must contain at least two numbers. Only one of `nimputations()`, `imputations()`, or `estimations()` may be specified.

`esample(`*varname*`)` restricts the prediction to the estimation sample identified by a binary variable *varname*. By default, predictions are obtained for all observations in the original data. Variable *varname* cannot be registered as imputed or passive and cannot vary across imputations.

`storecompleted` stores completed-data predictions in the newly created variables in each imputation. By default, the imputed data contain missing values in the newly created variables. The `store-completed` option may be specified only if the data are flong or flongsep; see [MI] **mi convert** to convert to one of those styles.

Reporting

`replay` replays estimation results from *miestfile*`.ster`, previously saved by `mi estimate, sav-ing(`*miestfile*`)`.

`cmdlegend` requests that the command line corresponding to the estimation command used to produce the estimation results saved in *miestfile*`.ster` be displayed.

Advanced

`iterate(#)`, `force`; see [R] **predictnl**.

The following options are available with `mi predict` and `mi predictnl` but are not shown in the dialog box:

`noupdate` in some cases suppresses the automatic `mi update` this command might perform; see [MI] **noupdate option**. This option is rarely used.

`noerrnotes` suppresses notes about failed estimation results. These notes appear when *miestfile*`.ster` contains estimation results, previously saved by `mi estimate, saving(`*miestfile*`)`, from imputations for which the estimation command used with `mi estimate` failed to estimate parameters.

`showimputations` displays imputation numbers corresponding to the estimation results saved in *miestfile*`.ster`.

Remarks

Remarks are presented under the following headings:

> *Introduction*
> *Using mi predict and mi predictnl*
> *Example 1: Obtain MI linear predictions and other statistics*
> *Example 2: Obtain MI linear predictions for the estimation sample*
> *Example 3: Obtain MI estimates of probabilities*
> *Example 4: Obtain other MI predictions*
> *Example 5: Obtain MI predictions after multiple-equation commands*

Introduction

Various predictions are often of interest after estimation. Within the MI framework, one must first decide what prediction means. There is no single dataset with respect to which prediction is made. Rather, there are multiple datasets in which values of imputed predictors vary from one dataset to another.

One definition is simply to consider an observation-specific prediction to be a parameter of interest and apply Rubin's combination rules to it as to any other estimand (White, Royston, and Wood 2011). The next thing to decide is what types of predictions are appropriate for pooling. For any parameter, the applicability of combination rules is subject to a number of conditions that the parameter must satisfy. One of them is asymptotic normality of the completed-data estimates of the parameter; see, for example, *Theory underlying multiple imputation* under *Remarks* of [MI] **intro substantive** for a full set of conditions.

It is safe to apply combination rules to the linear predictor, as computed by `mi predict`. It is also safe to apply combination rules to functions, possibly nonlinear, of the linear predictor, provided the sampling distribution of that function is asymptotically normal. This can be done by using `mi predictnl`. `mi predictnl` also provides, with the `predict()` specification, a way of obtaining MI estimates for various types of predictions specific to each estimation command used with `mi estimate`. Care should be taken when using this functionality. Some predictions may require preliminary transformation to a scale that improves normality, which is more appropriate for pooling. The obtained MI estimates of predictions may then be back-transformed to obtain final predictions in the original metric. For example, one can obtain MI estimates of probabilities of a positive outcome after logistic estimation by pooling the completed-data estimates of the actual probabilities. A better approach is to pool the completed-data estimates of the linear predictor and then apply an inverse-logit transformation to obtain the probability of a positive outcome. Other available predictions, such as standard errors, may not even be applicable for pooling.

The MI predictions should be treated as a final result; they should not be used as intermediate results in computations. For example, MI estimates of the linear predictor cannot be used to compute residuals as is done in non-MI analysis. Instead, completed-data residuals should be calculated for each imputed dataset, and these can be obtained by using the `mi xeq:` command. For example,

```
. mi xeq: regress ...; predict resid, r
```

Because completed-data predictions are super varying, they should only be computed in the flong or flongsep styles.

Using mi predict and mi predictnl

`mi predict` and `mi predictnl` require that completed-data estimation results saved by `mi estimate, saving()` are supplied with the `using` specification and that the mi data used to obtain these results are in memory. Apart from this, the use of these commands is similar to that of their non-mi counterparts, `predict` and `predictnl` (see [R] **predict** and [R] **predictnl**).

By default, `mi predict` computes MI linear predictions. If the `stdp` option is specified, `mi predict` computes standard errors of the MI linear predictions. As with `predict`, the `equation()` option can be used with `mi predict` after multiple-equation commands to obtain linear predictions or their standard errors from a specific equation.

Similarly to `predictnl`, a number of statistics associated with predictions can be obtained with `mi predictnl`, such as confidence intervals and p-values. Additionally, a number of MI statistics, such as relative variance increases and fractions of missing information, are available with `mi predictnl`. As we mentioned in *Introduction*, the `predict()` function of `mi predictnl` offers a variety of predictions. However, you should carefully consider whether the requested prediction is applicable for pooling or, perhaps, needs a preliminary transformation to improve normality.

Unlike `predict`, `mi predict` always defaults to the linear prediction. It supports only the linear prediction or its standard error and does not support any other command-specific predictions. Command-specific predictions appropriate for pooling may be obtained with the `predict()` function of `mi predictnl`. Also unlike `predict` after some multiple-equation commands, `mi predict` does not allow specification of multiple new variables to store predictions from all equations. For each equation *eqno*, you should use `mi predict, equation(`*eqno*`)` to obtain predictions from equation *eqno*.

To obtain estimation-sample predictions, the `if e(sample)` restriction is usually used with `predict` and `predictnl`. This restriction is not allowed with `mi predict` and `mi predictnl`. `mi estimate` does not set an estimation sample. There is no single estimation sample within the MI framework; there are M of them, and they may vary across imputed datasets. To obtain estimation-sample predictions with `mi predict` and `mi predictnl`, you must first store the estimation sample in a variable and then specify this variable in the `esample()` option. For example, you may use `mi estimate`'s `esample(`*newvar*`)` option to store the estimation sample in *newvar*. To use `mi estimate, esample()`, you must be in flong or flongsep style; use [MI] **mi convert** to convert to one of these styles.

`mi predict` and `mi predictnl` store MI predictions and statistics associated with them in the original data ($m = 0$). If your data are flong or flongsep, you may additionally store the corresponding completed-data estimates in the imputed data ($m > 0$) by specifying the `storecompleted` option. This option only affects results for which completed-data counterparts are available, such as predictions, standard errors, and confidence intervals. It has no effect on statistics specific to MI, such as relative variance increases and fractions of missing information.

When you restrict predictions to a subsample, `mi predict` and `mi predictnl` verify that the prediction samples are the same across imputed datasets. If varying prediction samples are detected, the commands terminate with an error. If such a situation occurs, you may consider modifying your restriction to define a sample common to all imputations. If there are a few imputations violating the consistency of the prediction sample, you may obtain MI predictions over a selected set of imputations using, for example, the `imputations()` option.

Example 1: Obtain MI linear predictions and other statistics

Recall the analysis of house resale prices from *Example 2: Completed-data linear regression analysis* in [MI] **mi estimate**:

```
. use http://www.stata-press.com/data/r12/mhouses1993s30
(Albuquerque Home Prices Feb15-Apr30, 1993)

. mi estimate, saving(miest): regress price tax sqft age nfeatures ne custom corner
```

Multiple-imputation estimates			Imputations	=	30
Linear regression			Number of obs	=	117
			Average RVI	=	0.0648
			Largest FMI	=	0.2533
			Complete DF	=	109
DF adjustment:	Small sample		DF: min	=	69.12
			avg	=	94.02
			max	=	105.51
Model F test:	Equal FMI		F(7, 106.5)	=	67.18
Within VCE type:	OLS		Prob > F	=	0.0000

price	Coef.	Std. Err.	t	P>|t|	[95% Conf. Interval]	
tax	.6768015	.1241568	5.45	0.000	.4301777	.9234253
sqft	.2118129	.069177	3.06	0.003	.0745091	.3491168
age	.2471445	1.653669	0.15	0.882	-3.051732	3.546021
nfeatures	9.288033	13.30469	0.70	0.487	-17.12017	35.69623
ne	2.518996	36.99365	0.07	0.946	-70.90416	75.94215
custom	134.2193	43.29755	3.10	0.002	48.35674	220.0818
corner	-68.58686	39.9488	-1.72	0.089	-147.7934	10.61972
_cons	123.9118	71.05816	1.74	0.085	-17.19932	265.0229

We saved complete-data estimation results to miest.ster using mi estimate's saving() option.

We store MI linear predictions in variable xb_mi:

```
. mi predict xb_mi using miest
(option xb assumed; linear prediction)
. mi xeq 0: summarize price xb_mi
m=0 data:
-> summarize price xb_mi
```

Variable	Obs	Mean	Std. Dev.	Min	Max
price	117	1062.735	380.437	540	2150
xb_mi	117	1062.735	344.2862	523.0295	2042.396

MI predictions are stored in the original data ($m = 0$). The predictions of price seem reasonable.

We compute standard errors of MI linear predictions by using the stdp option:

```
. mi predict stdp_mi using miest, stdp
```

To obtain other statistics, such as confidence intervals and Wald test statistics, we can use mi predictnl. For example, we compute linear predictions, 95% confidence intervals, and fractions of missing information of the linear predictions as follows:

```
. mi predictnl xb1_mi = predict(xb) using miest, ci(cil_mi ciu_mi) fmi(fmi)
```

Unlike confidence intervals produced by predictnl, confidence intervals from mi predictnl are based on observation-specific degrees of freedom. Recall from [MI] **mi estimate** that the degrees of freedom used for MI inference is inversely related to relative variance increases due to missing data, which are parameter-specific. The prediction for each observation is viewed as a separate parameter, so it has its own degrees of freedom. If desired, you may obtain observation-specific MI degrees of freedom by specifying the df() option with mi predictnl.

Example 2: Obtain MI linear predictions for the estimation sample

To obtain MI linear predictions for the estimation sample, we must first store the estimation sample in a variable. To store the estimation sample with mi estimate, the mi data must be flong or flongsep.

Continuing our house resale prices example, the data are mlong:

```
. use http://www.stata-press.com/data/r12/mhouses1993s30
(Albuquerque Home Prices Feb15-Apr30, 1993)
. mi query
data mi set mlong, M = 30
last mi update 03may2011 15:21:36, 1 day ago
```

We switch to the flong style by using the mi convert command (see [MI] **mi convert**) and store the estimation sample in variable touse by using mi estimate, esample():

```
. mi convert flong
. mi estimate, esample(touse): regress price tax sqft age nfeatures ne custom corner
```

Multiple-imputation estimates
Linear regression

Imputations	=	30
Number of obs	=	117
Average RVI	=	0.0648
Largest FMI	=	0.2533
Complete DF	=	109

DF adjustment: Small sample

DF:	min	=	69.12
	avg	=	94.02
	max	=	105.51

Model F test: Equal FMI
Within VCE type: OLS

F(7, 106.5)	=	67.18
Prob > F	=	0.0000

price	Coef.	Std. Err.	t	P>|t|	[95% Conf. Interval]	
tax	.6768015	.1241568	5.45	0.000	.4301777	.9234253
sqft	.2118129	.069177	3.06	0.003	.0745091	.3491168
age	.2471445	1.653669	0.15	0.882	-3.051732	3.546021
nfeatures	9.288033	13.30469	0.70	0.487	-17.12017	35.69623
ne	2.518996	36.99365	0.07	0.946	-70.90416	75.94215
custom	134.2193	43.29755	3.10	0.002	48.35674	220.0818
corner	-68.58686	39.9488	-1.72	0.089	-147.7934	10.61972
_cons	123.9118	71.05816	1.74	0.085	-17.19932	265.0229

Because we use the same regression model, we do not need to resave estimation results and we can use the previously saved `miest.ster` from *Example 1: Obtain MI linear predictions and other statistics* with `mi predict`.

To restrict the linear prediction to the estimation sample identified by the `touse` variable, we use `esample(touse)` with `mi predict`:

```
. mi predict xb_mi using miest, esample(touse)
(option xb assumed; linear prediction)
. mi xeq 0: summarize xb_mi
```

m=0 data:
-> summarize xb_mi

Variable	Obs	Mean	Std. Dev.	Min	Max
xb_mi	117	1062.735	344.2862	523.0295	2042.396

The estimation sample includes all observations, so we obtain the same predictions as we did in example 1.

We could simply use an `if` restriction instead of the `esample()` option to obtain the same results:

```
. mi predict xb_mi if touse using miest
  (output omitted )
```

But if you use the `esample()` option, `mi predict` and `mi predictnl` perform additional checks to verify that the supplied variable is a proper estimation-sample variable.

By default, the MI linear prediction is only stored in the original data ($m = 0$) and the imputed data contain missing values in the corresponding variable. In the flong and flongsep styles, we can also store completed-data predictions in the imputed data ($m > 0$) by specifying the `storecompleted` option:

```
. mi predict xb_mi_all using miest, esample(touse) storecompleted
(option xb assumed; linear prediction)
. mi xeq 0 1 2: summarize xb_mi_all
```

m=0 data:
```
-> summarize xb_mi_all
```

Variable	Obs	Mean	Std. Dev.	Min	Max
xb_mi_all	117	1062.735	344.2862	523.0295	2042.396

m=1 data:
```
-> summarize xb_mi_all
```

Variable	Obs	Mean	Std. Dev.	Min	Max
xb_mi_all	117	1062.735	346.1095	529.5227	2042.942

m=2 data:
```
-> summarize xb_mi_all
```

Variable	Obs	Mean	Std. Dev.	Min	Max
xb_mi_all	117	1062.735	344.8446	515.5598	2040.374

Variable xb_mi_all contains MI linear predictions in $m = 0$; completed-data linear predictions from imputation 1 in $m = 1$; completed-data linear predictions from imputation 2 in $m = 2$; and so on.

Example 3: Obtain MI estimates of probabilities

Recall the analysis of heart attacks from *Example 1: Completed-data logistic analysis* in [MI] **mi estimate**:

```
. use http://www.stata-press.com/data/r12/mheart1s20
(Fictional heart attack data; bmi missing)
. mi estimate, saving(miest, replace): logit attack smokes age bmi hsgrad female
```

Multiple-imputation estimates		Imputations	=	20
Logistic regression		Number of obs	=	154
		Average RVI	=	0.0312
		Largest FMI	=	0.1355
DF adjustment:	Large sample	DF: min	=	1060.38
		avg	=	223362.56
		max	=	493335.88
Model F test:	Equal FMI	F(5,71379.3)	=	3.59
Within VCE type:	OIM	Prob > F	=	0.0030

attack	Coef.	Std. Err.	t	P>|t|	[95% Conf. Interval]	
smokes	1.198595	.3578195	3.35	0.001	.4972789	1.899911
age	.0360159	.0154399	2.33	0.020	.0057541	.0662776
bmi	.1039416	.0476136	2.18	0.029	.010514	.1973692
hsgrad	.1578992	.4049257	0.39	0.697	-.6357464	.9515449
female	-.1067433	.4164735	-0.26	0.798	-.9230191	.7095326
_cons	-5.478143	1.685075	-3.25	0.001	-8.782394	-2.173892

We could have used a different estimation file to store the completed-data estimation results from logit. Instead, we replaced the existing estimation file miest.ster with new results by specifying saving()'s replace option.

Following the discussion in *Introduction*, we first obtain MI estimates of the probabilities of a positive outcome by using the transformation-based approach. We obtain MI estimates of linear predictions and apply the inverse-logit transformation to obtain the probabilities:

```
. mi predict xb_mi using miest
(option xb assumed; linear prediction)
. qui mi xeq: generate phat = invlogit(xb_mi)
```

Unlike predict after logit, mi predict after mi estimate: logit defaults to the linear prediction and not to the probability of a positive outcome. mi predict always assumes the linear prediction.

Alternatively, we can apply Rubin's combination rules directly to probabilities. Unlike predict, mi predict does not allow the pr option. You can obtain only linear predictions or standard errors using mi predict. We can use the predict() function of mi predictnl to obtain MI estimates of the probabilities by directly pooling completed-data probabilities:

```
. mi predictnl phat_mi = predict(pr) using miest
. mi xeq 0: summarize phat phat_mi
```

m=0 data:
-> summarize phat phat_mi

Variable	Obs	Mean	Std. Dev.	Min	Max
phat	154	.4478198	.1820425	.1410432	.8923041
phat_mi	154	.4480519	.1812098	.141361	.8912111

Although the first approach is preferable, we can see that we obtain similar estimates of the probabilities of a positive outcome with both approaches.

Example 4: Obtain other MI predictions

Consider the cancer data from *Example 3: Completed-data survival analysis* in [MI] **mi estimate**:

```
. use http://www.stata-press.com/data/r12/mdrugtrs25
(Patient Survival in Drug Trial)
. mi stset studytime, failure(died)

     failure event:  died != 0 & died < .
obs. time interval:  (0, studytime]
 exit on or before:  failure
```

```
      48  total obs.
       0  exclusions

      48  obs. remaining, representing
      31  failures in single record/single failure data
     744  total analysis time at risk, at risk from t =          0
                           earliest observed entry t =          0
                               last observed exit t =         39
```

In this example, we fit a parametric Weibull regression to the survival data and as before replace the estimation results in miest.ster with new ones from mi estimate: streg:

```
. mi estimate, saving(miest, replace): streg drug age, dist(weibull)
```

Multiple-imputation estimates   Imputations   =   25
Weibull regression: Log relative-hazard form   Number of obs   =   48
                                                 Average RVI   =   0.0927
                                                 Largest FMI   =   0.1847
DF adjustment: Large sample   DF: min   =   721.15
                                                           avg   =   6014.48
                                                           max   =   11383.09
Model F test: Equal FMI   F(2, 2910.0) =   14.94
Within VCE type: OIM   Prob > F   =   0.0000

_t	Coef.	Std. Err.	t	P>|t|	[95% Conf. Interval]	
drug	-2.093333	.4091925	-5.12	0.000	-2.895422	-1.291243
age	.126931	.0403526	3.15	0.002	.0477084	.2061536
_cons	-11.14588	2.584909	-4.31	0.000	-16.22013	-6.071634
/ln_p	.5524239	.1434973	3.85	0.000	.2711445	.8337033
p	1.737459	.2493207			1.311465	2.301827
1/p	.575553	.0825903			.4344374	.7625063

Suppose that we want to estimate median survival time. After `streg`, median survival time can be obtained by using `predict, median time`. `mi predict` does not support these options, but we can use the `predict(median time)` function with `mi predictnl` to obtain MI estimates of the median survival time.

To improve normality, we perform pooling in a log scale and then exponentiate results back to the original scale:

```
. mi predictnl p50_lntime_mi = ln(predict(median time)) using miest
. qui mi xeq: generate p50_time_mi = exp(p50_lntime_mi)
```

Above, we demonstrated the use of expressions with the `predict()` function by computing median log-survival time by using `ln(predict(median time))`. Alternatively, we can compute median log-survival time directly with `predict(median lntime)`:

```
. mi predictnl p50_lntime1_mi = predict(median lntime) using miest
. qui mi xeq: generate p50_time1_mi = exp(p50_lntime1_mi)
```

We verify that we obtain identical results:

```
. mi xeq 0: summarize p50_time_mi p50_time1_mi
m=0 data:
-> summarize p50_time_mi p50_time1_mi
```

Variable	Obs	Mean	Std. Dev.	Min	Max
p50_time_mi	48	21.74607	14.60662	3.707896	53.10997
p50_time1_mi	48	21.74607	14.60662	3.707896	53.10997

Example 5: Obtain MI predictions after multiple-equation commands

For illustrative purposes, let's use `mlogit` instead of `logit` to analyze the heart-attack data from *Example 3: Obtain MI estimates of probabilities*:

```
. use http://www.stata-press.com/data/r12/mheart1s20
(Fictional heart attack data; bmi missing)

. mi estimate, saving(miest, replace): mlogit attack smokes age bmi hsgrad female
```

Multiple-imputation estimates   Imputations   =   20
Multinomial logistic regression   Number of obs   =   154
                                           Average RVI   =   0.0312
                                           Largest FMI   =   0.1355
DF adjustment: Large sample   DF: min   =   1060.38
                                                    avg   = 223362.56
                                                    max   = 493335.88
Model F test: Equal FMI   F(5,71379.3) =   3.59
Within VCE type: OIM   Prob > F   =   0.0030

attack	Coef.	Std. Err.	t	P>|t|	[95% Conf. Interval]
0	(base outcome)					
1						
smokes	1.198595	.3578195	3.35	0.001	.4972789	1.899911
age	.0360159	.0154399	2.33	0.020	.0057541	.0662776
bmi	.1039416	.0476136	2.18	0.029	.010514	.1973692
hsgrad	.1578992	.4049257	0.39	0.697	-.6357464	.9515449
female	-.1067433	.4164735	-0.26	0.798	-.9230191	.7095326
_cons	-5.478143	1.685075	-3.25	0.001	-8.782394	-2.173892

We obtain the same results as with `mi estimate: logit`.

To obtain predictions after multiple-equation commands such as `mlogit`, we need to use the `equation()` option of `mi predict` or `mi predictnl` to obtain a prediction from a specific equation. By default, the first equation is assumed:

```
. mi predict xb_0_mi using miest
(option xb assumed; linear prediction)

. mi xeq 0: summarize xb_0_mi

m=0 data:
-> summarize xb_0_mi
```

Variable	Obs	Mean	Std. Dev.	Min	Max
xb_0_mi	154	0	0	0	0

In our example, the first equation corresponds to the base category, so the linear prediction is zero for this equation.

To obtain the linear prediction from the second equation, we specify the equation(*eqno*) option. *eqno* can refer to the equation number, #2, or to the equation name, 1. For example,

```
. mi predict xb_1_mi using miest, equation(#2)
(option xb assumed; linear prediction)
```

Suppose we want to compute observation-specific odds of a heart attack. Knowing that the odds of a disease is the exponentiated linear predictor, we can compute the odds simply as

```
. qui mi xeq: generate odds_mi = exp(xb_1_mi)
```

Instead, to illustrate a more advanced syntax of `mi predictnl`, we compute the odds using their definition as the ratio of a probability of a heart attack (attack==1) to the probability of no heart attack (attack==0). Log odds are asymptotically normally distributed, so we apply combination rules to log odds and then exponentiate the result to obtain odds:

```
. mi predictnl lnodds_mi = ln(predict(pr equation(1))/predict(pr equation(0))) using
> miest

. qui mi xeq: generate odds_mi = exp(lnodds_mi)
```

In the above, we used the names of the equations, 0 and 1, within equation() to obtain probabilities of no heart attack and a heart attack, respectively.

We can see, for example, that for older subjects or subjects who smoke, the odds of having a heart attack are noticeably higher:

```
. qui mi xeq: generate byte atrisk = smokes==1 | age>50

. mi xeq 0: by atrisk, sort: summ odds_mi
m=0 data:
-> by atrisk, sort: summ odds_mi
```

-> atrisk = 0

Variable	Obs	Mean	Std. Dev.	Min	Max
odds_mi	30	.3472545	.1451144	.1642029	.818259

-> atrisk = 1

Variable	Obs	Mean	Std. Dev.	Min	Max
odds_mi	124	1.327598	1.228176	.2198672	8.285403

Methods and formulas

Multiple-imputation predictions are obtained by considering an observation-specific prediction as an estimand and by applying Rubin's combination rules to it (White, Royston, and Wood 2011).

Let $\eta_i(\cdot)$ be a prediction of interest for subject i and $\widehat{\eta}_{i,m}(\cdot)$ be a completed-data estimate of the prediction for subject i, $i = 1, \ldots, N$, from imputation m, $m = 1, \ldots, M$. In what follows, we omit the functional argument of $\eta_i(\cdot)$ for brevity.

The MI estimate of prediction η_i is

$$\overline{\eta}_{i,M} = \frac{1}{M} \sum_{m=1}^{M} \widehat{\eta}_{i,m}, \ i = 1, \ldots, N$$

Let $\widehat{\mathrm{Var}}(\widehat{\eta}_{i,m})$ be the completed-data variance of the completed-data prediction $\widehat{\eta}_{i,m}$ for subject i from imputation m. The standard error of the MI prediction $\overline{\eta}_{i,M}$ is the square root of the total MI variance $T_{\overline{\eta}_{i,M}}$,

$$T_{\overline{\eta}_{i,M}} = \overline{U}_i + \left(1 + \frac{1}{M}\right) B_i, \ i = 1, \ldots, N$$

where $\overline{U}_i = \sum_{m=1}^{M} \widehat{\mathrm{Var}}(\widehat{\eta}_{i,m})/M$ is the within-imputation variance and $B_i = \sum_{m=1}^{M} (\widehat{\eta}_{i,m} - \overline{\eta}_{i,M})^2/(M-1)$ is the between-imputation variance.

Other statistics such as test statistics, confidence intervals, and relative variance increases are obtained by applying to η_i the same formulas as described in *Univariate case* under *Methods and formulas* of [MI] **mi estimate** for parameter Q. Also see Rubin (1987, 76–77).

As for any other parameter, the validity of applying Rubin's combination rules to η_i is subject to η_i satisfying a set of conditions as described, for example, in *Theory underlying multiple imputation* under *Remarks* of [MI] **intro substantive**. In particular, the combination rules should be applied to η_i in the metric for which the sampling distribution is closer to the normal distribution.

References

Barnard, J., and D. B. Rubin. 1999. Small-sample degrees of freedom with multiple imputation. *Biometrika* 86: 948–955.

Rubin, D. B. 1987. *Multiple Imputation for Nonresponse in Surveys.* New York: Wiley.

White, I. R., P. Royston, and A. M. Wood. 2011. Multiple imputation using chained equations: Issues and guidance for practice. *Statistics in Medicine* 30: 377–399.

Also see

Title

---
mi ptrace — Load parameter-trace file into Stata

Syntax

    mi ptrace <u>d</u>escribe [using] *filename*

    mi ptrace use *filename* [, *use_options*]

use_options	Description
clear	okay to replace existing data in memory
double	load variables as doubles (default is floats)
<u>sel</u>ect(*selections*)	what to load (default is all)

where *selections* is a space-separated list of individual selections. Individual selections are of the form

    b[*yname*, *xname*]
    v[*yname*, *yname*]

where *ynames* and *xnames* are displayed by mi ptrace describe. You may also specify

    b[#_y, #_x]
    v[#_y, #_y]

where #_y and #_x are the variable numbers associated with *yname* and *xname*, and those too are shown by mi ptrace describe.

For b, you may also specify * to mean all possible index elements. For instance,

    b[*,*]   all elements of b
    b[*yname*,*]   row corresponding to *yname*
    b[*,*xname*]   column corresponding to *xname*

Similarly, b[#_y,*] and b[*,#_x] are allowed. The same is allowed for v, and also, the second element can be specified as <, <=, =, >=, or >. For instance,

    v[*yname*,=]   variance of *yname*
    v[*,=]   all variances (diagonal elements)
    v[*,<]   lower triangle
    v[*,<=]   lower triangle and diagonal
    v[*,>=]   upper triangle and diagonal
    v[*,>]   upper triangle

In mi ptrace describe and in mi ptrace use, *filename* must be specified in quotes if it contains special characters or blanks. *filename* is assumed to be *filename*.stptrace if the suffix is not specified.

Description

Parameter-trace files, files with suffix `.stptrace`, are created by the `saveptrace()` option of `mi impute mvn`; see [MI] **mi impute mvn**. These are not Stata datasets, but they can be loaded as if they were by using `mi ptrace use`. Their contents can be described without loading them by using `mi ptrace describe`.

Options

`clear` specifies that it is okay to clear the dataset in memory, even if it has not been saved to disk since it was last changed.

`double` specifies that elements of b and v are to be loaded as doubles; they are loaded as floats by default.

`select(`*selections*`)` allows you to load subsets of b and v. If the option is not specified, all of b and v are loaded. That result is equivalent to specifying `select(b[*,*] v[*,<=])`. The `<=` specifies that just the diagonal and lower triangle of symmetric matrix v be loaded.

Specifying `select(b[*,*])` would load just b.

Specifying `select(v[*,<=])` would load just v.

Specifying `select(b[*,*] v[*,=])` would load b and the diagonal elements of v.

Remarks

Say that we impute the values of y_1 and y_2 assuming that they are multivariate normal distributed, with their means determined by a linear combination of x_1, x_2, and x_3, and their variance constant. Writing this more concisely, $\mathbf{y} = (y_1, y_2)'$ is distributed MVN($\mathbf{XB}, \mathbf{V}$), where $\mathbf{B}$: 2×3 and $\mathbf{V}$: 2×2. If we use MCMC or EM procedures to produce values of $\mathbf{B}$ and $\mathbf{V}$ to be used to generate values for $\mathbf{y}$, we must ensure that we use sufficient iterations so that the iterative procedure stabilizes. `mi impute mvn` (see [MI] **mi impute mvn**) provides the worst linear combination (WLC) of the elements of $\mathbf{B}$ and $\mathbf{V}$. If we want to perform other checks, we can specify `mi impute mvn`'s `saveptrace(`*filename*`)` option. `mi impute` then produces a file containing m (imputation number), iter (overall iteration number), and the corresponding B and V. The last iter for each m is the B and V that `mi impute mvn` used to impute the missing values.

When we used `mi impute mvn`, we specified burn-in and burn-between numbers, say, `burnin(300)` and `burnbetween(100)`. If we also specified `saveptrace()`, the file produced is organized as follows:

record #	m	iter	B	V	
1	1	−299	...	...	
2	1	−298	...	...	
.	.	.	.	.	
.	.	.	.	.	
299	1	−1	...	...	
300	1	0	...	...	<- used to impute *m*=1
301	2	1	.	.	
302	2	2	.	.	
.	.	.	.	.	
.	.	.	.	.	
399.	1	99	...	...	
400.	1	100	...	...	<- used to impute *m*=2
401.	2	101	...	...	
.	.	.	.	.	
.	.	.	.	.	

The file is not a Stata dataset, but mi ptrace use can load the file and convert it into Stata format, and then it will look just like the above except for the following:

- The record number will become the Stata observation number.

- B will become variables b_y1x1, b_y1x2, and b_y1x3; and b_y2x1, b_y2x2, and b_y2x3. (Remember, we had 2 y variables and 3 x variables.)

- V will become variables v_y1y1, v_y2y1, and v_y2y2. (This is the diagonal and lower triangle of **V**; variable v_y1y2 is not created because it would be equal to v_y2y1.)

- Variable labels will be filled in with the underlying names of the variables. For instance, the variable label for b_y1x1 might be "experience, age", and that would remind us that b_y1x1 contains the coefficient on age used to predict experience. v_y2y1 might be "education, experience", and that would remind us that v_y2y1 contains the covariance between education and experience.

Saved results

mi ptrace describe saves the following in r():

Scalars
r(tc)   %tc date-and-time file created
r(nx)   number of x variables (columns of B)
r(ny)   number of y variables (rows of B)
Macros
r(x)   space-separated [op.]varname of x
r(y)   space-separated [op.]varname of y
r(id)   name of file creator

Also see

[MI] **intro** — Introduction to mi

[MI] **mi impute mvn** — Impute using multivariate normal regression

Title

mi rename — Rename variable

Syntax

mi <u>ren</u>ame *oldname newname* [, <u>noup</u>date]

Menu

Statistics > Multiple imputation

Description

mi rename renames variables.

Option

noupdate in some cases suppresses the automatic mi update this command might perform; see [MI] **noupdate option**.

Remarks

Remarks are presented under the following headings:

> *Specifying the noupdate option*
> *What to do if you accidentally use rename*
> *What to do if you accidentally use rename on wide data*
> *What to do if you accidentally use rename on mlong data*
> *What to do if you accidentally use rename on flong data*
> *What to do if you accidentally use rename on flongsep data*

Specifying the noupdate option

If you are renaming more than one variable, you can speed execution with no loss of safety by specifying the noupdate option after the first mi rename:

. mi rename ageyears age

. mi rename timeinstudy studytime, noupdate

. mi rename personid id, noupdate

The above is generally good advice. When giving one mi command after another, you may specify noupdate after the first command to speed execution.

What to do if you accidentally use rename

Assume that you just typed

    . rename ageyears age

rather than typing

    . mi rename ageyears age

as you should have. No damage has been done yet, but if you give another mi command and it runs mi update (see [MI] **mi update**), real damage will be done. We will discuss that and what to do about it in the sections that follow, but first, if you have given no additional mi commands, use rename (not mi rename) to rename the variable back to how it was:

    . rename age ageyears

Then use mi rename as you should have in the first place:

    . mi rename ageyears age

The sections below handle the case where mi update has run. You will know that mi update has run because since the rename, you gave some mi command—perhaps even mi update itself—and you saw a message like one of these:

> (variable ageyears dropped in $m > 0$)
>
> (imputed variable ageyears unregistered because not in $m = 0$)
>
> (passive variable ageyears unregistered because not in $m = 0$)
>
> (regular variable ageyears unregistered because not in $m = 0$)

What to do if you accidentally use rename on wide data

If ageyears was unregistered, no damage was done, and no additional action needs to be taken.

If ageyears was registered as regular, no damage was done. However, your renamed variable is no longer registered. Reregister the variable under its new name by typing mi register regular age; see [MI] **mi set**.

If ageyears was registered as imputed or passive, you just lost all values for $m > 0$. Passive variables are usually not too difficult to re-create; see [MI] **mi passive**. If the variable was imputed, well, hope that you will have saved your data recently when you make this error and, before that, learn good computing habits.

What to do if you accidentally use rename on mlong data

If ageyears was unregistered, no damage was done, and no additional action needs to be taken.

If ageyears was registered as regular, no damage was done. However, your renamed variable is no longer registered. Reregister the variable under its new name by typing mi register regular age; see [MI] **mi set**.

If ageyears was registered as imputed or passive, you just lost all values for $m > 0$. We offer the same advice as we offered when the data were wide: Passive variables are usually not too difficult to re-create—see [MI] **mi passive**—and otherwise hope that you will have saved your data recently when you make this error. It is always a good idea to save your data periodically.

What to do if you accidentally use rename on flong data

The news is better in this case; no matter how your variables were registered, you have not lost data.

If ageyears was unregistered, no further action is required.

If ageyears was registered as regular, you need to reregister the variable under its new name by typing mi register regular age; see [MI] **mi set**.

If ageyears was registered as passive or imputed, you need to reregister the variable under its new name by typing mi register passive age or mi register imputed age.

What to do if you accidentally use rename on flongsep data

The news is not as good in this case.

If ageyears was unregistered, no damage was done. When mi update ran, it noticed that old variable ageyears no longer appeared in $m > 0$ and that new variable age now appeared in $m = 0$, so mi update dropped the first and added the second to $m > 0$, thus undoing any damage. There is nothing more that needs to be done.

If ageyears was registered as regular, no damage was done, but you need to reregister the variable by typing mi register regular age; see [MI] **mi set**.

If ageyears was registered as passive or imputed, you have lost the values in $m > 0$. Now would probably be a good time for us to mention how you should work with a copy of your flongsep data; see [MI] **mi copy**.

Also see

[MI] **intro** — Introduction to mi

Title

---
mi replace0 — Replace original data

Syntax

mi replace0 using *filename*, id(*varlist*)

Typical use is

. mi extract 0

. (*perform data-management commands*)

. mi replace0 using *origfile*, id(*idvar*)

Menu

Statistics > Multiple imputation

Description

mi replace0 replaces $m = 0$ of an mi dataset with the non-mi data of another and then carries out whatever changes are necessary in $m > 0$ of the former to make the resulting mi data consistent. mi replace0 starts with one of the datasets in memory and the other on disk (it does not matter which is which) and leaves in memory the mi data with $m = 0$ replaced.

Option

id(*varlist*) is required; it specifies the variable or variables to use to match the observations in $m = 0$ of the mi data to the observations of the non-mi dataset. The ID variables must uniquely identify the observations in each dataset, and equal values across datasets must indicate corresponding observations, but one or both datasets can have observations found (or not found) in the other.

Remarks

It is often easier to perform data management on $m = 0$ and then let mi replace0 duplicate the results for $m = 1$, $m = 2$, ..., $m = M$ rather than perform the data management on all m's simultaneously. It is easier because $m = 0$ by itself is a non-mi dataset, so you can use any of the general Stata commands (that is, non-mi commands) with it.

You use mi extract to extract $m = 0$; see [MI] **mi extract**. The extracted dataset is just a regular Stata dataset; it is not mi set, nor does it have any secret complexities.

You use mi replace0 to recombine the datasets after you have modified the $m = 0$ data. mi replace0 can deal with the following changes to $m = 0$:

- changes to the values of existing variables,
- removal of variables,

288

- addition of new variables,
- dropped observations, and
- added observations.

For instance, you could use `mi extract` and `mi replace0` to do the following:

```
. use my_midata, clear
. mi extract 0
. replace age = 26 if age==6
. replace age = 32 if pid==2088
. merge 1:1 pid using newvars, keep(match) nogen
. by location: egen avgrate = mean(rate)
. drop proxyrate
. mi replace0 using my_midata, id(pid)
```

In the above,

1. we extract $m = 0$;

2. we update existing variable `age` (we fix a typo and the age of `pid` 2088);

3. we merge $m = 0$ with `newvars.dta` to obtain some new variables and, in the process, keep only the observations that were found in both $m = 0$ and `newvars.dta`;

4. we create new variable `avgrate` equal to the mean rate by location; and

5. we drop previously existing variable `proxyrate`.

We then take that result and use it to replace $m = 0$ in our original `mi` dataset. We leave it to `mi replace0` to carry out the changes to $m = 1$, $m = 2$, ..., $m = M$ to account for what we did to $m = 0$.

By the way, it turns out that `age` in `my_midata.dta` is registered as imputed. We changed one nonmissing value to another nonmissing value and changed one missing value to a nonmissing value. `mi replace0` will deal with the implications of that. It would even deal with us having changed a nonmissing value to a missing value.

There is no easier way to do data management than by using `mi extract` and `mi replace0`.

Also see

[MI] **intro** — Introduction to mi

[MI] **mi extract** — Extract original or imputed data from mi data

Title

> **mi reset** — Reset imputed or passive variables

Syntax

mi reset *varlist* [= *exp*] [*if*] [, *options*]

options	Description
Main	
m(*numlist*)	*m* to reset; default all
noupdate	see [MI] **noupdate option**

Menu

Statistics > Multiple imputation

Description

mi reset resets the imputed or passive variables specified. Values are reset to the values in $m = 0$, which are typically missing, but if you specify = *exp*, they are reset to the value specified.

Options

⌐ Main ⌐

m(*numlist*) specifies the values of m that are to be reset; the default is to update all values of m. If M were equal to 3, the default would be equivalent to specifying m(1/3) or m(1 2 3). If you wished to update the specified variable(s) in just $m = 2$, you could specify m(2).

noupdate in some cases suppresses the automatic mi update this command might perform; see [MI] **noupdate option**.

Remarks

Remarks are presented under the following headings:

Using mi reset
Technical notes and relation to mi update

Using mi reset

Resetting an imputed or passive variable means setting its values in $m > 0$ equal to the values recorded in $m = 0$. For instance, if variable inc were imputed, typing

```
. mi reset inc
(15 values reset)
```

would reset its incomplete values back to missing in all m. In the sample output shown, we happen to have $M = 5$ and reset back to missing the three previously imputed values in each imputation.

It is rare that you would want to reset an imputed variable, but one can imagine cases. Your coworker Joe sent you the data and just buzzed you on the telephone. "There is one value wrong in the data I sent you," he says. "There is an imputed value for `inc` that is 15,000, which is obviously absurd. Just reset it back to missing until I find out what happened." So you type

```
. mi reset inc if inc==15000
(1 value reset)
```

Later Joe calls back. "It is a long and very stupid story," he begins, and you can hear him settling into his chair to tell it. As you finish your second cup of coffee, he is wrapping up. "So the value of `inc` for `pid` 1433 should be 0.725." You type

```
. mi reset inc = .725 if pid=1433
(1 value reset)
```

It is common to need to reset passive variables if imputed values change. For instance, you have variables `age` and `lnage` in your data. You imputed `lnage`; `age` is passive. You recently updated the imputed values for `lnage`. One way to update the values for `age` would be to type

```
. mi passive: replace age = exp(lnage)
m=0:
m=1:
(10 real changes made)
m=2:
(10 real changes made)
m=3:
(8 real changes made)
```

Alternatively, you could type

```
. mi reset age = exp(lnage)
(28 values reset)
```

Technical notes and relation to mi update

`mi reset`, used with an imputed variable, changes only the values for which the variable contains hard missing (.) in $m = 0$. The other values are, by definition, already equal to their $m = 0$ values.

`mi reset`, used with a passive variable, changes only the values in incomplete observations, observations in which any imputed variable contains hard missing. The other values of the passive variable are, by definition, already equal to their $m = 0$ values.

`mi update` can be used to ensure that values that are supposed to be equal to their $m = 0$ values in fact are equal to them; see [MI] **mi update**.

Also see

[MI] **intro** — Introduction to mi

[MI] **mi update** — Ensure that mi data are consistent

Title

> **mi reshape** — Reshape mi data

Syntax

Overview

(The words long *and* wide *in what follows have nothing to do with mi styles mlong, flong, flongsep, and wide; they have to do with reshape's concepts.)*

	long	
i	*j*	*stub*
1	1	4.1
1	2	4.5
2	1	3.3
2	2	3.0

reshape ⟵⟶

	wide	
i	*stub1*	*stub2*
1	4.1	4.5
2	3.3	3.0

To go from long to wide:

j existing variable

        mi reshape wide *stub*, i(*i*) j(*j*)

To go from wide to long:

        mi reshape long *stub*, i(*i*) j(*j*)

j new variable

Basic syntax

Convert mi data from long form to wide form

        mi reshape wide *stubnames*, i(*varlist*) j(*varname*) [*options*]

Convert mi data from wide form to long form

        mi reshape long *stubnames*, i(*varlist*) j(*varname*) [*options*]

options	Description
i(*varlist*)	*i* variable(s)
j(*varname* [*values*])	long→wide: j, existing variable
	wide→long: j, new variable
	optionally specify values to subset j
<u>string</u>	j is string variable (default is numeric)

where *values* is $\#[-\#][\dots]$ if j is numeric (the default)
               "*string*" ["*string*" ...]   if j is string

and where *stubnames* are variable names (long→wide), or stubs of variable names (wide→long), and either way, may contain @, denoting where j appears or is to appear in the name.

In the example above, when we wrote "`mi reshape wide` *stub*", we could have written "`mi reshape wide` *stub*@" because j by default is used as a suffix. Had we written *stu*@*b*, then the wide variables would have been named *stu1b* and *stu2b*.

Menu

Statistics > Multiple imputation

Description

`mi reshape` is Stata's `reshape` for `mi` data; see [D] **reshape**.

Options

`noupdate` in some cases suppresses the automatic `mi update` this command might perform; see [MI] **noupdate option**.

See [D] **reshape** for descriptions of the other options.

Remarks

The `reshape` command you specify is carried out on the $m = 0$ data, and then the result is duplicated in $m = 1$, $m = 2$, ..., $m = M$.

Also see

[MI] **intro** — Introduction to mi

[D] **reshape** — Convert data from wide to long form and vice versa

[MI] **mi replace0** — Replace original data

Title

> **mi select** — Programmer's alternative to mi extract

Syntax

```
mi select init [ , fast ]

mi select #
```

where $0 \leq \# \leq M$, and where typical usage is

```
quietly mi query
local M = r(M)

preserve
mi select init
local priorcmd "'r(priorcmd)'"

forvalues m=1(1)'M' {
        mi select 'm'
        ...
        'priorcmd'
}

restore
```

Description

mi select is a programmer's command. It is a faster, more dangerous version of mi extract; see [MI] **mi extract**.

Before using mi select, the mi data must be preserved; see [P] **preserve**.

mi select init initializes mi select. mi select returns macro r(priorcmd), which you are to issue as a command between each mi select # call. r(priorcmd) is not required to be issued before the first call to mi select #, although you may issue it if that is convenient.

mi select # replaces the data in memory with a copy of the data for $m = \#$. The data are not mi set. Changes to the selected data will not be posted back to the underlying mi data. mi select # calls can be made in any order, and the same m may be selected repeatedly.

Option

fast, specified with mi select init, specifies that the data delivered by mi select # commands not be changed except for sort order. Then mi select can operate more quickly. fast is allowed with all styles but currently affects the performance with the wide style only.

If fast is not specified, the data delivered by mi select # may be modified freely before the next mi select # call. However, the data may not be dropped. mi select uses characteristics (see [P] **char**) stored in _dta[] to know its state.

294

Remarks

The data delivered by mi select # differ from those delivered by mi extract in that there may be extra variables in the dataset.

One of the extra variables, _mi_id, is a unique observation identifier that you can use. If you want to post changes made in the selected data back to the mi data, you can write a file containing mi_id and the updated variables and then use _mi_id to match that to the mi data after your final restore.

In the case of wide data, the mi data have no _mi_id variable. _mi_id in the selected data is reflected in the current order of the mi data.

Saved results

mi select init returns the following in r():

Macros
    r(priorcmd)   command to be issued prior to calling mi select #; this command will be either restore, preserve or nothing

Also see

[MI] **intro** — Introduction to mi

[MI] **mi extract** — Extract original or imputed data from mi data

[MI] **technical** — Details for programmers

Title

mi set — Declare multiple-imputation data

Syntax

mi set *style*

>   where *style* is <u>w</u>ide
>   <u>ml</u>ong
>   <u>fl</u>ong
>   <u>fl</u>ongsep *name*

mi <u>reg</u>ister { <u>imp</u>uted | <u>pass</u>ive | <u>reg</u>ular } *varlist*

mi <u>unr</u>egister *varlist*

mi set M { = | += | -= } #

mi set m -= (*numlist*)

mi unset [, asis]

Menu

Statistics > Multiple imputation

Description

mi set is used to set a regular Stata dataset to be an mi dataset. mi set is also used to modify the attributes of an already set dataset. An mi set dataset has the following attributes:

- The data are recorded in a *style*: wide, mlong, flong, or flongsep; see [MI] **styles**.
- Variables are registered as imputed, passive, or regular, or they are left unregistered.
- In addition to $m = 0$, the data with missing values, the data include $M \geq 0$ imputations of the imputed variables.

mi set *style* begins the setting process by setting the desired style. mi set *style* sets all variables as unregistered and sets $M = 0$.

mi register registers variables as imputed, passive, or regular. Variables can be registered one at a time or in groups and can be registered and reregistered.

mi unregister unregisters registered variables, which is useful if you make a mistake. Exercise caution. Unregistering an imputed or passive variable can cause loss of the filled-in missing values in $m > 0$ if your data are recorded in the wide or mlong styles. In such cases, just mi register the variable correctly without mi unregistering it first.

296

mi set M modifies M, the total number of imputations. M may be increased or decreased. M may be set before or after imputed variables are registered.

mi set m drops selected imputations from the data.

mi unset is a rarely used command to unset the data. Better alternatives include mi extract and mi export (see [MI] **mi extract** and [MI] **mi export**, respectively).

Remarks

Data must be mi set before they can be used with the other mi commands. There are two ways data can become mi set: direct use of mi set *style* or use of mi import (see [MI] **mi import**).

The mi register, mi set M, and mi set m commands are for use with already set data and are useful even with imported data.

Remarks are presented under the following headings:

> *mi set style*
> *mi register and mi unregister*
> *mi set M and mi set m*
> *mi unset*

mi set style

mi set *style* begins the setting process. mi set *style* has the following forms:

```
mi set wide
mi set mlong
mi set flong
mi set flongsep name
```

It does not matter which style you choose because you can always use mi convert (see [MI] **mi convert**) to change the style later. We typically choose wide to begin.

If your data are large, you may have to use flongsep. mi set flongsep requires you to specify a name for the flongsep dataset collection. See *Advice for using flongsep* in [MI] **styles**.

If you intend to have super-varying variables, you need to choose either flong or flongsep, or you will need to mi convert to flong or flongsep style later.

The current style of the data is shown by the mi query and mi describe commands; see [MI] **mi describe**.

mi register and mi unregister

mi register has three forms:

```
mi register imputed varlist
mi register passive varlist
mi register regular varlist
```

See [MI] **Glossary** for a definition of imputed, passive, and regular variables.

You are required to register imputed variables. If you intend to use mi impute (see [MI] **mi impute**) to impute missing values, you must still register the variables first.

Concerning passive variables, we recommend that you register them, and if your data are style wide, you are required to register them. If you create passive variables by using mi passive (see [MI] **mi passive**), that command automatically registers them for you.

Whether you register regular variables is up to you. Registering them is safer in all styles except wide, where it does not matter. We say registering is safer because regular variables should not vary across m, and in the long styles, you can unintentionally create variables that vary. If variables are registered, mi will detect and fix mistakes for you.

Super-varying variables—see [MI] **Glossary**—rarely occur, but if you have them, be aware that they can be stored only in flong and flongsep data and that they never should be registered.

The registration status of variables is listed by mi describe (see [MI] **mi describe**).

Use mi unregister if you accidentally register a variable incorrectly, with one exception: if you mistakenly register a variable as imputed but intended to register it as passive, or vice versa, use mi register directly to reregister the variable. The mere act of unregistering a passive or imputed variable can cause values in $m > 0$ to be replaced with those from $m = 0$ if the data are wide or mlong.

That exception aside, you first mi unregister variables before reregistering them.

mi set M and mi set m

mi set M is seldom used, and mi set m is sometimes used.

mi set M sets M, the total number of imputations. The syntax is

```
mi set M  = #
mi set M += #
mi set M -= #
```

mi set M = # sets $M = \#$. Imputations are added or deleted as necessary. If imputations are added, the new imputations obtain their values of imputed and passive variables from $m = 0$, which means that the missing values are not yet replaced in the new imputations. It is not necessary to increase M if you intend to use mi impute to impute values; see [MI] **mi impute**.

mi set M += # increments M by #.

mi set M -= # decrements M by #.

mi set m -= (*numlist*) deletes the specified imputations. For instance, if you had $M = 5$ imputations and wanted to delete imputation 2, leaving you with $M = 4$, you would type mi set m -= (2).

mi unset

If you wish to unset your data, your best choices are mi extract and mi export; see [MI] **mi extract** and [MI] **mi export**. The mi extract 0 command replaces the data in memory with the data from $m = 0$, unset. The mi export command replaces the data in memory with unset data in a form that can be sent to a non–Stata user.

mi unset is included for completeness, and if it has any use at all, it would be by programmers.

Also see

[MI] **intro** — Introduction to mi

[MI] **mi convert** — Change style of mi data

[MI] **mi describe** — Describe mi data

[MI] **mi extract** — Extract original or imputed data from mi data

[MI] **mi export** — Export mi data

[MI] **mi import** — Import data into mi

[MI] **mi XXXset** — Declare mi data to be svy, st, ts, xt, etc.

[MI] **styles** — Dataset styles

Title

> **mi stsplit** — Stsplit and stjoin mi data

Syntax

To split at designated times

> mi stsplit *newvar* $[$ *if* $]$, { at(*numlist*) | <u>eve</u>ry(#) } $[$ *options* $]$

options	Description
Main	
* at(*numlist*)	split at specified analysis times
* <u>eve</u>ry(#)	split when analysis time is a multiple of #
<u>af</u>ter(*spec*)	use time since *spec* instead of analysis time for at() or every()
trim	exclude observations outside of range
<u>noup</u>date	see [MI] **noupdate option**
<u>nopre</u>serve	programmer's option

* at() or every() is required.

nopreserve is not included in the dialog box.

To split at failure times

> mi stsplit $[$ *if* $]$, at(<u>f</u>ailures) $[$ *options* $]$

options	Description
Main	
* at(<u>f</u>ailures)	split at times of observed failures
<u>st</u>rata(*varlist*)	perform splitting by failures within stratum, strata defined by *varlist*
<u>ri</u>skset(*newvar*)	create risk-set ID variable
<u>noup</u>date	see [MI] **noupdate option**
<u>nopre</u>serve	programmer's option

* at() is required.

nopreserve is not included in the dialog box.

To join episodes

   mi stjoin [, *options*]

options	Description
Main	
<u>c</u>ensored(*numlist*)	values of failure that indicate no event
<u>noup</u>date	see [MI] **noupdate option**

Menu

Statistics > Multiple imputation

Description

mi stsplit and mi stjoin are stsplit and stjoin for mi data; see [ST] **stsplit**. Except for the addition of the noupdate option, the syntax is identical. Except for generalization across m, the results are identical.

Your mi data must be stset to use these commands. If your data are not already stset, use mi stset rather than the standard stset; see [MI] **mi XXXset**.

Options

noupdate in some cases suppresses the automatic mi update this command might perform; see [MI] **noupdate option**.

See [ST] **stsplit** for documentation on the remaining options.

Remarks

One should never use any heavyweight data-management commands with mi data. Heavyweight commands are commands that make sweeping changes to the data rather than simply deleting some observations, adding or dropping some variables, or changing some values of existing variables. stsplit and stjoin are examples of heavyweight commands (see [ST] **stsplit**).

Also see

[MI] **intro** — Introduction to mi

[ST] **stsplit** — Split and join time-span records

[MI] **mi XXXset** — Declare mi data to be svy, st, ts, xt, etc.

Title

mi test — Test hypotheses after mi estimate

Syntax

Test that coefficients are zero

> mi test *coeflist*

Test that coefficients within a single equation are zero

> mi test $\big[$ *eqno* $\big]$ $\big[$: *coeflist* $\big]$

Test that subsets of coefficients are zero (full syntax)

> mi test (*spec*) $\big[$(*spec*) ...$\big]$ $\big[$, *test_options* $\big]$

Test that subsets of transformed coefficients are zero

> mi testtransform *name* $\big[$ (*name*) ...$\big]$ $\big[$, *transform_options* $\big]$

test_options	Description
Test	
<u>ufmitest</u>	perform unrestricted FMI model test
<u>nosmall</u>	do not apply small-sample correction to degrees of freedom
<u>cons</u>tant	include the constant in coefficients to be tested

transform_options	Description
Test	
<u>ufmitest</u>	perform unrestricted FMI model test
<u>nosmall</u>	do not apply small-sample correction to degrees of freedom
<u>nolegend</u>	suppress transformation legend

coeflist may contain factor variables and time-series operators; see [U] **11.4.3 Factor variables** and [U] **11.4.4 Time-series varlists**.

coeflist is

> *coef* $\big[$ *coef* ...$\big]$
> [*eqno*]*coef* $\big[$ [*eqno*]*coef*...$\big]$
> [*eqno*] _b[*coef*] $\big[$ [*eqno*] _b[*coef*]...$\big]$

eqno is

> # #
> *eqname*

spec is

> *coeflist*
>
> [*eqno*] [: *coeflist*]

coef identifies a coefficient in the model; see [R] **test** for details. *eqname* is an equation name.

name is an expression name as specified with mi estimate or mi estimate using (see [MI] **mi estimate** or [MI] **mi estimate using**).

Menu

Statistics > Multiple imputation

Description

mi test performs joint tests of coefficients.

mi testtransform performs joint tests of transformed coefficients as specified with mi estimate or mi estimate using (see [MI] **mi estimate** or [MI] **mi estimate using**).

Options

⌐ Test ⌐

ufmitest specifies that the unrestricted fraction missing information (FMI) model test be used. The default test performed assumes equal fractions of information missing due to nonresponse for all coefficients. This is equivalent to the assumption that the between-imputation and within-imputation variances are proportional. The unrestricted test may be preferable when this assumption is suspect provided that the number of imputations is large relative to the number of estimated coefficients.

nosmall specifies that no small-sample adjustment be made to the degrees of freedom. By default, individual tests of coefficients (and transformed coefficients) use the small-sample adjustment of Barnard and Rubin (1999), and the overall model test uses the small-sample adjustment of Reiter (2007).

constant specifies that _cons be included in the list of coefficients to be tested when using the [*eqno*] form of *spec* with mi test. The default is to not include _cons.

nolegend, specified with mi testtransform, suppresses the transformation legend.

Remarks

Remarks are presented under the following headings:

> *Introduction*
> *Overview*
> *Example 1: Testing subsets of coefficients equal to zero*
> *Example 2: Testing linear hypotheses*
> *Example 3: Testing nonlinear hypotheses*

Introduction

The major issue arising when performing tests after MI estimation is the validity of the variance–covariance estimator (VCE) of the MI estimates. MI variance consists of two sources of variation: within-imputation variation and between-imputation variation. With a small number of imputations, the estimate of the between-imputation variance–covariance matrix is imprecise. In fact, when the number of imputations is less than or equal to the number of estimated parameters, the between-imputation matrix does not even have a full rank. As such, the estimated VCE may not be a valid variance–covariance matrix and thus not suitable for joint inference.

One solution to this problem was proposed by Rubin (1987) and Li et al. (1991). The idea is to assume that the between-imputation variance is proportional to the within-imputation variance. This assumption implies equal FMIs for all jointly tested parameters. Li et al. (1991) found that the procedure performs well in terms of power and maintaining the significance level even with moderately variable FMIs. mi test and mi testtransform, by default, perform tests using this procedure.

When the number of imputations is large enough relative to the number of tested parameters so that the corresponding VCE is trustworthy, you can request the unrestricted FMI test by specifying the ufmitest option. The unrestricted FMI test is the conventional test described by Rubin (1987, 77).

For testing nonlinear hypotheses, direct application of the conventional delta method to the estimated coefficients may not be feasible when the number of imputations is small enough that the VCE of the MI estimates cannot be used for inference. To test these hypotheses, one can first obtain MI estimates of the transformed coefficients by applying Rubin's combination rules to the transformed completed-data estimates and then apply the above MI-specific hypotheses tests to the combined transformed estimates. The first step can be done by specifying expressions with mi estimate (or mi estimate using). The second step is performed with mi testtransform. mi testtransform uses the same method to test transformed coefficients as mi test uses to test coefficients.

Overview

Use mi test to perform joint tests that coefficients are equal to zero:

```
. mi estimate: y x1 x2 x3 x4
. mi test x2 x3 x4
```

Use mi testtransform, however, to perform tests of more general linear hypotheses, such as _b[x1]=_b[x2], or _b[x1]=_b[x2] and _b[x1]=_b[x3]. Testing general linear hypotheses requires estimation of between and within variances corresponding to the specific hypotheses and requires recombining the imputation-specific estimation results. One way you could do that would be to refit the model and include the additional parameters during the estimation step. To test _b[x1]=_b[x2], you could type

```
. mi estimate (diff: _b[x1]-_b[x2]): regress y x1 x3 x3 x4
. mi testtransform diff
```

A better approach, however, is to save each of the imputation-specific results at the time the original model is fit and then later recombine results using mi estimate using. To save the imputation-specific results, specify mi estimate's saving() option when the model is originally fit:

```
. mi estimate, saving(myresults): regress y x1 x2 x3 x4
```

To test _b[x1]=_b[x2], you type

```
. mi estimate (diff: _b[x1]-_b[x2]) using myresults
. mi testtransform diff
```

The advantage of this approach is that you can test additional hypotheses without refitting the model. For instance, if we now wanted to test _b[x1]=_b[x2] and _b[x1]=_b[x3], we could type

```
. mi estimate (diff1: _b[x1]-_b[x2]) (diff2: _b[x1]=_b[x3]) using myresults
. mi testtransform diff1 diff2
```

To test nonlinear hypotheses, such as _b[x1]/_b[x2]=_b[x3]/_b[x4], we could then type

```
. mi estimate (diff: _b[x1]/_b[x2]-_b[x3]/_b[x4]) using myresults
. mi testtransform diff
```

Example 1: Testing subsets of coefficients equal to zero

We are going to test that tax, sqft, age, nfeatures, ne, custom, and corner are in the regression analysis of house resale prices we performed in *Example 1: Completed-data logistic analysis* of [MI] **mi estimate**. Following the advice above, when we fit the model, we are going to save the imputation-specific results even though we will not need them in this example; we will need them in the following examples.

```
. use http://www.stata-press.com/data/r12/mhouses1993s30
(Albuquerque Home Prices Feb15-Apr30, 1993)
. mi estimate, saving(miest): regress price tax sqft age nfeatures ne custom corner
```

Multiple-imputation estimates				Imputations	=	30
Linear regression				Number of obs	=	117
				Average RVI	=	0.0648
				Largest FMI	=	0.2533
				Complete DF	=	109
DF adjustment:	Small sample			DF: min	=	69.12
				avg	=	94.02
				max	=	105.51
Model F test:	Equal FMI			F(7, 106.5)	=	67.18
Within VCE type:	OLS			Prob > F	=	0.0000

price	Coef.	Std. Err.	t	P>|t|	[95% Conf. Interval]	
tax	.6768015	.1241568	5.45	0.000	.4301777	.9234253
sqft	.2118129	.069177	3.06	0.003	.0745091	.3491168
age	.2471445	1.653669	0.15	0.882	-3.051732	3.546021
nfeatures	9.288033	13.30469	0.70	0.487	-17.12017	35.69623
ne	2.518996	36.99365	0.07	0.946	-70.90416	75.94215
custom	134.2193	43.29755	3.10	0.002	48.35674	220.0818
corner	-68.58686	39.9488	-1.72	0.089	-147.7934	10.61972
_cons	123.9118	71.05816	1.74	0.085	-17.19932	265.0229

In the above mi estimate command, we use the saving() option to create a Stata estimation file called miest.ster, which contains imputation-specific estimation results.

mi estimate reports the joint test of all coefficients equal to zero in the header. We can reproduce this test with mi test by typing

```
. mi test tax sqft age nfeatures ne custom corner
note: assuming equal fractions of missing information
 ( 1)  tax = 0
 ( 2)  sqft = 0
 ( 3)  age = 0
 ( 4)  nfeatures = 0
 ( 5)  ne = 0
 ( 6)  custom = 0
 ( 7)  corner = 0
       F(  7, 106.5) =   67.18
            Prob > F =    0.0000
```

We obtain results identical to those from mi estimate.

We can test that a subset of coefficients, say, sqft and tax, are equal to zero by typing

```
. mi test sqft tax
note: assuming equal fractions of missing information
 ( 1)  sqft = 0
 ( 2)  tax = 0
       F(  2, 105.7) =  114.75
            Prob > F =    0.0000
```

Example 2: Testing linear hypotheses

Now we want to test the equality of the coefficients for sqft and tax. Following our earlier suggestion, we use mi estimate using to estimate the difference between coefficients (and avoid refitting the models) and then use mi testtransform to test that the difference is zero:

```
. mi estimate (diff: _b[tax]-_b[sqft]) using miest, nocoef
Multiple-imputation estimates          Imputations       =         30
Linear regression                      Number of obs     =        117
                                       Average RVI       =     0.1200
                                       Largest FMI       =     0.1100
                                       Complete DF       =        109
DF adjustment:    Small sample         DF:       min     =      92.10
                                                 avg     =      92.10
Within VCE type:          OLS                    max     =      92.10

        command: regress price tax sqft age nfeatures ne custom corner
           diff: _b[tax]-_b[sqft]
```

price	Coef.	Std. Err.	t	P>|t|	[95% Conf. Interval]
diff	.4649885	.1863919	2.49	0.014	.0948037	.8351733

```
. mi testtransform diff
note: assuming equal fractions of missing information
           diff: _b[tax]-_b[sqft]
 ( 1)  diff = 0
       F(  1,  92.1) =    6.22
            Prob > F =    0.0144
```

We suppress the display of the coefficient table by specifying the nocoef option with mi estimate using. We obtain the same results from the F test as those of the t test reported in the transformation table.

Similarly, we can test whether three coefficients are jointly equal:

```
. mi estimate (diff1: _b[tax]-_b[sqft]) (diff2: _b[custom]-_b[tax]) using miest, nocoef
Multiple-imputation estimates              Imputations     =        30
Linear regression                          Number of obs   =       117
                                           Average RVI     =    0.0748
                                           Largest FMI     =    0.1100
                                           Complete DF     =       109
DF adjustment:    Small sample             DF:      min    =     92.10
                                                    avg    =     97.95
Within VCE type:          OLS                       max    =    103.80

        command: regress price tax sqft age nfeatures ne custom corner
          diff1: _b[tax]-_b[sqft]
          diff2: _b[custom]-_b[tax]
```

price	Coef.	Std. Err.	t	P>|t|	[95% Conf. Interval]	
diff1	.4649885	.1863919	2.49	0.014	.0948037	.8351733
diff2	133.5425	43.30262	3.08	0.003	47.66984	219.4151

```
. mi testtr diff1 diff2
note: assuming equal fractions of missing information
          diff1: _b[tax]-_b[sqft]
          diff2: _b[custom]-_b[tax]
 ( 1)  diff1 = 0
 ( 2)  diff2 = 0
        F(  2, 105.6) =     7.34
            Prob > F =    0.0010
```

We estimate two differences, _b[tax]-_b[sqft] and _b[custom]-_b[tax], using mi estimate using and test whether they are jointly equal to zero by using mi testtransform.

We can perform tests of other hypotheses similarly by reformulating the hypotheses of interest such that we are testing equality to zero.

Example 3: Testing nonlinear hypotheses

In the examples above, we tested linear hypotheses. Testing nonlinear hypotheses is no different. We simply replace the specification of linear expressions in mi estimate using with the nonlinear expressions corresponding to the tests of interest.

For example, let's test that the ratio of the coefficients for tax and sqft is one, an equivalent but less efficient way of testing whether the two coefficients are the same. Similarly to the earlier example, we specify the corresponding nonlinear expression with mi estimate using and then use mi testtransform to test that the ratio is one:

```
. mi estimate (rdiff: _b[tax]/_b[sqft] - 1) using miest, nocoef
Multiple-imputation estimates              Imputations      =         30
Linear regression                          Number of obs    =        117
                                           Average RVI      =     0.0951
                                           Largest FMI      =     0.0892
                                           Complete DF      =        109
DF adjustment:    Small sample             DF:     min      =      95.33
                                                   avg      =      95.33
Within VCE type:           OLS                     max      =      95.33
        command: regress price tax sqft age nfeatures ne custom corner
          rdiff: _b[tax]/_b[sqft] - 1
```

price	Coef.	Std. Err.	t	P>|t|	[95% Conf. Interval]	
rdiff	2.2359	1.624546	1.38	0.172	-.9890876	5.460888

```
. mi testtr rdiff
note: assuming equal fractions of missing information
        rdiff: _b[tax]/_b[sqft] - 1
 ( 1)  rdiff = 0
       F(  1,  95.3) =    1.89
            Prob > F =    0.1719
```

We do not need to use `mi testtransform` (or `mi test`) to test one transformation (or coefficient) because the corresponding test is provided in the output from `mi estimate using`.

Saved results

`mi test` and `mi testtransform` save the following in `r()`:

Scalars
    `r(df)`   test constraints degrees of freedom
    `r(df_r)`   residual degrees of freedom
    `r(p)`   two-sided p-value
    `r(F)` F statistic
    `r(drop)`   1 if constraints were dropped, 0 otherwise
    `r(dropped_i)` index of ith constraint dropped

Methods and formulas

`mi test` and `mi testtransform` use the methodology described in *Multivariate case* under *Methods and formulas* of [MI] **mi estimate**, where we replace $\mathbf{q}$ with $\mathbf{Rq} - \mathbf{r}$ and $\mathbf{q}_0 = \mathbf{0}$ for the test H_0: $\mathbf{Rq} = \mathbf{r}$.

References

Barnard, J., and D. B. Rubin. 1999. Small-sample degrees of freedom with multiple imputation. *Biometrika* 86: 948–955.

Li, K.-H., X.-L. Meng, T. E. Raghunathan, and D. B. Rubin. 1991. Significance levels from repeated p-values with multiply-imputed data. *Statistica Sinica* 1: 65–92.

Reiter, J. P. 2007. Small-sample degrees of freedom for multi-component significance tests with multiple imputation for missing data. *Biometrika* 94: 502–508.

Rubin, D. B. 1987. *Multiple Imputation for Nonresponse in Surveys*. New York: Wiley.

Also see

Title

> **mi update** — Ensure that mi data are consistent

Syntax

```
mi update
```

Menu

Statistics > Multiple imputation

Description

mi update verifies that mi data are consistent. If the data are not consistent, mi update reports the inconsistencies and makes the necessary changes to make the data consistent.

mi update can change the sort order of the data.

Remarks

Remarks are presented under the following headings:

> *Purpose of mi update*
> *What mi update does*
> *mi update is run automatically*

Purpose of mi update

mi update allows you to

- change the values of existing variables, whether imputed, passive, regular, or unregistered;
- add or remove missing values from imputed variables (or from any variables);
- drop variables;
- create new variables;
- drop observations; and
- duplicate observations (but not add observations in other ways).

You can make any or all of the above changes and then type

```
. mi update
```

and mi update will handle making whatever additional changes are required to keep the data consistent. For instance,

```
. drop if sex==1
(75 observations deleted)
. mi update
(375 m>0 obs. dropped due to dropped obs. in m=0)
```

In this example, we happen to have five imputations and are working with flongsep data. We dropped 75 observations in $m = 0$, and that still left $5 \times 75 = 375$ observations to be dropped in $m > 0$.

310

The messages `mi update` produces vary according to the style of the data because the changes required to make the data consistent are determined by the style. Had we been working with flong data, we might have seen

```
. drop if sex==1
(450 observations deleted)
. mi update
(system variable _mi_id updated due to change in number of obs.)
```

With flong data in memory, when we dropped `if sex==1`, we dropped all $75 + 5 \times 75 = 450$ observations, so no more observations needed to be dropped; but here `mi update` needed to update one of its system variables because of the change we made.

Had we been working with mlong data, we might have seen

```
. drop if sex==1
(90 observations deleted)
. mi update
(system variable _mi_id updated due to change in number of obs.)
```

The story here is very much like the story in the flong case. In mlong data, dropping `if sex==1` drops the 75 observations in $m = 0$ and also drops the incomplete observations among the 75 in $m = 1$, $m = 2$, ..., $m = 5$. In this example, there are three such observations, so a total of $75 + 5 \times 3 = 90$ were dropped, and because of the change, `mi update` needed to update its system variable.

Had we been using wide data, we might have seen

```
. drop if sex==1
(75 observations deleted)
. mi update
```

`mi update`'s silence indicates that `mi update` did nothing, because after dropping observations in wide data, nothing more needs to be done. We could have skipped typing `mi update` here, but do not think that way because changing values, dropping variables, creating new variables, dropping observations, or creating new observations can have unanticipated consequences.

For instance, in our data is variable `farmincome`, and it seems obvious that `farmincome` should be 0 if the person does not have a farm, so we type

```
. replace farmincome = 0 if !farm
(15 real changes made)
```

After changing values, you should type `mi update` even if you do not suspect that it is necessary. Here is what happens when we do that with these data:

```
. mi update
(12 m=0 obs. now marked as complete)
```

Typing `mi update` was indeed necessary! We forgot that the `farmincome` variable was imputed, and it turns out that the variable contained missing in 12 nonfarm observations; `mi` needed to deal with that.

Running `mi update` is so important that `mi` itself is constantly running it just in case you forget. For instance, let's "forget" to type `mi update` and then convert our data to wide:

```
. replace farmincome = 0 if !farm
(15 real changes made)
. mi convert wide, clear
(12 m=0 obs. now marked as complete)
```

The parenthetical message was produced because mi convert ran mi update for us. For more information on this, see [MI] **noupdate option**.

What mi update does

- mi update checks whether you have changed N, the number of observations in $m = 0$, and resets N if necessary.

- mi update checks whether you have changed M, the number of imputations, and adjusts the data if necessary.

- mi update checks whether you have added, dropped, registered, or unregistered any variables and takes the appropriate action.

- mi update checks whether you have added or deleted any observations. If you have, it then checks whether you carried out the operation consistently for $m = 0$, $m = 1$, ..., $m = M$. If you have not carried it out consistently, mi update carries it out consistently for you.

- In the mlong, flong, and flongsep styles, mi update checks system variable _mi_id, which links observations across m, and reconstructs the variable if necessary.

- mi update checks that the system variable _mi_miss, which marks the incomplete observations, is correct and, if not, updates it and makes any other changes required by the change.

- mi update verifies that the values recorded in imputed variables in $m > 0$ are equal to the values in $m = 0$ when they are nonmissing and updates any that differ.

- mi update verifies that the values recorded in passive variables in $m > 0$ are equal to the values recorded in $m = 0$'s complete observations and updates any that differ.

- mi update verifies that the values recorded in regular variables in $m > 0$ equal the values in $m = 0$ and updates any that differ.

- mi update adds any new variables in $m = 0$ to $m > 0$.

- mi update drops any variables from $m > 0$ that do not appear in $m = 0$.

mi update is run automatically

As we mentioned before, running mi update is so important that many mi commands simply run it as a matter of course. This is discussed in [MI] **noupdate option**. In a nutshell, the mi commands that run mi update automatically have a noupdate option, so you can identify them, and you can specify the option to skip running the update and so speed execution, but only with the adrenaline rush caused by a small amount of danger.

Whether you specify noupdate or not, we advise you to run mi update periodically and to always run mi update after dropping or adding variables or observations, or changing values.

Also see

[MI] **intro** — Introduction to mi

[MI] **noupdate option** — The noupdate option

Title

mi varying — Identify variables that vary across imputations

Syntax

mi <u>vary</u>ing [*varlist*] [, <u>noup</u>date]

mi <u>vary</u>ing, <u>unregis</u>tered [<u>noup</u>date]

Menu

Statistics > Multiple imputation

Description

mi varying lists the names of variables that are unexpectedly varying and super varying; see [MI] **Glossary** for a definition of varying and super-varying variables.

Options

<u>unregistered</u> specifies that the listing be made only for unregistered variables. Specifying this option saves time, especially when the data are flongsep.

<u>noupdate</u> in some cases suppresses the automatic mi update this command might perform; see [MI] **noupdate option**.

Remarks

A variable is said to be varying if it varies over m in the complete observations. A variable is said to be super varying if it varies over m in the incomplete observations.

Remarks are presented under the following headings:

> *Detecting problems*
> *Fixing problems*

Detecting problems

mi varying looks for five potential problems:

1. *Imputed nonvarying*. Variables that are registered as imputed and are nonvarying either

   a. do not have their missing values in $m > 0$ filled in yet, in which case you should use mi impute (see [MI] **mi impute**) to impute them, or

   b. have no missing values in $m = 0$, in which case you should mi unregister the variables and perhaps use mi register to register the variables as regular (see [MI] **mi set**).

2. *Passive nonvarying.* Variables that are registered as passive and are nonvarying either

    a. have missing values in the incomplete observations in $m > 0$, in which case after you have filled in the missing values of your imputed variables, you should use `mi passive` (see [MI] **mi passive**) to update the values of these variables, or

    b. have no missing values in $m = 0$, in which case you should `mi unregister` the variables and perhaps use `mi register` to register the variables as regular (see [MI] **mi set**).

3. *Unregistered varying.*

    a. It is most likely that such variables should be registered as imputed or as passive.

    b. If the variables are varying but should not be, use `mi register` to register them as regular. That will fix the problem; values from $m = 0$ will be copied to $m > 0$.

    c. It is possible that this is just like potential problem 5, below, and it just randomly turned out that the only observations in which variation occurred were the incomplete observations. In that case, leave the variable unregistered.

4. *Unregistered super/varying.* These are variables that are super varying but would have been categorized as varying if they were registered as imputed. This is to say that while they have varying values in the complete observations as complete is defined this instant—which is based on the variables currently registered as imputed—these variables merely vary in observations for which they themselves contain missing in $m = 0$, and thus they could be registered as imputed without loss of information. Such variables should be registered as imputed.

5. *Unregistered super varying.* These variables really do super vary and could not be registered as imputed without loss of information. These variables either contain true errors or they are passive variables that are functions of groups of observations. Fix the errors by registering the variables as regular and leave unregistered those intended to be super varying. If you intentionally have super-varying variables in your data, remember never to convert to the wide or mlong styles. Super-varying variables can appear only in the flong and flongsep styles.

`mi varying` output looks like this:

```
              Possible problem   Variable names

          imputed nonvarying:    (none)
          passive nonvarying:    (none)
       unregistered varying:     (none)
*unregistered super/varying:     (none)
  unregistered super varying:    (none)

  * super/varying means super varying but would be varying if registered as
    imputed; variables vary only where equal to soft missing in m=0.
```

If there are possible problems, variable names are listed in the table.

Super-varying variables can arise only in flong and flongsep data, so the last two categories are omitted when `mi varying` is run on wide or mlong data. If there are no imputed variables, or no passive variables, or no unregistered variables, the corresponding categories are omitted from the table.

Fixing problems

If mi varying detects problems, register all imputed variables before registering passive variables. Rerun mi varying as you register new imputed variables. Registering new variables as imputed can change which observations are classified as complete and incomplete, and that classification in turn can change the categories to which the other variables are assigned. After registering a variable as imputed, another variable previously listed as super varying might now be merely varying.

Saved results

mi varying saves the following in r():

Macros
r(ivars)	nonvarying imputed variables
r(pvars)	nonvarying passive variables
r(uvars_v)	varying unregistered variables
r(uvars_s_v)	(super) varying unregistered variables
r(uvars_s_s)	super-varying unregistered variables

Also see

[MI] **intro** — Introduction to mi

[MI] **mi misstable** — Tabulate pattern of missing values

Title

---
mi xeq — Execute command(s) on individual imputations

Syntax

mi xeq [*numlist*]: *command* [; *command* [; ...]]

Description

mi xeq: *XXX* executes *XXX* on $m = 0$, $m = 1$, ..., $m = M$.

mi xeq *numlist*: *XXX* executes *XXX* on $m = numlist$.

XXX can be any single command or it can be multiple commands separated by a semicolon. If specifying multiple commands, the delimiter must not be set to semicolon; see [P] **#delimit**.

Remarks

Remarks are presented under the following headings:

> *Using mi xeq with reporting commands*
> *Using mi xeq with data-modification commands*
> *Using mi xeq with data-modification commands on flongsep data*

Using mi xeq with reporting commands

By reporting commands, we mean any general Stata command that reports results but leaves the data unchanged. summarize (see [R] **summarize**) is an example. mi xeq is especially useful with such commands. If you wanted to see the summary statistics for variables outcome and age among the females in your mi data, you could type

```
. mi xeq: summarize outcome age if sex=="female"
m=0 data:
-> summarize outcome age if sex=="female"
    (output omitted)

m=1 data:
-> summarize outcome age if sex=="female"
    (output omitted)

m=2 data:
-> summarize outcome age if sex=="female"
    (output omitted)
```

$M = 2$ in the data above.

If you wanted to see a particular regression run on the $m = 2$ data, you could type

```
. mi xeq 2: regress outcome age bp
m=2 data:
-> regress outcome age bp
    (output omitted)
```

In both cases, once the command executes, the entire mi dataset is brought back into memory.

Using mi xeq with data-modification commands

You can use data-modification commands with mi xeq but doing that is not especially useful unless you are using flongsep data.

If variable lnage were registered as passive and you wanted to update its values, you could type

```
. mi xeq: replace lnage = ln(age)
  (output omitted )
```

That would work regardless of style, although it is just as easy to update the variable using mi passive (see [MI] **mi passive**):

```
. mi passive: replace lnage = ln(age)
  (output omitted )
```

If what you are doing depends on the sort order of the data, include the sort command among the commands to be executed; do not assume that the individual datasets will be sorted the way the data in memory are sorted. For instance, if you have passive variable totalx, do not type

```
. sort id time
. mi xeq: by id: replace totalx = sum(x)
```

That will not work. Instead, type

```
. mi xeq: sort id time;  by id: replace totalx = sum(x)
m=0 data:
-> sort id time
-> by id: replace total x = sum(x)
(8 changes made)

m=1 data:
-> sort id time
-> by id: replace total x = sum(x)
(8 changes made)

m=2 data:
-> sort id time
-> by id: replace total x = sum(x)
(8 changes made)
```

Again we note that it would be just as easy to update this variable with mi passive:

```
. mi passive: by id (time): replace totalx = sum(x)
m=0:
(8 changes made)
m=1:
(8 changes made)
m=2:
(8 changes made)
```

With the wide, mlong, and flong styles, there is always another way to proceed, and often the other way is easier.

Using mi xeq with data-modification commands on flongsep data

With flongsep data, mi xeq is especially useful. Consider the case where you want to add new variable lnage = ln(age) to your data, and age is just a regular or unregistered variable. With flong, mlong, or wide data, you would just type

```
. generate lnage = ln(age)
```

and be done with it.

With flongsep data, you have multiple datasets to update. Of course, you could mi convert (see [MI] **mi convert**) your data to one of the other styles, but we will assume that if you had sufficient memory to do that, you would have done that long ago and so would not be using flongsep data.

The easy way to create lnage with flongsep data is by typing

```
. mi xeq: gen lnage = ln(age)
  (output omitted)
```

You could use the mi xeq approach with any of the styles, but with flong, mlong, or wide data, it is not necessary. With flongsep, it is.

Saved results

mi xeq saves in r() whatever the last command run on the last imputation or specified imputation returns. For instance,

```
. mi xeq: tabulate g ;  summarize x
```

returns the saved results for summarize x run on $m = M$.

```
. mi xeq 1 2: tabulate g ;  summarize x
```

returns the saved results for summarize x run on $m = 2$.

```
. mi xeq 0: summarize x
```

returns the saved results for summarize x run on $m = 0$.

Also see

[MI] **intro** — Introduction to mi

[MI] **mi passive** — Generate/replace and register passive variables

Title

> **mi XXXset** — Declare mi data to be svy, st, ts, xt, etc.

Syntax

mi fvset ...	see [R] **fvset**
mi svyset ...	see [SVY] **svyset**
mi stset ...	see [ST] **stset**
mi streset ...	
mi st ...	
mi tsset ...	see [TS] **tsset**
mi xtset ...	see [XT] **xtset**

Description

Using some features of Stata requires setting your data. The commands listed above allow you to do that with mi data. The mi variants have the same syntax and work the same way as the original commands.

Remarks

If you have set your data with any of the above commands before you mi set them, there is no problem; the settings were automatically imported. Once you mi set your data, however, you will discover that Stata's other set commands no longer work. For instance, here is the result of typing stset on an mi set dataset:

```
. stset ...
no; data are mi set
    Use mi stset to set or query these data; mi stset has the same
    syntax as stset.
r(119);
```

The solution is to use mi stset:

```
. mi stset ...
(usual output appears)
```

After mi setting your data, put mi in front of Stata's other set commands.

Also, you might sometimes see an error like the one above when you give a command that depends on the data being set by one of Stata's other set commands. In general, it is odd that you would be running such a command directly on mi data because what you will get will depend on the mi style of data. Perhaps, however, you are using mi wide data, where the structure of the data more or less corresponds to the structure of non-mi data, or perhaps you have smartly specified the appropriate if statement to account for the mi style of data you are using. In any case, the result might be

319

```
. some_other_command
no; data are mi set
    Use mi XXXset to set or query these data; mi XXXset has the same
    syntax as XXXset.
r(119);
```

Substitute one of the set commands listed above for XXXset, and then understand what just happened. You correctly used `mi XXXset` to set your data, you thought your data were set, yet when you tried to use a command that depended on the data being XXXset, you received this error.

If this happens to you, the solution is to use `mi extract` (see [MI] **mi extract**) to obtain the data on which you want to run the command—which is probably $m = 0$, so you would type `mi extract 0`—and then run the command.

Also see

[MI] **intro** — Introduction to mi

Title

noupdate option — The noupdate option

Syntax

mi ... [, ... <u>noupdate</u> ...]

Description

Many mi commands allow the noupdate option. This entry describes the purpose of that option.

Option

noupdate specifies that the mi command in question need not perform an mi update because you are certain that there are no inconsistencies that need fixing; see [MI] **mi update**. noupdate is taken as a suggestion; mi update will still be performed if the command sees evidence that it needs to be. Not specifying the option does not mean that an mi update will be performed.

Remarks

Some mi commands perform modifications to the data, and those modifications will go very poorly—even to the point of corrupting your data—if certain assumptions about your data are not true. Usually, those assumptions are true, but to be safe, the commands check the assumptions. They do this by calling mi update; see [MI] **mi update**. mi update checks the assumptions and, if they are not true, corrects the data so that the assumptions are true. mi update always reports the data corrections it makes.

All of this reflects an abundance of caution, with the result that some commands spend more time running mi update than they spend performing their intended task.

Commands that use mi update to verify assumptions have a noupdate option. When you specify that option, the command skips checking the assumptions, which is to say it skips calling mi update. More correctly, the command skips calling mi update if the command sees no obvious evidence that mi update needs to be called.

You can make commands run faster by specifying noupdate. Should you? Unless you are noticing poor performance, we would say no. It is, however, absolutely safe to specify noupdate if the only commands executed since the last mi update are mi commands. The following would be perfectly safe:

```
. mi update
. mi passive, noupdate: gen agesq = age*age
. mi rename age age_at_admission, noupdate
. mi ...
```

The following would be safe, too:

```
. mi update
. mi passive, noupdate: gen agesq = age*age
. summarize agesq
. mi rename age age_at_admission, noupdate
. mi ...
```

321

It would be safe because `summarize` is a reporting command that does not change the data; see [R] **summarize**.

The problem `mi` has is that it is not in control of your session and data. Between `mi` commands, `mi` does not know what you have done to the data. The following would not be recommended and has the potential to go very poorly:

```
. mi update
. mi passive, noupdate: gen agesq = age*age
. drop if female
. drop agesq
. mi ..., noupdate            // do not do this
```

By the rules for using `mi`, you should perform an `mi update` yourself after a `drop` command, or any other command that changes the data, but it usually does not matter whether you follow that rule because `mi` will check eventually, when it matters. That is, `mi` will check if you do not specify the `noupdate` option.

The `noupdate` option is recommended for use by programmers in programs that code a sequence of `mi` commands.

Also see

[MI] **intro** — Introduction to mi

[MI] **mi update** — Ensure that mi data are consistent

Title

styles — Dataset styles

Syntax

There are four dataset styles available for storing mi data:

wide

mlong

flong

flongsep

Description

The purpose of this entry is to familiarize you with the four styles in which mi data can be stored.

Remarks

Remarks are presented under the following headings:

The four styles
  Style wide
  Style flong
  Style mlong
  Style flongsep
  How we constructed this example
Using mi system variables
Advice for using flongsep

The four styles

We have highly artificial data, which we will first describe verbally and then show to you in each of the styles. The original data have two observations on two variables:

a	b
1	2
4	.

Variable b has a missing value. We have two imputed values for b, namely, 4.5 and 5.5. There will also be a third variable, c, in our dataset, where $c = a + b$.

Thus, in the jargon of mi, we have $M = 2$ imputations, and the datasets $m = 0$, $m = 1$, and $m = 2$ are

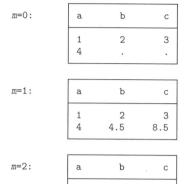

m=0:	a	b	c
	1	2	3
	4	.	.

m=1:	a	b	c
	1	2	3
	4	4.5	8.5

m=2:	a	b	c
	1	2	3
	4	5.5	9.5

Continuing with jargon, a is a regular variable, b is an imputed variable, and c is a passive variable.

Style wide

The above data have been stored in `miproto.dta` in the wide style.

```
. use http://www.stata-press.com/data/r12/miproto
(mi prototype)
. list
```

	a	b	c	_1_b	_2_b	_1_c	_2_c	_mi_miss
1.	1	2	3	2	2	3	3	0
2.	4	.	.	4.5	5.5	8.5	9.5	1

There is no significance to the order in which the variables appear.

On the left, under variables a, b, and c, you can see the original data.

The imputed values for b appear under the variables named _1_b and _2_b; $m = 1$ appears under _1_b, and $m = 2$ appears under _2_b. Note that in the first observation, the observed value of b is simply repeated in _1_b and _2_b. In the second observation, however, _1_b and _2_b show the replacement values for the missing value of b.

The passive values for c appear under the variables named _1_c and _2_c in the same way that the imputed values appeared under the variables named _1_b and _2_b.

Finally, one extra variable appears: _mi_miss. This is an example of an mi system variable. You are never to change mi system variables; they take care of themselves. The wide style has only one system variable. _mi_miss contains 0 for complete observations and 1 for incomplete observations.

Style flong

Let's convert this dataset to style flong:

```
. mi convert flong, clear
. list, separator(2)
```

	a	b	c	_mi_miss	_mi_m	_mi_id
1.	1	2	3	0	0	1
2.	4	.	.	1	0	2
3.	1	2	3	.	1	1
4.	4	4.5	8.5	.	1	2
5.	1	2	3	.	2	1
6.	4	5.5	9.5	.	2	2

We listed these data with a separator line after every two rows so that they would be easier to understand. Ignore the mi system variables and focus on variables a, b, and c. Observations 1 and 2 contain $m = 0$; observations 3 and 4 contain $m = 1$; observations 5 and 6 contain $m = 2$.

We will now explain the system variables, but you do not need to remember this.

1. We again see _mi_miss, just as we did in the wide style. It marks the incomplete observations in $m = 0$. It contains missing in $m > 0$.

2. _mi_m records m. The first two observations are $m = 0$; the next two, $m = 1$; and the last two, $m = 2$.

3. _mi_id records an arbitrarily coded observation-identification variable. It is 1 and 2 in $m = 0$, and then repeats in $m = 1$ and $m = 2$. Observations _mi_id = 1 correspond to each other for all m. The same applies to _mi_id = 2.

   Warning: Do not use _mi_id as your own ID variable. You might look one time, see that a particular observation has _mi_id = 8, and look a little later, and see that the observation has changed from _mi_id = 8 to _mi_id = 5. _mi_id belongs to mi. If you want your own ID variable, make your own. All that is true of _mi_id is that, at any instant, it uniquely identifies, and ties together, the observations.

There is no significance to the order of the variables or, for that matter, to the order of the observations.

Style mlong

Let's convert this dataset to the mlong style:

```
. mi convert mlong, clear
. list
```

	a	b	c	_mi_miss	_mi_m	_mi_id
1.	1	2	3	0	0	1
2.	4	.	.	1	0	2
3.	4	4.5	8.5	.	1	2
4.	4	5.5	9.5	.	2	2

This listing will be easier to read if we add some carefully chosen blank lines:

	a	b	c	_mi_miss	_mi_m	_mi_id
1.	1	2	3	0	0	1
2.	4	.	.	1	0	2
3.	4	4.5	8.5	.	1	2
4.	4	5.5	9.5	.	2	2

The mlong style is just like flong except that the complete observations—observations for which $\_mi\_miss = 0$ in $m = 0$—are omitted in $m > 0$.

Observations 1 and 2 are the original, $m = 0$ data.

Observation 3 is the $m = 1$ replacement observation for observation 2.

Observation 4 is the $m = 2$ replacement observation for observation 2.

Style flongsep

Let's look at these data in the flongsep style:

```
. mi convert flongsep example, clear
(files example.dta _1_example.dta _2_example.dta created)

. list
```

	a	b	c	_mi_miss	_mi_id
1.	1	2	3	0	1
2.	4	.	.	1	2

The flongsep style stores $m = 0$, $m = 1$, and $m = 2$ in separate files. When we converted to the flongsep style, we had to specify a name for these files, and we chose example. This resulted in $m = 0$ being stored in example.dta, $m = 1$ being stored in _1_example.dta, and $m = 2$ being stored in _2_example.dta.

In the listing above, we see the original, $m = 0$ data.

After conversion, $m = 0$ (example.dta) was left in memory. When working with flongsep data, you always work with $m = 0$ in memory. Nothing can stop us, however, from taking a brief peek:

```
. save example, replace
file example.dta saved

. use _1_example, clear
(mi prototype)

. list
```

	a	b	c	_mi_id
1.	1	2	3	1
2.	4	4.5	8.5	2

There are the data for $m = 1$. As previously, system variable _mi_id ties together observations. In the $m = 1$ data, however, _mi_miss is not repeated.

Let's now look at _2_example.dta:

```
. use _2_example, clear
(mi prototype)
. list
```

	a	b	c	_mi_id
1.	1	2	3	1
2.	4	5.5	9.5	2

And there are the data for $m = 2$.

We have an aside, but an important one. Review the commands we just gave, stripped of their output:

```
. mi convert flongsep example, clear
. list
. save example, replace
. use _1_example, clear
. list
. use _2_example, clear
. list
```

What we want you to notice is the line save example, replace. After converting to flongsep, for some reason we felt obligated to save the dataset. We will explain below. Now look farther down the history. After using _1_example.dta, we did not feel obligated to resave that dataset before using _2_example.dta. We will explain that below, too.

The flongsep style data are a matched set of datasets. You work with the $m = 0$ dataset in memory. It is your responsibility to save that dataset. Sometimes mi will have already saved the dataset for you. That was true here after mi convert, but it is impossible for you to know that in general, and it is your responsibility to save the dataset just as you would save any other dataset.

The $m > 0$ datasets, _#_name.dta, are mi's responsibility. We do not have to concern ourselves with saving them. Obviously, it was not necessary to save them here because we had just used the data and made no changes. The point is that, in general, the $m > 0$ datasets are not our responsibility. The $m = 0$ dataset, however, is our responsibility.

We are done with the demonstration:

```
. drop _all
. mi erase example
(files example.dta _1_example.dta _2_example.dta erased)
```

How we constructed this example

You might be curious as to how we constructed miproto.dta. Here is what we did:

```
. drop _all
. input a b
            a          b
  1. 1 2
  2. 4 .
  3. end
. mi set wide
. mi set M = 2
(2 imputations added; M = 2)
```

```
. mi register regular a
. mi register imputed b
. replace _1_b = 4.5 in 2
(1 real change made)
. replace _2_b = 5.5 in 2
(1 real change made)
. mi passive: gen c = a + b
m=0:
(1 missing value generated)
m=1:
m=2:
. order a b c _1_b _2_b _1_c _2_c _mi_miss
```

Using mi system variables

You can use mi's system variables to make some tasks easier. For instance, if you wanted to know the overall number of complete and incomplete observations, you could type

```
. tabulate _mi_miss
```

because in all styles, the _mi_miss variable is created in $m = 0$ containing 0 if complete and 1 if incomplete.

If you wanted to know the summary statistics for weight in $m = 1$, the general solution is

```
. mi xeq 1: summarize weight
```

If you were using wide data, however, you could instead type

```
. summarize _1_weight
```

If you were using flong data, you could type

```
. summarize weight if _mi_m==1
```

If you were using mlong data, you could type

```
. summarize weight if (_mi_m==0 & !_mi_miss) | _mi_m==1
```

Well, that last is not so convenient.

What is convenient to do directly depends on the style you are using. Remember, however, you can always switch between styles by using mi convert (see [MI] **mi convert**). If you were using mlong data and wanted to compare summary statistics of the weight variable in the original data and in all imputations, you could type

```
. mi convert wide
. summarize *weight
```

Advice for using flongsep

Use the flongsep style when your data are too big to fit into any of the other styles. If you already have flongsep data, you can try to convert it to another style. If you get the error "no room to add more observations" or "no room to add more variables", then you need to increase the amount of memory Stata is allowed to use (see [D] **memory**) or resign yourself to using the flongsep style.

There is nothing wrong with the flongsep style except that you need to learn some new habits. Usually, in Stata, you work with a copy of the data in memory, and the changes you make are not reflected in the underlying disk file until and unless you explicitly save the data. If you want to change the name of the data, you merely save them in a file of a different name. None of that is true when working with flongsep data. Flongsep data are a collection of datasets; you work with the one corresponding to $m = 0$ in memory, and `mi` handles keeping the others in sync. As you make changes, the datasets on disk change.

Think of the collection of datasets as having one name. That name is established when the flongsep data are created. There are three ways that can happen. You might start with a non-`mi` dataset in memory and `mi set` it; you might import a dataset into Stata and the result be flongsep; or you might convert another `mi` dataset to flongsep. Here are all the corresponding commands:

. mi set flongsep *name*	(1)
. mi import flongsep *name*	(2)
. mi import nhanes1 *name*	
. mi convert flongsep *name*	(3)

In each command, you specify a name and that name becomes the name of the flongsep dataset collection. In particular, *name*.dta becomes $m = 0$, _1_*name*.dta becomes $m = 1$, _2_*name*.dta becomes $m = 2$, and so on. You use flongsep data by typing use *name*, just as you would any other dataset. As we said, you work with $m = 0$ in memory and `mi` handles the rest.

Flongsep data are stored in the current (working) directory. Learn about `pwd` to find out where you are and about `cd` to change that; see [D] **cd**.

As you work with flongsep data, it is your responsibility to save *name*.dta almost as it would be with any Stata dataset. The difference is that `mi` might and probably has saved *name*.dta along the way without mentioning the fact, and `mi` has doubtlessly updated the _#_*name*.dta datasets, too. Nevertheless, it is still your responsibility to save *name*.dta when you are done because you do not know whether `mi` has saved *name*.dta recently enough. It is not your responsibility to worry about _#_*name*.dta.

It is a wonderful feature of Stata that you can usually work with a dataset in memory without modifying the original copy on disk except when you intend to update it. It is a unpleasant feature of flongsep that the same is not true. We therefore recommend working with a copy of the data, and `mi` provides an `mi copy` command (see [MI] **mi copy**) for just that purpose:

    . mi copy *newname*

With flongsep data in memory, when you type `mi copy` *newname*, the current flongsep files are saved in their existing name (this is one case where you are not responsible for saving *name*.dta), and then the files are copied to *newname*, meaning that $m = 0$ is copied to *newname*.dta, $m = 1$ is copied to _1_*newname*.dta, and so on. You are now working with the same data, but with the new name *newname*.

As you work, you may reach a point where you would like to save the data collection under *name* and continue working with *newname*. Do the following:

    . mi copy *name*, replace
    . use *newname*

When you are done for the day, if you want your data saved, do not forget to save them by using `mi copy`. It is also a good idea to erase the flongsep *newname* dataset collection:

    . mi copy *name*, replace
    . mi erase *newname*

By the way, *name*.dta, _1_*name*.dta, ... are just ordinary Stata datasets. By using general (non-mi) Stata commands, you can look at them and even make changes to them. Be careful about doing the latter; see [MI] **technical**.

See [MI] **mi copy** to learn more about mi copy.

Also see

[MI] **intro** — Introduction to mi

[MI] **mi copy** — Copy mi flongsep data

[MI] **mi erase** — Erase mi datasets

[MI] **technical** — Details for programmers

Title

Description

Technical information for programmers who wish to extend `mi` is provided below.

Remarks

Remarks are presented under the following headings:

Notation

$$M = \text{\# of imputations}$$

$$m = \text{imputation number}$$

       0. original data with missing values

       1. first imputation dataset

       .

       .

$$M. \text{ last imputation dataset}$$

$$N = \text{number of observations in } m = 0$$

Definition of styles

Style describes how the mi data are stored. There are four styles: wide, mlong, flong, and flongsep.

Style all

Characteristics:

    _dta[_mi_marker]   "_mi_ds_1"

Description: _dta[_mi_marker] is set with all styles, including flongsep_sub. The definitions below apply only if "`_dta[_mi_marker]'" = "_mi_ds_1".

Style wide

Characteristics:

_dta[_mi_style]	"wide"
_dta[_mi_M]	M
_dta[_mi_ivars]	imputed variables; variable list
_dta[_mi_pvars]	passive variables; variable list
_dta[_mi_rvars]	regular variables; variable list
_dta[_mi_update]	time last updated; %tc_value/1000

Variables:

_mi_miss	whether incomplete; 0 or 1
_#_varname	*varname* for $m = \#$, defined for each
	'_dta[_mi_ivars]' and '_dta[_mi_pvars]'

Description: $m = 0$, $m = 1$, ..., $m = M$ are stored in one dataset with $\_N = N$ observations. Each imputed and passive variable has M additional variables associated with it. If variable bp contains the values in $m = 0$, then values for $m = 1$ are contained in variable _1_bp, values for $m = 2$ in _2_bp, and so on. wide stands for *wide*.

Style mlong

Characteristics:

_dta[_mi_style]	"mlong"
_dta[_mi_M]	M
_dta[_mi_N]	N
_dta[_mi_n]	# of observations in marginal
_dta[_mi_ivars]	imputed variables; variable list
_dta[_mi_pvars]	passive variables; variable list
_dta[_mi_rvars]	regular variables; variable list
_dta[_mi_update]	time last updated; %tc_value/1000

Variables:

_mi_m	m; 0, 1, ..., M
_mi_id	ID; 1, ..., N
_mi_miss	whether incomplete; 0 or 1 if _mi_m = 0, else .

Description: $m = 0$, $m = 1$, ..., $m = M$ are stored in one dataset with $\_N = N + M \times n$ observations, where n is the number of incomplete observations in $m = 0$. mlong stands for *marginal long*.

Style flong

Characteristics:

_dta[_mi_style]	"flong"
_dta[_mi_M]	M
_dta[_mi_N]	N
_dta[_mi_ivars]	imputed variables; variable list
_dta[_mi_pvars]	passive variables; variable list
_dta[_mi_rvars]	regular variables; variable list
_dta[_mi_update]	time last updated; %tc_value/1000

Variables:

_mi_m	m; 0, 1, ..., M
_mi_id	ID; 1, ..., N
_mi_miss	whether incomplete; 0 or 1 if _mi_m = 0, else .

Description: $m = 0$, $m = 1$, ..., $m = M$ are stored in one dataset with $\_N = N + M \times N$ observations, where N is the number of observations in $m = 0$. flong stands for *full long*.

Style flongsep

Characteristics:

_dta[_mi_style]	"flongsep"
_dta[_mi_name]	*name*
_dta[_mi_M]	M
_dta[_mi_N]	N
_dta[_mi_ivars]	imputed variables; variable list
_dta[_mi_pvars]	passive variables; variable list
_dta[_mi_rvars]	regular variables; variable list
_dta[_mi_update]	time last updated; %tc_value/1000

Variables:

_mi_id	ID; 1, ..., N
_mi_miss	whether incomplete; 0 or 1

Description: $m = 0$, $m = 1$, ..., $m = M$ are each separate .dta datasets. If $m = 0$ data are stored in pat.dta, then $m = 1$ data are stored in _1_pat.dta, $m = 2$ in _2_pat.dta, and so on.

The definitions above apply only to $m = 0$, the dataset named '_dta[_mi_name]'.dta. See *Style flongsep_sub* directly below for $m > 0$. flongsep stands for *full long and separate*.

Style flongsep_sub

Characteristics:

_dta[_mi_style]	"flongsep_sub"
_dta[_mi_name]	*name*
_dta[_mi_m]	m; 0, 1, ..., M

Variables:

_mi_id	ID; 1, ..., N

Description: The description above applies to the _'_dta[_mi_m]'_'_dta[_mi_name]'.dta datasets. There are M such datasets recording $m = 1$, ..., M used by the flongsep style directly above.

Adding new commands to mi

New commands are written in ado. Name the new command mi_cmd_*newcmd* and store it in mi_cmd_*newcmd*.ado. When the user types mi *newcmd* ..., mi_cmd_*newcmd*.ado will be executed.

See *Writing programs for use with mi* of [P] **program properties** for details on how to write estimation commands for use with the mi estimate prefix.

Outline for new commands

```
program mi_cmd_newcmd, rclass                              (1)
        version 12
        u_mi_assert_set                                    (2)
        syntax ... [, ... noUPdate ...]                    (3)
        ...
        u_mi_certify_data, acceptable                      (4)
        ...
        if ("`update'"=="") {
                u_mi_certify_data, proper                  (5)
        }
        ...
end
```

Notes:

1. The command may be rclass; that is not required. It may be eclass instead if you wish.

2. u_mi_assert_set verifies that the data are mi data; see *u_mi_assert_set* below.

3. If you intend for your command to use mi update to update the data before performing its intended task, include a noupdate option; see [MI] **noupdate option**. Some commands instead or in addition run mi update to perform cleanup after performing their task. Such use does not require a noupdate option.

4. u_mi_certify_data is the internal routine that performs mi update. An update is divided into two parts, called acceptable and proper. All commands should verify that the data are acceptable; see *u_mi_certify_data* below.

5. u_mi_certify_data, proper performs the second step of mi update; it verifies that acceptable data are proper. Whether you verify properness is up to you, but if you do, you are supposed to include a noupdate option to skip running the check.

Utility routines

The only information you absolutely need to know is that already revealed. Using the utility routines described below, however, will simplify your programming task and make your code appear more professional to the end user.

As you read what follows, remember that you may review the source code for the routines by using viewsource; see [P] **viewsource**. If you wanted to see the source for u_mi_assert_set, you would type viewsource u_mi_assert_set.ado. If you do this, you will sometimes see that the routines allow options not documented below. Ignore those options; they may not appear in future releases.

Using viewsource, you may also review examples of the utility commands being used by viewing the source of the mi commands we have written. Each mi command appears in the file mi_cmd_*command*.ado. Also remember that other mi commands make useful utility routines. For instance, if your new command makes passive variables, use mi register to register them. Always call existing mi commands through mi; code mi passive and not mi_cmd_passive.

u_mi_assert_set

u_mi_assert_set $\big[$ *desired_style* $\big]$

This utility verifies that data are mi and optionally of the desired style; it issues the appropriate error message and stops execution if not. The optional argument *desired_style* can be wide, mlong, flong, or flongsep, but is seldom specified. When not specified, any style is allowed.

u_mi_certify_data

u_mi_certify_data $\big[$, acceptable proper noupdate sortok $\big]$

This command performs mi update. mi update is equivalent to u_mi_certify_data, acceptable proper sortok.

Specify one or both of acceptable and proper. If the noupdate option is specified, then proper is specified. The sortok option specifies that u_mi_certify_data need not spend extra time to preserve and restore the original sort order of the data.

An update is divided into two parts. In the first part, called acceptable, $m = 0$ and the _dta[_mi_*] characteristics are certified. Your program will use the information recorded in those characteristics, and before that information can be trusted, the data must be certified as acceptable. Do not trust any _dta[_mi_*] characteristics until you have run u_mi_certify_data, acceptable.

u_mi_certify_data, proper verifies that data known to be acceptable are proper. In practice, this means that in addition to trusting $m = 0$, you can trust $m > 0$.

Running u_mi_certify_data, acceptable might actually result in the data being certified as proper, although you cannot depend on that. When you run u_mi_certify_data, acceptable and certain problems are observed in $m = 0$, they are fixed in all m, which can lead to other problems being detected, and by the time the whole process is through, the data are proper.

u_mi_no_sys_vars and u_mi_no_wide_vars

u_mi_no_sys_vars "*variable_list*" $\big[$ "*word*" $\big]$

u_mi_no_wide_vars "*variable_list*" $\big[$ "*word*" $\big]$

These routines are for use in parsing user input.

u_mi_no_sys_vars verifies that the specified list of variable names does not include any mi system variables such as _mi_m, _mi_id, _mi_miss, etc.

u_mi_no_wide_vars verifies that the specified list of variable names does not include any style wide $m > 0$ variables of the form _#_varname. u_mi_no_wide_vars may be called with any style of data but does nothing if the style is not wide.

Both functions issue appropriate error messages if problems are found. If *word* is specified, the error message will be "*word* may not include ...". Otherwise, the error message is "may not specify ...".

u_mi_zap_chars

u_mi_zap_chars

u_mi_zap_chars deletes all _dta[_mi_*] characteristics from the data in memory.

u_mi_xeq_on_tmp_flongsep

> u_mi_xeq_on_tmp_flongsep [, nopreserve]: *command*

u_mi_xeq_on_tmp_flongsep executes *command* on the data in memory, said data converted to style flongsep, and then converts the flongsep result back to the original style. If the data already are flongsep, a temporary copy is made and, at the end, posted back to the original. Either way, *command* is run on a temporary copy of the data. If anything goes wrong, the user's original data are restored; that is, they are restored unless nopreserve is specified. If *command* completes without error, the flongsep data in memory are converted back to the original style and the original data are discarded.

It is not uncommon to write commands that can deal only with flongsep data, and yet these seem to users as if they work with all styles. That is because the routines use u_mi_xeq_on_tmp_flongsep. They start by allowing any style, but the guts of the routine are written assuming flongsep. mi stjoin is implemented in this way. There are two parts to mi stjoin: mi_cmd_stjoin.ado and mi_sub_stjoin_flongsep.ado. mi_cmd_stjoin.ado ends with

> u_mi_xeq_on_tmp_flongsep: mi_sub_stjoin_flongsep 'if', 'options'

mi_sub_stjoin_flongsep does all the work, while u_mi_xeq_on_tmp_flongsep handles the issue of converting to flongsep and back again. The mi_sub_stjoin_flongsep subroutine must appear in its own ado-file because u_mi_xeq_on_tmp_flongsep is itself implemented as an ado-file. u_mi_xeq_on_tmp_flongsep would be unable to find the subroutine otherwise.

u_mi_get_flongsep_tmpname

> u_mi_get_flongsep_tmpname *macname* : *basename*

u_mi_get_flongsep_tmpname creates a temporary flongsep name based on *basename* and stores it in the local macro *macname*. u_mi_xeq_on_tmp_flongsep, for your information, obtains the temporary name it uses from this routine.

u_mi_get_flongsep_tmpname is seldom used directly because u_mi_xeq_on_tmp_flongsep works well for shifting temporarily into flongsep mode, and u_mi_xeq_on_tmp_flongsep does a lot more than just getting a name under which the data should be temporarily stored. There are instances, however, when one needs to be more involved in the conversion. For examples, see the source mi_cmd_append.ado and mi_cmd_merge.ado. The issue these two routines face is that they need to shift two input datasets to flongsep, then they create a third from them, and that is the only one that needs to be shifted back to the original style. So these two commands handle the conversions themselves using u_mi_get_flongsep_tmpname and mi convert (see [MI] **mi convert**).

For instance, they start with something like

> u_mi_get_flongsep_tmpname master : __mimaster

That creates a temporary name suitable for use with mi convert and stores it in 'master'. The suggested name is __mimaster, but if that name is in use, then u_mi_get_flongsep_tmpname will form from it __mimaster1, or __mimaster2, etc. We recommend that you specify a *basename* that begins with __mi, which is to say, two underscores followed by mi.

Next you must appreciate that it is your responsibility to eliminate the temporary files. You do that by coding something like

```
        ...
        local origstyle "'_dta[_mi_style]'"
        if ("'origstyle'"=="flongsep") {
                local origstyle "'origstyle' '_dta[_mi_name]'"
        }
        u_mi_get_flongsep_tmpname master : __mimaster
        capture {
                quietly mi convert flongsep 'master'
                ...
                ...
                quietly mi convert 'origstyle', clear replace
        {
        nobreak {
                local rc = _rc
                mata: u_mi_flongsep_erase("'master'", 0, 0)
                if ('rc') {
                        exit 'rc'
                }
        }
```

The other thing to note above is our use of mi convert 'master' to convert our data to flongsep under the name 'master'. What, you might wonder, happens if our data already is flongsep? A nice feature of mi convert is that when run on data that are already flongsep, it performs an mi copy; see [MI] **mi copy**.

mata: u_mi_flongsep_erase()

> mata: u_mi_flongsep_erase("*name*", *from* $\left[\, , \textit{output}\right]$)

where

name	*string*; flongsep name
from	#; where to begin erasing
output	0|1; whether to produce output

mata: u_mi_flongsep_erase() is the internal version of mi erase (see [MI] **mi erase**); use whichever is more convenient.

Input *from* is usually specified as 0 and then mata: u_mi_flongsep_erase() erases *name*.dta, _1_*name*.dta, _2_*name*.dta, and so on. *from* may be specified as a number greater than zero, however, and then erased are _<*from*>_*name*.dta, _<*from+1*>_*name*.dta, _<*from+2*>_*name*.dta, ...

If *output* is 0, no output is produced; otherwise, the erased files are also listed. If *output* is not specified, files are listed.

See viewsource u_mi.mata for the source code for this routine.

u_mi_sortback

> u_mi_sortback *varlist*

u_mi_sortback removes dropped variables from *varlist* and sorts the data on the remaining variables. The routine is for dealing with sort-preserve problems when program *name*, sortpreserve is not adequate, such as when the data might be subjected to substantial editing between the preserving of the sort order and the restoring of it. To use u_mi_sortback, first record the order of the data:

```
local sortedby : sortedby
tempvar recnum
gen long 'recnum' = _n
quietly compress 'recnum'
```

Later, when you want to restore the sort order, you code

```
u_mi_sortback 'sortedby' 'recnum'
```

u_mi_save and u_mi_use

u_mi_save *macname* : *filename* [, *save_options*]

u_mi_use '"'*macname*'"' *filename* [, clear nolabel]

save_options are as described in [D] **save**. clear and nolabel are as described in [D] **use**. In both commands, *filename* must be specified in quotes if it contains any special characters or blanks.

It is sometimes necessary to save data in a temporary file and reload them later. In such cases, when the data are reloaded, you would like to have the original c(filename), c(filedate), and c(changed) restored. u_mi_save saves that information in *macname*. u_mi_use restores the information from the information saved in *macname*. Note the use of compound quotes around '*macname*' in u_mi_use; they are not optional.

mata: u_mi_wide_swapvars()

mata: u_mi_wide_swapvars(*m*, *tmpvarname*)

where

m	#; $1 \le \# \le M$
tmpvarname	*string*; name from tempvar

This utility is for use with wide data only. For each variable name contained in _dta[_mi_ivars] and _dta[_mi_pvars], mata: u_mi_wide_swapvars() swaps the contents of *varname* with _*m*_varname. Argument *tmpvarname* must be the name of a temporary variable obtained from command tempvar, and the variable must not exist. mata: u_mi_wide_swapvars() will use this variable while swapping. See [P] **macro** for more information on tempvar.

This function is its own inverse, assuming _dta[_mi_ivars] and _dta[_mi_pvars] have not changed.

See viewsource u_mi.mata for the source code for this routine.

u_mi_fixchars

u_mi_fixchars [, acceptable proper]

u_mi_fixchars makes the data and variable characteristics the same in $m = 1$, $m = 2$, ..., $m = M$ as they are in $m = 0$. The options specify what is already known to be true about the data, that the data are known to be acceptable or known to be proper. If neither is specified, you are stating that you do not know whether the data are even acceptable. That is okay. u_mi_fixchars handles performing whatever certification is required. Specifying the options makes u_mi_fixchars run faster.

This stabilizing of the characteristics is not about mi's characteristics; that is handled by u_mi_certify_data. Other commands of Stata set and use characteristics, while u_mi_fixchars ensures that those characteristics are the same across all m.

mata: u_mi_cpchars_get() and mata: u_mi_cpchars_put()

mata: u_mi_cpchars_get(*matavar*)

mata: u_mi_cpchars_put(*matavar*, {0|1|2})

where *matavar* is a Mata transmorphic variable. Obtain *matavar* from u_mi_get_mata_instanced_var() when using these functions from Stata.

These routines replace the characteristics in one dataset with those of another. They are used to implement u_mi_fixchars.

mata: u_mi_cpchars_get(*matavar*) stores in *matavar* the characteristics of the data in memory. The data in memory remain unchanged.

mata: u_mi_cpchars_put(*matavar*, #) replaces the characteristics of the data in memory with those previously recorded in *matavar*. The second argument specifies the treatment of _dta[_mi_*] characteristics:

0	delete them in the destination data
1	copy them from the source just like any other characteristic
2	retain them as-is from the destination data.

mata: u_mi_get_mata_instanced_var()

mata: u_mi_get_mata_instanced_var("*macname*", "*basename*" [, *i_value*])

where

macname	name of local macro
basename	suggested name for instanced variable
i_value	initial value for instanced variable

mata: u_mi_get_mata_instanced_var() creates a new Mata global variable, initializes it with *i_value* or as a 0×0 real, and places its name in local macro *macname*. Typical usage is

```
local var
capture noisily {
        mata: u_mi_get_mata_instanced_var("var", "myvar")
        ...
        ... use 'var' however you wish ...
        ...
}
nobreak {
        local rc = _rc
        capture mata: mata drop 'var'
        if ('rc') {
                exit 'rc'
        }
}
```

mata: u_mi_ptrace_*()

h = u_mi_ptrace_open(*"filename"*, {"r"|"w"} [, {0|1}])

u_mi_ptrace_write_stripes(h, *id*, *ynames*, *xnames*)

u_mi_ptrace_write_iter(h, *m*, *iter*, *B*, *V*)

u_mi_ptrace_close(h)

u_mi_ptrace_safeclose(h)

The above are Mata functions, where

> h, if it is declared, should be declared transmorphic
> *id* is a string scalar
> *ynames* and *xnames* are string scalars
> *m* and *iter* are real scalars
> B and V are real matrices; V must be symmetric

These routines write parameter-trace files; see [MI] **mi ptrace**. The procedure is 1) open the file; 2) write the stripes; 3) repeatedly write iteration information; and 4) close the file.

1. Open the file: *filename* may be specified with or without a file suffix. Specify the second argument as "w". The third argument should be 1 if the file may be replaced when it exists, and 0 otherwise.

2. Write the stripes: Specify *id* as the name of your routine or as ""; mi ptrace describe will show this string as the creator of the file if the string is not "". *ynames* and *xnames* are both string scalars containing space-separated names or, possibly, *op.names*.

3. Repeatedly write iteration information: Written are m, the imputation number; *iter*, the iteration number; B, the matrix of coefficients; and V, the variance matrix. B must be $ny \times nx$ and V must be $ny \times ny$ and symmetric, where $nx =$ length(tokens(*xnames*)) and $ny =$ length(tokens(*ynames*)).

4. Close the file: In Mata, use u_mi_ptrace_close(h). It is highly recommended that, before step 1, h be obtained from inside Stata (not Mata) using mata: u_mi_get_mata_instanced_var("h", "*myvar*"). If you follow this advice, include a mata: u_mi_ptrace_safeclose('h') in the ado-file cleanup code. This will ensure that open files are closed if the user presses *Break* or something else causes your routine to exit before the file is closed. A correctly written program will have two closes, one in Mata and another in the ado-file, although you could omit the one in Mata. See *mata: u_mi_get_mata_instanced_var()* directly above.

Also included in u_mi_ptrace_*() are routines to read parameter-trace files. You should not need these routines because users will use Stata command mi ptrace use to load the file you have written. If you are interested, however, then type viewsource u_mi_ptrace.mata.

How to write other set commands to work with mi

This section concerns the writing of other set commands such as [ST] **stset** or [XT] **xtset**—set commands having nothing to do with mi—so that they properly work with mi.

The definition of a set command is any command that creates characteristics in the data, and possibly creates variables in the data, that other commands in the suite will subsequently access. Making such set commands work with mi is mostly mi's responsibility, but there is a little you need to do to assist mi. Before dealing with that, however, write and debug your set command ignoring mi. Once that is done, go back and add a few lines to your code. We will pretend your set command is named mynewset and your original code looks something like this:

```
program mynewset
    ...
    syntax ... [, ... ]
    ...
end
```

Our goal is to make it so that mynewset will not run on mi data while simultaneously making it so that mi can call it (the user types mi mynewset). When the user types mi mynewset, mi will 1) give mynewset a clean, $m = 0$ dataset on which it can run and 2) duplicate whatever mynewset does to $m = 0$ on $m = 1$, $m = 2$, ..., $m = M$.

To achieve this, modify your code to look like this:

```
program mynewset
    ...
    syntax ... [, ... MI]                               (1)
    if ("'mi'"=="") {                                   (2)
            u_mi_not_mi_set "mynewset"
            local checkvars "*"                         (3)
    }
    else {
            local checkvars "u_mi_check_setvars settime" (3)
    }
    ...
    'checkvars' 'varlist'                               (4)
    ...
end
```

That is,

1. Add the mi option to any options you already have.

2. If the mi option is not specified, execute u_mi_not_mi_set, passing to it the name of your set command. If the data are not mi, then u_mi_not_mi_set will do nothing. If the data are mi, then u_mi_not_mi_set will issue an error telling the user to run mi mynewset.

3. Set new local macro checkvars to * if the mi option is not specified, and otherwise to u_mi_check_setvars. We should mention that the mi option will be specified when mi mynewset calls mynewset.

4. Run 'checkvars' on any input variables mynewset uses that must not vary across m. mi does not care about other variables or even about new variables mynewset might create; it cares only about existing variables that should not vary across m.

Let's understand what "'checkvars' *varlist*" does. If the mi option was not specified, the line expands to "* *varlist*", which is a comment, and does nothing. If the mi option was specified, the line expands to "u_mi_check_setvars settime *varlist*". We are calling mi routine u_mi_check_setvars, telling it that we are calling at set time, and passing along *varlist*. u_mi_check_setvars will verify that *varlist* does not contain mi system variables or variables that vary across m. Within mynewset, you may call 'checkvars' repeatedly if that is convenient.

You have completed the changes to mynewset. You finally need to write one short program that reads

```
program mi_cmd_mynewset
        version 12
        mi_cmd_genericset '"mynewset '0'"' "_mynewset_x _mynewset_y"
end
```

In the above, we assume that mynewset might add one or two variables to the data named _mynewset_x and _mynewset_y. List in the second argument all variables mynewset might create. If mynewset never creates new variables, then the program should read

```
program mi_cmd_mynewset
        version 12
        mi_cmd_genericset '"mynewset '0'"'
end
```

You are done.

Also see

Title

workflow — Suggested workflow

Description

Provided below are suggested workflows for working with original data and for working with data that already have imputations.

Remarks

Remarks are presented under the following headings:

Suggested workflow for original data
Suggested workflow for data that already have imputations
Example

Suggested workflow for original data

By original data, we mean data with missing values for which you do not already have imputations. Your task is to identify the missing values, impute values for them, and perform estimation.

mi does not have a fixed order in which you must perform tasks except that you must mi set the data first.

1. mi set your data; see [MI] **mi set**.

   Set the data to be wide, mlong, flong, or flongsep. Choose flongsep only if your data are bumping up against the constraints of memory. Choose flong or flongsep if you will need super-varying variables.

   Memory is not usually a problem, and super-varying variables are seldom necessary, so we generally start with the data as wide:

   . use *originaldata*

   . mi set wide

   If you need to use flongsep, you also need to specify a name for the flongsep dataset collection. Choose a name different from the current name of the dataset:

   . use *originaldata*

   . mi set flongsep *newname*

   If the original dataset is chd.dta, you might choose chdm for *newname*. *newname* does not include the .dta suffix. If you choose chdm, the data will then be stored in chdm.dta, _1_chdm.dta, and so on. It is important that you choose a name different from *originaldata* because you do not want your mi data to overwrite the original. Stata users are used to working with a copy of the data in memory, meaning that the changes made to the data are not reflected in the .dta dataset until the user saves them. With flongsep data, however, changes are made to the mi .dta dataset collection as you work. See *Advice for using flongsep* in [MI] **styles**.

2. Use mi describe often; see [MI] **mi describe**.

   mi describe will not tell you anything useful yet, but as you set more about the data, mi describe will be more informative.

   . mi describe

3. Use mi misstable to identify missing values; see [MI] **mi misstable**.

   mi misstable is the standard misstable (see [R] **misstable**) but tailored for mi data. Several Stata commands have mi variants—become familiar with them. If there is no mi variant, then it is generally safe to use the standard command directly, although it may not be appropriate. For instance, typing misstable rather than mi misstable would produce appropriate results right now, but it would not produce appropriate results later. If mi datasets $m = 0$, $m = 1$, ..., $m = M$ exist and you run misstable, you might end up running the command on a strange combination of the m's. We recommend the wide style because general Stata commands will do what you expect. The same is true for the flongsep style. It is your responsibility to get this right.

   So what is the difference between mi misstable and misstable? mi misstable amounts to mi xeq 0: misstable, exok, which is to say it runs on $m = 0$ and specifies the exok option so that extended missing values are treated as hard missings.

   In general, you need to become familiar with all the mi commands, use the mi variant of regular Stata commands whenever one exists, and think twice before using a command without an mi prefix. Doing the right thing will become automatic once you gain familiarity with the styles; see [MI] **styles**.

   To learn about the missing values in your data, type

   . mi misstable summarize

4. Use mi register imputed to register the variables you wish to impute; see [MI] **mi set**.

   The only variables that mi will impute are those registered as imputed. You can register variables one at a time or all at once. If you register a variable mistakenly, use mi unregister to unregister it.

   . mi register imputed varname [varname ...]

5. Use mi impute to impute (fill in) the missing values; see [MI] **mi impute**.

   There is a lot to be said here. For instance, in a dataset where variables age and bmi contain missing, you might type

   . mi register imputed age bmi
   . mi impute mvn age bmi = attack smokes hsgrad, add(10)

   mi impute's add(#) option specifies the number of imputations to be added. We currently have 0 imputations, so after imputation, we will have 10. We usually start with a small number of imputations and add more later.

6. Use mi describe to verify that all missing values are filled in; see [MI] **mi describe**.

   . mi describe

   You might also want to use mi xeq (see [MI] **mi xeq**) to look at summary statistics in each of the imputation datasets:

   . mi xeq: summarize

7. Generate passive variables; see [MI] **mi passive**.

   Passive variables are variables that are functions of imputed variables, such as `lnage` when some values of `age` are imputed. The values of passive variables differ across m just as the values of imputed variables do. The official way to generate imputed values is by using `mi passive`:

   ```
   . mi passive: generate lnage = ln(age)
   ```

   Rather than use the official way, however, we often switch our data to mlong and just generate the passive variables directly:

   ```
   . mi convert mlong
   . generate lnage = ln(age)
   . mi register passive lnage
   ```

   If you work as we do, remember to register any passive variables you create. When you are done, you may `mi convert` your data back to wide, but there is no reason to do that.

8. Use `mi estimate` (see [MI] **mi estimate**) to fit models:

   ```
   . mi estimate: logistic attack smokes age bmi hsgrad
   ```

   You fit your model just as you would ordinarily except that you add `mi estimate:` in front of the command.

To see an example of the advice applied to a simple dataset, see *Example* below.

In theory, you should get your data cleaning and data management out of the way before `mi` setting your data. In practice that will not happen, so you will want to become familiar with the other `mi` commands. Among the data-management commands available are `mi append` (see [MI] **mi append**), `mi merge` (see [MI] **mi merge**), `mi expand` (see [MI] **mi expand**), and `mi reshape` (see [MI] **mi reshape**). If you are working with survival-time data, also see [MI] **mi stsplit**. To `stset` your data, or `svyset`, or `xtset`, see [MI] **mi set** and [MI] **mi XXXset**.

Suggested workflow for data that already have imputations

Data sometimes come with imputations included. The data might be made by another researcher for you or the data might come from an official source. Either way, we will assume that the data are not in Stata format, because if they were, you would just use the data and would type `mi describe`.

`mi` can import officially produced datasets created by the National Health and Nutrition Examination Survey (NHANES) with the `mi import nhanes1` command, and `mi` can import more informally created datasets that are wide-, flong-, or flongsep-like with `mi import wide`, `mi import flong`, or `mi import flongsep`; see [MI] **mi import**.

The required workflow is hardly different from *Suggested workflow for original data*, presented above. The differences are that you will use `mi import` rather than `mi set` and you will skip using `mi impute` to generate the imputations. In this sense, your job is easier.

On the other hand, you need to verify that you have imported your data correctly, and we have a lot to say about that. Basically, after importing, you need to be careful about which `mi` commands you use until you have verified that you have the variables registered correctly. That is discussed in [MI] **mi import**.

Example

We are going to repeat *A simple example* from [MI] **intro**, but this time we are going to follow the advice given above in *Suggested workflow for original data*.

We have fictional data on 154 patients and want to examine the relationship between binary outcome attack, recording heart attacks, and variables smokes, age, bmi, hsgrad, and female. We will use logistic regression. Below we load our original data and show you a little about it using the standard commands describe and summarize. We emphasize that mheart5.dta is just a standard Stata dataset; it has not been mi set.

```
. use http://www.stata-press.com/data/r12/mheart5
(Fictional heart attack data; bmi and age missing)

. describe
Contains data from http://www.stata-press.com/data/r12/mheart5.dta
  obs:           154                        Fictional heart attack data;
                                            bmi and age missing
  vars:            6                        19 Jun 2011 10:50
  size:        1,848

              storage   display    value
variable name   type    format     label      variable label

attack          byte    %9.0g                 Outcome (heart attack)
smokes          byte    %9.0g                 Current smoker
age             float   %9.0g                 Age, in years
bmi             float   %9.0g                 Body Mass Index, kg/m^2
female          byte    %9.0g                 Gender
hsgrad          byte    %9.0g                 High school graduate

Sorted by:
. summarize
    Variable |        Obs        Mean    Std. Dev.       Min        Max

      attack |        154    .4480519    .4989166          0          1
      smokes |        154    .4155844    .4944304          0          1
         age |        142    56.43324    11.59131   20.73613   83.78423
         bmi |        126    25.23523    4.029325   17.22643   38.24214
      female |        154    .2467532    .4325285          0          1

      hsgrad |        154    .7532468    .4325285          0          1
```

The first guideline is

1. mi set your data; see [MI] **mi set**.

We will set the data to be flong even though in *A simple example* we set the data to be mlong. mi provides four styles—flong, mlong, wide, and flongsep—and at this point it does not matter which we choose. mi commands work the same way regardless of style. Four styles are provided because, should we decide to step outside of mi and attack the data with standard Stata commands, we will find different styles more convenient depending on what we want to do. It is easy to switch styles.

Below we type mi set flong and then, to show you what that command did to the data, we show you the output from a standard describe:

```
. mi set flong

. describe
Contains data from http://www.stata-press.com/data/r12/mheart5.dta
  obs:            154                          Fictional heart attack data;
                                                 bmi and age missing
  vars:             9                          19 Jun 2011 10:50
  size:         2,618
```

variable name	storage type	display format	value label	variable label
attack	byte	%9.0g		Outcome (heart attack)
smokes	byte	%9.0g		Current smoker
age	float	%9.0g		Age, in years
bmi	float	%9.0g		Body Mass Index, kg/m^2
female	byte	%9.0g		Gender
hsgrad	byte	%9.0g		High school graduate
_mi_miss	byte	%8.0g		
_mi_m	int	%8.0g		
_mi_id	int	%12.0g		

```
Sorted by:
```

Typing mi set flong added three variables to our data: _mi_miss, _mi_m, and _mi_id. Those variables belong to mi. If you are curious about them, see [MI] **styles**. Advanced users can even use them. No matter how advanced you are, however, you must never change their contents.

Except for the three added variables, the data are unchanged, and we would see that if we typed summarize. The three added variables are due to the style we chose. When you mi set your data, different styles will change the data differently, but the changes will be just around the edges.

The second guideline is

2. Use mi describe often; see [MI] **mi describe**.

The guideline is to use mi describe, not describe as we just did. Here is the result:

```
. mi describe
  Style: flong
         last mi update 02apr2011 11:07:59, 0 seconds ago
  Obs.:  complete          154
         incomplete          0   (M = 0 imputations)
         ─────────────────────
         total             154
  Vars.: imputed:  0
         passive:  0
         regular:  0
         system:   3; _mi_m _mi_id _mi_miss
         (there are 6 unregistered variables)
```

As the guideline warned us, "mi describe will not tell you anything useful yet."

The third guideline is

3. Use `mi misstable` to identify missing values; see [MI] **mi misstable**.

Below we type `mi misstable summarize` and `mi misstable nested`:

```
. mi misstable summarize
                                                             Obs<.
                                                +------------------------------
                |                               | Unique
       Variable |     Obs=.     Obs>.     Obs<. | values          Min         Max
  --------------+-------------------------------+------------------------------
            age |        12                 142 |    142     20.73613    83.78423
            bmi |        28                 126 |    126     17.22643    38.24214
  -----------------------------------------------------------------------------

. mi misstable nested
     1.  age(12) -> bmi(28)
```

`mi misstable summarize` reports the variables containing missing values. Those variables in our data are `age` and `bmi`. Notice that `mi misstable summarize` draws a distinction between, as it puts it, "Obs=." and "Obs>.", which is to say between standard missing (`.`) and extended missing (`.a, .b, ..., .z`). That is because `mi` has a concept of soft and hard missing, and it associates soft missing with system missing and hard missing with extended missing. Hard missing values—extended missings—are taken to mean missing values that are not to be imputed. Our data have no missing values like that.

After typing `mi misstable summarize`, we typed `mi misstable nested` because we were curious whether the missing values were nested or, to use the jargon, monotone. We discovered that they were. That is, `age` has 12 missing values in the data, and in every observation in which `age` is missing, so is `bmi`, although `bmi` has another 16 missing values scattered around the data. That means we can use a monotone imputation method, and that is good news because monotone methods are more flexible and faster. We will discuss the implications of that shortly. There is a mechanical detail we must handle first.

The fourth guideline is

4. Use `mi register imputed` to register the variables you wish to impute; see [MI] **mi set**.

We know that `age` and `bmi` have missing values, and before we can impute replacements for those missing values, we must register the variables as to-be-imputed, which we do by typing

```
. mi register imputed age bmi
(28 m=0 obs. now marked as incomplete)
```

Guideline 2 suggested that we type `mi describe` often. Perhaps now would be a good time:

```
. mi describe
  Style:  flong
          last mi update 02apr2011 11:07:59, 0 seconds ago
  Obs.:   complete          126
          incomplete         28  (M = 0 imputations)
          ---------------------
          total             154
  Vars.:  imputed:  2; age(12) bmi(28)
          passive:  0
          regular:  0
          system:   3; _mi_m _mi_id _mi_miss
         (there are 4 unregistered variables; attack smokes female hsgrad)
```

The output has indeed changed. mi knows just as it did before that we have 154 observations, and it now knows that 126 of them are complete and 28 of them are incomplete. It also knows that age and bmi are to be imputed. The numbers in parentheses are the number of missing values.

The fifth guideline is

5. Use mi impute to impute (fill in) the missing values; see [MI] **mi impute**.

In *A simple example* from [MI] **intro**, we imputed values for age and bmi by typing

. mi impute mvn age bmi = attack smokes hsgrad female, add(10)

This time, we will impute values by typing

. mi impute monotone (regress) age bmi = attack smokes hsgrad female, add(20)

We changed add(10) to add(20) for no other reason than to show that we could, although we admit to a preference for more imputations whenever possible. add() specifies the number of imputations to be added to the data. For every missing value, we will impute 20 nonmissing replacements.

We switched from mi impute mvn to mi impute monotone because our data are monotone. Here mi impute monotone will be faster than mi impute mvn but will offer no statistical advantage. In other cases, there might be statistical advantages. All of which is to say that when you get to the imputation step, you have important decisions to make and you need to become knowledgeable about the subject. You can start by reading [MI] **mi impute**.

```
. set seed 20039
. mi impute monotone (regress) age bmi = attack smokes hsgrad female, add(20)
Conditional models:
            age: regress age attack smokes hsgrad female
            bmi: regress bmi age attack smokes hsgrad female

Multivariate imputation                    Imputations =         20
Monotone method                                  added =         20
Imputed: m=1 through m=20                       updated =          0
            age: linear regression
            bmi: linear regression
```

	Observations per m			
Variable	Complete	Incomplete	Imputed	Total
age	142	12	12	154
bmi	126	28	28	154

```
(complete + incomplete = total; imputed is the minimum across m
 of the number of filled-in observations.)
```

Note that we typed set seed 20039 before issuing the mi impute command. Doing that made our results reproducible. We could have specified mi impute's rseed(20039) option instead. Or we could have skipped setting the random-number seed altogether, and then we would not be able to reproduce our results.

The sixth guideline is

6. Use mi describe to verify that all missing values are filled in; see [MI] **mi describe**.

```
. mi describe, detail
  Style:    flong
            last mi update 02apr2011 11:07:59, 0 seconds ago
  Obs.:     complete        126
            incomplete       28   (M = 20 imputations)
            ─────────────────────
            total           154
  Vars.:    imputed:  2; age(12; 20*0) bmi(28; 20*0)
            passive:  0
            regular:  0
            system:   3; _mi_m _mi_id _mi_miss
            (there are 4 unregistered variables; attack smokes female hsgrad)
```

This time, we specified mi describe's detail option, although you have to look closely at the output to see the effect. When you do not specify detail, mi describe reports results for the original, unimputed data only, what we call $m = 0$ throughout this documentation. When you specify detail, mi describe also includes information about the imputation data, what we call $m > 0$ and is $m = 1$, $m = 2$, ..., $m = 20$ here. Previously, mi describe reported "age(12)", meaning that age in $m = 0$ has 12 missing values. This time, it reports "age(12; 20*0)", meaning that age still has 12 missing values in $m = 0$, and it has 0 missing values in the 20 imputations. bmi also has 0 missing values in the imputations. Success!

Let's take a detour to see how our data really look. Let's type Stata's standard describe command. The last time we looked, our data had three extra variables.

```
. describe
Contains data from http://www.stata-press.com/data/r12/mheart5.dta
  obs:        3,234                        Fictional heart attack data;
                                             bmi and age missing
  vars:           9                        21 Jun 2011 13:36
  size:      54,978
─────────────────────────────────────────────────────────────────────
              storage   display    value
variable name    type    format    label    variable label
─────────────────────────────────────────────────────────────────────
attack           byte    %9.0g               Outcome (heart attack)
smokes           byte    %9.0g               Current smoker
age              float   %9.0g               Age, in years
bmi              float   %9.0g               Body Mass Index, kg/m^2
female           byte    %9.0g               Gender
hsgrad           byte    %9.0g               High school graduate
_mi_id           int     %12.0g
_mi_miss         byte    %8.0g
_mi_m            int     %8.0g
─────────────────────────────────────────────────────────────────────
Sorted by: _mi_m _mi_id
```

Nothing has changed as far as variables are concerned, but notice the number of observations. Previously, we had 154 observations. Now we have 3,234! That works out to 21*154. Stored is our original data plus 20 imputations. The flong style makes extra copies of the data.

We chose style flong only because it is so easy to explain. In *A simple example* from [MI] **intro** using this same data, we choose style mlong. It is not too late:

```
. mi convert mlong
```

All that is required to change styles is typing mi convert. The style of the data changes, but not the contents. Let's see what describe has to report:

```
. describe
Contains data from http://www.stata-press.com/data/r12/mheart5.dta
  obs:             714                          Fictional heart attack data;
                                                  bmi and age missing
  vars:              9                          21 Jun 2011 13:36
  size:         12,138

              storage   display    value
variable name   type    format     label        variable label

attack          byte    %9.0g                   Outcome (heart attack)
smokes          byte    %9.0g                   Current smoker
age             float   %9.0g                   Age, in years
bmi             float   %9.0g                   Body Mass Index, kg/m^2
female          byte    %9.0g                   Gender
hsgrad          byte    %9.0g                   High school graduate
_mi_id          int     %12.0g
_mi_miss        byte    %8.0g
_mi_m           int     %8.0g

Sorted by:  _mi_m  _mi_id
```

The data look much like they did when they were flong, except that the number of observations has fallen from 3,234 to 714! Style mlong is an efficient style in that rather than storing the full data for every imputation, it stores only the changes. Back when the data were flong, mi describe reported that we had 28 incomplete observations. We get 714 from the 154 original observations plus 20×28 replacement observations for the incomplete observations.

We recommend style mlong. Style wide is also recommended. Below we type mi convert to convert our mlong data to wide, and then we run the standard describe command:

```
. mi convert wide
. describe
Contains data from http://www.stata-press.com/data/r12/mheart5.dta
  obs:             154                          Fictional heart attack data;
                                                  bmi and age missing
  vars:             47                          21 Jun 2011 13:43
  size:         26,642

              storage   display    value
variable name   type    format     label        variable label

attack          byte    %9.0g                   Outcome (heart attack)
smokes          byte    %9.0g                   Current smoker
age             float   %9.0g                   Age, in years
bmi             float   %9.0g                   Body Mass Index, kg/m^2
female          byte    %9.0g                   Gender
hsgrad          byte    %9.0g                   High school graduate
_mi_miss        byte    %8.0g
_1_age          float   %9.0g                   Age, in years
_1_bmi          float   %9.0g                   Body Mass Index, kg/m^2
_2_age          float   %9.0g                   Age, in years
_2_bmi          float   %9.0g                   Body Mass Index, kg/m^2
  (output omitted )
_20_age         float   %9.0g                   Age, in years
_20_bmi         float   %9.0g                   Body Mass Index, kg/m^2

Sorted by:
```

In the wide style, our data are back to having 154 observations, but now we have 47 variables!

Variable _1_age contains age for $m = 1$, _1_bmi contains bmi for $m = 1$, _2_age contains age for $m = 2$, and so on.

Guideline 7 is

7. Generate passive variables.

Passive variables are variables derived from imputed variables. For instance, if we needed lnage = ln(age), variable lnage would be passive. Passive variables are easy to create; see [MI] **mi passive**. We are not going to need any passive variables in this example.

Guideline 8 is

8. Use mi estimate to fit models; see [MI] **mi estimate**.

Our data are wide right now, but that does not matter. We fit our model:

```
. mi estimate: logistic attack smokes age bmi hsgrad female
Multiple-imputation estimates              Imputations     =          20
Logistic regression                        Number of obs   =         154
                                           Average RVI     =      0.0547
                                           Largest FMI     =      0.1377
DF adjustment:   Large sample              DF:      min     =     1027.48
                                                    avg     =    55394.62
                                                    max     =   168501.59
Model F test:       Equal FMI              F(  5,25165.6)   =        3.35
Within VCE type:         OIM               Prob > F         =      0.0050
```

attack	Coef.	Std. Err.	t	P>|t|	[95% Conf. Interval]
smokes	1.186791	.359663	3.30	0.001	.4818394 1.891743
age	.0297742	.0164346	1.81	0.070	-.0024699 .0620184
bmi	.1033297	.0468362	2.21	0.028	.0114494 .1952101
hsgrad	.1529883	.4033788	0.38	0.704	-.6376254 .943602
female	-.079329	.4145832	-0.19	0.848	-.8919049 .7332468
_cons	-5.100976	1.685697	-3.03	0.003	-8.408779 -1.793173

Those familiar with the logistic command will be surprised that mi estimate: logistic reported coefficients rather than odds ratios. That is because the estimation command is not logistic using mi estimate, it is mi estimate using logistic. If we wanted to see odds ratios at estimation time, we could have typed

```
. mi estimate, or:  logistic ...
```

By the same token, if we wanted to replay results, we would not type logistic, we would type mi estimate:

```
. mi estimate
  (output omitted )
```

If we wanted to replay results with odds ratios, we would type

```
. mi estimate, or
```

And that concludes the guidelines.

Also see

[MI] **intro** — Introduction to mi

[MI] **Glossary**

Title

> **Glossary**

Description

Please read. The terms defined below are used throughout the documentation, sometimes without explanation.

Glossary

arbitrary missing pattern. Any missing-value pattern. Some imputation methods are suitable only when the pattern of missing values is special, such as a monotone-missing pattern. An imputation method suitable for use with an arbitrary missing pattern may be used regardless of the pattern.

augmented regression. Regression performed on the augmented data, the data with a few extra observations with small weights. The data are augmented in a way that prevents perfect prediction, which may arise during estimation of categorical data. See *The issue of perfect prediction during imputation of categorical data* under *Remarks* of [MI] **mi impute**.

burn-between period. The number of iterations between two draws of an MCMC sequence such that these draws may be regarded as independent.

burn-in period. The number of iterations it takes for an MCMC sequence to reach stationarity.

casewise deletion. See *listwise deletion*.

chained equations. See *FCS*.

complete and incomplete observations. An observation in the $m = 0$ data is said to be complete if no imputed variable in the observation contains soft missing (.). Observations that are not complete are said to be incomplete.

complete-cases analysis. See *listwise deletion*.

complete data. Data that do not contain any missing values.

complete-data analysis. The analysis or estimation performed on the complete data, the data for which all values are observed. This term does not refer to analysis or estimation performed on the subset of complete observations. Do not confuse this with completed-data analysis.

complete DF, complete degrees of freedom. The degrees of freedom that would have been used for inference if the data were complete.

completed data. See *imputed data*.

completed-data analysis. The analysis or estimation performed on the made-to-be completed (imputed) data. This term does not refer to analysis or estimation performed on the subset of complete observations.

conditional imputation. Imputation performed using a conditional sample, a restricted part of the sample. Missing values outside the conditional sample are replaced with a conditional constant, the constant value of the imputed variable in the nonmissing observations outside the conditional sample. See *Conditional imputation* under *Remarks* of [MI] **mi impute**.

DA, data augmentation. An MCMC method used for the imputation of missing data.

EM, expectation-maximization algorithm. In the context of MI, an iterative procedure for obtaining maximum likelihood or posterior-mode estimates in the presence of missing data.

FCS, fully conditional specification. Consider imputation variables $X_1, X_2, \ldots, X_p$. Fully conditional specification of the prediction equation for X_j includes all variables except X_j; that is, variables $\mathbf{X}_{-j} = (X_1, X_2, \ldots, X_{j-1}, X_{j+1}, \ldots, X_p)$.

flong data. See *style*.

flongsep data. See *style*.

FMI, fraction of missing information. The ratio of information lost due to the missing data to the total information that would be present if there were no missing data.

An equal FMI test is a test under the assumption that FMIs are equal across parameters.

An unrestricted FMI test is a test without the equal FMI assumption.

hard missing and soft missing. A hard missing value is a value of .a, .b, $\ldots$, .z in $m = 0$ in an imputed variable. Hard missing values are not replaced in $m > 0$.

A soft missing value is a value of . in $m = 0$ in an imputed variable. If an imputed variable contains soft missing, then that value is eligible to be imputed, and perhaps is imputed, in $m > 0$.

Although you can use the terms hard missing and soft missing for passive, regular, and unregistered variables, it has no special significance in terms of how the missing values are treated.

ignorable missing-data mechanism. The missing-data mechanism is said to be ignorable if missing data are missing at random and the parameters of the data model and the parameters of the missing-data mechanism are distinct; that is, the joint distribution of the model and the missing-data parameters can be factorized into two independent marginal distributions of model parameters and of missing-data parameters.

imputed, passive, and regular variables. An imputed variable is a variable that has missing values and for which you have or will have imputations.

A passive variable is a varying variable that is a function of imputed variables or of other passive variables. A passive variable will have missing values in $m = 0$ and varying values for observations in $m > 0$.

A regular variable is a variable that is neither imputed nor passive and that has the same values, whether missing or not, in all m.

Imputed, passive, and regular variables can be registered using the `mi register` command; see [MI] **mi set**. You are required to register imputed variables, and we recommend that you register passive variables. Regular variables can also be registered. See *registered and unregistered variables*.

imputed data. Data in which all missing values are imputed.

incomplete observations. See *complete and incomplete observations*.

ineligible missing value. An ineligible missing value is a missing value in a to-be-imputed variable that is due to inability to calculate a result rather than an underlying value being unobserved. For instance, assume that variable `income` had some missing values and so you wish to impute it. Because `income` is skewed, you decide to impute the log of income, and you begin by typing

```
. generate lnincome = log(income)
```

If `income` contained any zero values, the corresponding missing values in `lnincome` would be ineligible missing values. To ensure that values are subsequently imputed correctly, it is of vital importance that any ineligible missing values be recorded as hard missing. You would do that by typing

```
. replace lnincome = .a if lnincome==. & income!=.
```

As an aside, if after imputing `lnincome` using `mi impute` (see [MI] **mi impute**), you wanted to fill in `income`, `income` surprisingly would be a passive variable because `lnincome` is the imputed variable and `income` would be derived from it. You would type

> . mi register passive income
> . mi passive: replace income = cond(lnincome==.a, 0, exp(lnincome))

In general, you should avoid using transformations that produce ineligible missing values to avoid the loss of information contained in other variables in the corresponding observations. For example, in the above, for zero values of `income` we could have assigned the log of income, `lnincome`, to be the smallest value that can be stored as `double`, because the logarithm of zero is negative infinity:

> . generate lnincome = cond(income==0, mindouble(), log(income))

This way, all observations for which `income==0` will be used in the imputation model for `lnincome`.

jackknifed standard error. See *Monte Carlo error*.

listwise deletion, casewise deletion. Omitting from analysis observations containing missing values.

M, m. M is the number of imputations. m refers to a particular imputation, $m = 1, 2, \ldots, M$. In `mi`, $m = 0$ is used to refer to the original data, the data containing the missing values. Thus `mi` data in effect contain $M + 1$ datasets, corresponding to $m = 0$, $m = 1$, ..., and $m = M$.

MAR, missing at random. Missing data are said to be missing at random (MAR) if the probability that data are missing does not depend on unobserved data but may depend on observed data. Under MAR, the missing-data values do not contain any additional information given observed data about the missing-data mechanism. Thus the process that causes missing data can be ignored.

MCAR, missing completely at random. Missing data are said to be missing completely at random (MCAR) if the probability that data are missing does not depend on observed or unobserved data. Under MCAR, the missing data values are a simple random sample of all data values, so any analysis that discards the missing values remains consistent, albeit perhaps inefficient.

MCE, Monte Carlo error. Within the multiple-imputation context, a Monte Carlo error is defined as the standard deviation of the multiple-imputation results across repeated runs of the same imputation procedure using the same data. The Monte Carlo error is useful for evaluating the statistical reproducibility of multiple-imputation results. See *Example 6: Monte Carlo error estimates* under *Remarks* of [MI] **mi estimate**.

MCMC, Markov chain Monte Carlo. A class of methods for simulating random draws from otherwise intractable multivariate distributions. The Markov chain has the desired distribution as its equilibrium distribution.

mi data. Any data that have been `mi set` (see [MI] **mi set**), whether directly by `mi set` or indirectly by `mi import` (see [MI] **mi import**). The `mi` data might have no imputations (have $M = 0$) and no imputed variables, at least yet, or they might have $M > 0$ and no imputed variables, or vice versa. An `mi` dataset might have $M > 0$ and imputed variables, but the missing values have not yet been replaced with imputed values. Or `mi` data might have $M > 0$ and imputed variables and the missing values of the imputed variables filled in with imputed values.

mlong data. See *style*.

monotone-missing pattern, monotone missingness. A special pattern of missing values in which if the variables are ordered from least to most missing, then all observations of a variable contain missing in the observations in which the prior variable contains missing.

MNAR, missing not at random. Missing data are missing not at random (MNAR) if the probability that data are missing depends on unobserved data. Under MNAR, a missing-data mechanism (the process that causes missing data) must be modeled to obtain valid results.

original data. Original data are the data as originally collected, with missing values in place. In mi data, the original data are stored in $m = 0$. The original data can be extracted from mi data by using mi extract; see [MI] **mi extract**.

passive variable. See *imputed, passive, and regular variables*.

registered and unregistered variables. Variables in mi data can be registered as imputed, passive, or regular by using the mi register command; see [MI] **mi set**.

You are required to register imputed variables.

You should register passive variables; if your data are style wide, you are required to register them. The mi passive command (see [MI] **mi passive**) makes creating passive variables easy, and it automatically registers them for you.

Whether you register regular variables is up to you. Registering them is safer in all styles except wide, where it does not matter. By definition, regular variables should be the same across m. In the long styles, you can unintentionally create variables that vary. If the variable is registered, mi will detect and fix your mistakes.

Super-varying variables, which rarely occur and can be stored only in flong and flongsep data, should never be registered.

The registration status of variables is listed by the mi describe command; see [MI] **mi describe**.

regular variable. See *imputed, passive, and regular variables*.

relative efficiency. Ratio of variance of a parameter given estimation with finite M to the variance if M were infinite.

RVI, relative variance increase. The increase in variance of a parameter estimate due to nonresponse.

style. Style refers to the format in which the mi data are stored. There are four styles: flongsep, flong, mlong, and wide. You can ignore styles, except for making an original selection, because all mi commands work regardless of style. You will be able to work more efficiently, however, if you understand the details of the style you are using; see [MI] **styles**. Some tasks are easier in one style than another. You can switch between styles by using the mi convert command; see [MI] **mi convert**.

The flongsep style is best avoided unless your data are too big to fit into one of the other styles. In flongsep style, a separate .dta set is created for $m = 0$, for $m = 1$, ..., and for $m = M$. Flongsep is best avoided because mi commands work more slowly with it.

In all the other styles, the $M + 1$ datasets are stored in one .dta file. The other styles are both more convenient and more efficient.

The most easily described of these .dta styles is flong; however, flong is also best avoided because mlong style is every bit as convenient as flong, and mlong is memorywise more efficient. In flong, each observation in the original data is repeated M times in the .dta dataset, once for $m = 1$, again for $m = 2$, and so on. Variable _mi_m records m and takes on values 0, 1, 2, ..., M. Within each value of m, variable _mi_id takes on values 1, 2, ..., N and thus connects imputed with original observations.

The mlong style is recommended. It is efficient and easy to use. Mlong is much like flong except that complete observations are not repeated.

Equally recommended is the wide style. In wide, each imputed and passive variable has an additional M variables associated with it, one for the variable's value in $m = 1$, another for its value in $m = 2$, and so on. If an imputed or passive variable is named *vn*, then the values of *vn* in $m = 1$ are stored in variable _1_*vn*; the values for $m = 2$, in _2_*vn*; and so on.

What makes mlong and wide so convenient? In mlong, there is a one-to-one correspondence of your idea of a variable and Stata's idea of a variable—variable *vn* refers to *vn* for all values of m. In wide, there is a one-to-one correspondence of your idea of an observation and Stata's idea—physical observation 5 is observation 5 in all datasets.

Choose the style that matches the problem at hand. If you want to create new variables or modify existing ones, choose mlong. If you want to drop observations or create new ones, choose wide. You can switch styles with the `mi convert` command; see [MI] **mi convert**.

For instance, if you want to create new variable `ageXexp` equal to `age*exp` and your data are mlong, you can just type `generate ageXexp = age*exp`, and that will work even if `age` and `exp` are imputed, passive, or a mix. Theoretically, the right way to do that is to type `mi passive: generate agexExp = age*exp`, but concerning variables, if your data are mlong, you can work the usual Stata way.

If you want to drop observation 20 or drop `if sex==2`, if your data are wide, you can just type `drop in 20` or `drop if sex==2`. Here the "right" way to do the problem is to type the `drop` command and then remember to type `mi update` so that `mi` can perform whatever machinations are required to carry out the change throughout $m > 0$; however, in the wide form, there are no machinations required.

super-varying variables. See *varying and super-varying variables*.

unregistered variables. See *registered and unregistered variables*.

varying and super-varying variables. A variable is said to be varying if its values in the incomplete observations differ across m. Imputed and passive variables are varying. Regular variables are nonvarying. Unregistered variables can be either.

Imputed variables are supposed to vary because their incomplete values are filled in with different imputed values, although an imputed variable can be temporarily nonvarying if you have not imputed its values yet. Similarly, passive variables should vary because they are or will be filled in based on values of varying imputed variables.

A variable is said to be super varying if its values in the complete observations differ across m. The existence of super-varying variables is usually an indication of error. It makes no sense for a variable to have different values in, say, $m = 0$ and $m = 2$ in the complete observations—in observations that contain no missing values. That is, it makes no sense unless the values of the variable is a function of the values of other variables across multiple observations. If variable `sumx` is the sum of x across observations, and if x is imputed, then `sumx` will differ across m in all observations after the first observation in which x is imputed.

The `mi varying` command will identify varying and super-varying variables, as well as nonvarying imputed and passive variables. [MI] **mi varying** explains how to fix problems when they are due to error.

Some problems that theoretically could arise cannot arise because `mi` will not let them. For instance, an imputed variable could be super varying and that would obviously be a serious error. Or a regular variable could be varying and that, too, would be a serious error. When you register a variable, `mi` fixes any such problems and, from that point on, watches for problems and fixes them as they arise.

Use mi register to register variables; see [MI] **mi set**. You can perform the checks and fixes at any time by running mi update; see [MI] **mi update**. Among other things, mi update replaces values of regular variables in $m > 0$ with their values from $m = 0$; it replaces values of imputed variables in $m > 0$ with their nonmissing values from $m = 0$; and it replaces values of passive variables in incomplete observations of $m > 0$ with their $m = 0$ values. mi update follows a hands-off policy with respect to unregistered variables.

If you need super-varying variables, use flong or flongsep style and do not register the variable. You must use one of the flong styles because in the wide and mlong styles, there is simply no place to store super-varying values.

wide data. See *style*.

WLF, worst linear function. A linear combination of all parameters being estimated by an iterative procedure that is thought to converge slowly.

Also see

[MI] **intro** — Introduction to mi

Subject and author index

This is the subject and author index for the *Multiple-Imputation Reference Manual*. Readers interested in topics other than multiple imputation should see the combined subject index (and the combined author index) in the *Quick Reference and Index*.

Semicolons set off the most important entries from the rest. Sometimes no entry will be set off with semicolons, meaning that all entries are equally important.

A

Abayomi, K., [MI] **intro substantive**, [MI] **mi impute**
add, mi subcommand, [MI] **mi add**
Albert, A., [MI] **mi impute**
Alfaro, R., [MI] **intro**
Allison, P. D., [MI] **intro substantive**, [MI] **mi impute**
analysis step, [MI] **intro substantive**, [MI] **mi estimate**, *also see* estimation
Anderson, J. A., [MI] **mi impute**
Anderson, T. W., [MI] **intro substantive**
append, mi subcommand, [MI] **mi append**
appending data, [MI] **mi append**
arbitrary pattern of missing values, [MI] **Glossary**, [MI] **mi impute chained**, [MI] **mi impute mvn**, *also see* pattern of missingness
Arnold, B. C., [MI] **intro substantive**, [MI] **mi impute chained**
augmented regression, *see* imputation, perfect prediction
available-case analysis, [MI] **intro substantive**
average RVI, [MI] **Glossary**, [MI] **mi estimate**

B

Barnard, J., [MI] **intro substantive**, [MI] **mi estimate**, [MI] **mi estimate using**, [MI] **mi predict**, [MI] **mi test**
Bayesian concepts, [MI] **intro substantive**
between-imputation variability, [MI] **mi estimate**, [MI] **mi predict**
Bibby, J. M., [MI] **mi impute mvn**
binary variable imputation, *see* imputation, binary
Binder, D. A., [MI] **intro substantive**
Boshuizen, H. C., [MI] **intro substantive**, [MI] **mi impute**, [MI] **mi impute chained**, [MI] **mi impute monotone**
Brand, J. P. L., [MI] **intro substantive**, [MI] **mi impute chained**
Burkhauser, R. V., [MI] **intro substantive**
burn-between period, [MI] **Glossary**, [MI] **mi impute**, [MI] **mi impute chained**, [MI] **mi impute mvn**
burn-in period, [MI] **Glossary**, [MI] **mi impute**, [MI] **mi impute chained**, [MI] **mi impute mvn**

C

cancer data, [MI] **mi estimate**, [MI] **mi predict**

Carlin, J. B., [MI] **intro**, [MI] **intro substantive**, [MI] **mi estimate**, [MI] **mi impute**, [MI] **mi impute mvn**, [MI] **mi impute regress**
Carpenter, J. R., [MI] **intro**, [MI] **intro substantive**, [MI] **mi impute**
casewise deletion, *see* listwise deletion
Castillo, E., [MI] **intro substantive**, [MI] **mi impute chained**
categorical variable imputation, *see* imputation, categorical
censoring, *see* imputation, interval-censored data
certifying mi data are consistent, [MI] **mi update**
chained equations, *see* imputation, chained equations
Cleves, M. A., [MI] **mi estimate**
Coffey, C., [MI] **intro substantive**
combination step, [MI] **intro substantive**, [MI] **mi estimate**, [MI] **mi estimate using**, [MI] **mi predict**
combining data, [MI] **mi add**, [MI] **mi append**, [MI] **mi merge**
complete data, [MI] **Glossary**
complete degrees of freedom for coefficients, [MI] **Glossary**, [MI] **mi estimate**
complete observations, [MI] **Glossary**
complete-cases analysis, [MI] **Glossary**
complete-data analysis, [MI] **Glossary**
completed data, [MI] **Glossary**
completed-data analysis, [MI] **Glossary**, [MI] **intro substantive**, [MI] **mi estimate**
conditional imputation, [MI] **Glossary**, [MI] **mi impute**, *see* imputation, conditional
continuous variable imputation, *see* imputation, continuous
convergence of MCMC, *see* MCMC, convergence
convert, mi subcommand, [MI] **mi convert**
converting between styles, [MI] **mi convert**
copy, mi subcommand, [MI] **mi copy**, [MI] **styles**
count data, *see* imputation, count data
custom prediction equations, [MI] **mi impute chained**, [MI] **mi impute monotone**

D

DA, *see* data augmentation
Daniel, R., [MI] **intro substantive**, [MI] **mi impute**, [MI] **mi impute chained**, [MI] **mi impute monotone**
data augmentation, [MI] **Glossary**, [MI] **mi impute**, [MI] **mi impute mvn**
data management, [MI] **mi add**, [MI] **mi append**, [MI] **mi expand**, [MI] **mi extract**, [MI] **mi merge**, [MI] **mi rename**, [MI] **mi replace0**, [MI] **mi reset**, [MI] **mi reshape**
data,
   exporting, *see* exporting
   flong, *see* flong
   flongsep, *see* flongsep
   importing, *see* importing

J

K

L